Principles and Applications of GSM

ISBN 0-13-949124-4

90000

Prentice Hall Communications Engineering and Emerging Technologies Series

Theodore S. Rappaport, Series Editor

RAPPAPORT, *Wireless Communication: Principles & Practice*

RAZAVI, *RF Microelectronics*

GARG & WILKES, *Principles and Applications of GSM*

FORTHCOMING

LIBERTI & RAPPAPORT, *Smart Antennas for Wireless CDMA: IS-95 and Broadband CDMA Applications*

TRANTER, KOSBAR, RAPPAPORT, & SHANMUGAN, *Simulation of Modern Communications Systems with Wireless Applications*

Principles and Applications of GSM

Vijay K. Garg, Ph.D., PE, SE
Distinguished Member of Technical Staff, Lucent Technologies

and

Joseph E. Wilkes, Ph.D., PE
Senior Research Scientist, Bell Communications Research

Prentice Hall PTR
Upper Saddle River, NJ 07458
http://www.phptr.com

Library of Congress Cataloging-in-Publication Data

```
Garg, Vijay Kumar
     Principles and applications of GSM / Vijay K. Garg and Joseph E. Wilkes.
        p.cm. — (Prentice Hall communications engineering and emerging technologies
     series)
        Includes bibliographical references and index.
        ISBN 0-13-949124-4
        1. Global system for mobile communications.I. Wilkes, Joseph E.II. Title.
     III. Series.
     TK5103.483.G37 1999
     621.382—dc21                                            98-36519
                                                                 CIP
```

Editorial/production supervision: BooksCraft, Inc., Indianapolis, IN
Cover design director: Jerry Votta
Cover design: Design Source
Acquisition editor: Bernard M. Goodwin
Manufacturing manager: Alan Fischer

 © 1999 by Prentice Hall PTR
Prentice-Hall, Inc.
A Simon & Schuster Company
Upper Saddle River, NJ 07458

The publisher offers discounts on this book when ordered in bulk quantities. For more information contact:

Corporate Sales Department
Phone: 800-382-3419 Fax: 201-236-7141
E-mail: corpsales@prenhall.com
Or write:
Prentice Hall PTR
Corp. Sales Dept.
One Lake Street
Upper Saddle River, NJ 07458

Printed in the United States of America

10 9 8 7 6 5 4 3 2 1

ISBN 0-13-949124-4

Prentice-Hall International (UK) Limited, *London*
Prentice-Hall of Australia Pty. Limited, *Sydney*
Prentice-Hall Canada Inc., *Toronto*
Prentice-Hall Hispanoamericana, S.A., *Mexico*
Prentice-Hall of India Private Limited, *New Delhi*
Prentice-Hall of Japan, Inc., *Tokyo*
Simon & Schuster Asia Pte. Ltd., *Singapore*
Editora Prentice-Hall do Brasil, Ltda., *Rio de Janeiro*

This book is dedicated to our families: Pushpa Garg; Nina, Rajiv and Monica Taneja; Meena, Jeff and Adam Dorr; and Ravi and Mridu Garg.

Lois Wilkes; Joanne Wilkes; Tom, Mandy and Kylie Wilkes; and Peter Wilkes.

Vijay Garg
Joe Wilkes

Contents

Preface

During the last several decades, the world has seen phenomenal changes in the telecommunications industry. Communications that were formerly provided by wires are now provided by radio (wireless) means. Thus, wireless communication, which uncouples the telephone from its wires to the local telephone exchange, has exploded. During the early 1980s, six incompatible analog mobile systems were operational in Western Europe, which precluded interoperability between systems; thus, roaming between countries of Europe was not possible. With the growth of the European Common Market, roaming between these countries became important.

In 1985 the main governing body of the European Postal Telephone and Telegraph (PTTs)—Conference Europeenne des Postes et Telecommunications (CEPT)—set up a committee known as Groupe Special Mobile, later changed to Global System for Mobile Communications (GSM), under the auspices of its Committee on Harmonization to define a mobile system that could be introduced across Europe in the 1990s. This initiative gave the European mobile communications industry a home market of about 300 million subscribers, while at the same time providing it with a significant technical challenge.

The early years of the GSM were devoted mainly to the selection of the radio techniques for the air interface. In 1986 field trials of different candidate systems proposed for the GSM air interface were conducted in Paris. A set of criteria ranked in order of importance was established to assess these candidate systems. Some of the criteria to be met by the candidate system included

☞ Spectral efficiency

☞ Subjective voice quality

☞ Cost of mobile

☞ Hand portable feasibility

☞ Cost of base station

☞ Ability to support the new services

☞ Coexistence with existing systems

The performance of a cellular radio system depends primarily upon cochannel interference, and a given quality of voice can be achieved at much higher levels of cochannel interference when digital transmission instead of analog transmission is used for the system. After a considerable debate over the most suitable transmission mode (Frequency-Division Multiple Access [FDMA], Time-Division Multiple Access [TDMA], or Code-Division Multiple Access [CDMA]), the final decision in 1987 was to adopt TDMA for GSM. In 1989 the responsibility for generating specifications for GSM was passed from the CEPT to the newly formed European Telecommunications Standards Institute (ETSI). The specifications for GSM Phase 1 were completed by 1990 and are divided into 12 sets of recommendations covering different aspects of the GSM system. GSM Phase 1 is a self-contained version of the standard that supports only a subset of the services that were originally thought to be in GSM.

Based on these original specifications, GSM systems have been installed and mobile communications services have been offered in most continents of the world.

GSM Phase 2 is a full-fledged version of the standard. It differs from Phase 1 primarily in a number of added supplementary services. However, upon a close examination one finds that the signaling protocol Mobile Application Part (MAP) and the protocol between mobile station and the infrastructure have been modified in many areas. The GSM Phase 2+ activities are organized as a set of independent tasks so that each of them could be introduced with little or no impact on the others. So far, more than 80 tasks are identified in the Special Mobile Group (SMG). They cover aspects from radio transmission to call management. The challenge of GSM Phase 2+ is to gradually introduce important changes while maintaining upward compatibility. Interfaces, protocols, and protocol stacks in GSM are aligned with the Open System Interconnect (OSI) principles. GSM is an open architecture that provides maximum independence between network elements (such as the Base Station Controller [BSC], Home Location Register [HLR], and so on). This approach simplifies design, testing, and implementation of the system. It also favors an evolutionary growth path, since network element independence implies that modification to one network element can be made with minimum or no impact on the others. Also, a system operator has a choice of using network elements from different manufacturers.

GSM 900 has been adopted in many countries, including a major part of Europe, North Africa, Middle East, many East Asian countries, and Australia. In most of these cases, roaming agreements exist to make it possible for subscribers to travel in different parts of the world and enjoy continuity of their telecommunications services with a single number and a single bill. DCS 1800 is also being deployed in East Asia and some South American countries. PCS 1900, a derivative of GSM for North America, is planned to cover a substantial area of the United States. All these systems will also enjoy a form of roaming, referred to as Subscriber Identity Module (SIM) roaming, between them and with all other GSM-based systems. A subscriber from any of these systems could access telecommunication services by using the SIM card in a handset suitable to the network from which coverage is provided. If the subscriber has a multiband phone, then one phone could be used worldwide. This globalization is making GSM and its derivatives the leading contenders to offer digital cellular and Personal Communications Services (PCS) worldwide.

This book describes the emerging digital cellular communications and Personal Communications Networks (PCNs) being envisioned. It discusses the recent history of GSM technology that is being used to synthesize one version of PCN and delineates the alternative approaches being considered.

The primary focus of this book is to discuss the past, present, and future evolution of the technical aspects of the GSM 900 system and its derivatives such as DCS 1800 and PCS 1900. The book describes GSM Phase 1 and Phase 2 and identifies major future trends.

The book is divided in three parts. The first part covers the technical aspects of cellular communications that are used for GSM 900 and its derivatives. The second part describes the GSM system in detail. The third part of the book describes adjuncts to the GSM system and compares GSM with other cellular systems that use TDMA. Finally we examine the future trends of wireless systems and GSM.

This book can be used by telecommunication managers engaged in managing GSM cellular/PCS networks with little or no technical background in GSM technologies; by practicing communication engineers involved in the design of the GSM cellular/PCS systems; and by senior/graduate students in electrical, telecommunication, or computer engineering planning to pursue a career as a telecommunication engineer.

Since the deployment of GSM in Europe in the early 1990s, about a dozen books have been written to describe the GSM technology. Unfortunately, none of these books provide a total picture of the GSM technology and GSM networks. Some of these books focus on the radio aspects of the GSM, whereas others provide a simple high-level view of GSM networks but ignore important parameters of them. Also, most of these books are written with the assumption that a reader has been exposed to the telecommunications field and possesses adequate knowledge of fundamental disciplines such as traffic engineering and Signaling System 7 (SS7) upon which GSM is built.

We use a different approach in this book by presenting a comprehensive treatment of the subject. We start from ground zero and first introduce basic principles essential to understanding any wireless technology. Next we discuss the GSM technology in depth and provide GSM architecture; radio link operations; logical channel structure and framing; speech coding; the physical, data link, and network layers; and the message flows between network elements. Since there is a worldwide explosion in wireless data communications, we present material on the GSM solution for wireless data. Security is an important feature in digital systems; therefore we cover the security methods used in GSM without revealing details that would compromise the system. We then cover modulation and radio propagation and use the knowledge to show how a GSM system is planned. We then describe the network aspects of the GSM network and provide a step-by-step procedure to size GSM network elements. We also discuss management of the GSM network based on the Telecommunications Management Network (TMN) approach. Finally we introduce the reader to traffic engineering and SS7. To make the book fully self-sufficient, we include the comparison of various wireless systems based on TDMA technology, and the low-mobility systems adjunct to GSM called Digital Enhanced Cordless Telephone (DECT). Finally, we conclude the book by providing a chapter on the future of wireless technology with a focus on GSM.

We have written the book to satisfy the needs of the large technical community ranging from college students to practicing designers/engineers and managers. To satisfy the needs of students, we provide several numerical examples to illustrate the applications of formulas. On the other hand, we limit the use of advanced mathematical principles to make the book readable for practicing engineers and managers. For a GSM designer, we include enough design examples so that the book can be used as a reference during the early design and planning phases of GSM networks.

If this book is used as a textbook for teaching a two-semester course in wireless and GSM technology, we recommend covering chapters 1 through 8 during the first semester and chapters 9 to 15, and 20 during the second semester. For engineering managers, we recommend chapters 1–3, 5–7, 11, and 20, and for the practicing engineers/designers we suggest chapters 1–11, 14, 15, and 20. For those readers who are not exposed to the SS7 and teletraffic engineering, we recommend reading chapters 17 and 18.

Throughout the book, we present material on GSM and on other radio systems that have evolved in parallel with GSM. By comparing the various systems, the reader can understand the advantages and disadvantages of GSM compared to other systems.

In chapter 1 we describe first-generation analog cellular systems, second-generation digital cellular systems, emerging digital cellular communications, and PCNs being envisioned. Also discussed is the recent history of GSM technology.

In chapter 2 we present standards for wireless communications. We discuss the European cordless systems—CT2, DECT, and others—and discuss the GSM standard developed by ETSI. We describe Universal Personal Telecommunication (UPT) based on the use of a Personal Telecommunication Number (PTN) and concentrate on IMT-2000 and Universal Mobile Telecommunications System (UMTS) standards. We also discuss the North American standards IS-54 and IS-95 and include the Japanese standards.

Next we discuss, in chapter 3, the narrowband channelized and wideband nonchannelized systems for wireless communications. Our focus is on access technologies including FDMA, TDMA, and CDMA. We also present the concepts of the Frequency Division Duplex (FDD) and Time Division Duplex (TDD).

In chapter 4 the fundamentals of cellular communications are presented. The concept of cochannel interference for both omnidirectional and sectorized cell sites is studied. The cell splitting procedures used in cellular communications is also discussed.

In chapter 5 we present an overview of GSM as described in the ETSI's recommendations. We describe the architecture and network interfaces of a GSM system.

In chapter 6 we first describe radio link measurements in GSM and present the details of Adaptive Power Control (APC), Discontinuous Transmission (DTX), and Slow Frequency Hopping (SFH). The chapter also discusses future techniques (e.g., channel borrowing and smart antenna) that may be used in GSM to reduce interference and improve system performance.

GSM has a rich set of logical channels that are used on the radio link. We devote chapter 7 to the discussion of these logical channels. They are used to carry user information and control signaling data.

In chapter 8 we first discuss speech coding methods and attributes of speech codec. These are then followed by a brief discussion of the Linear-Prediction-based Analysis-by Synthesis (LPAS) and the ITU-T standards providing comparison of different codecs.

In chapter 9 we discuss messages that are passed on the open interfaces in a GSM network. These interfaces are: mobile station to base station, base station to MSC and MSC to HLR and Visitor Location Register (VLR). Using these messages, we construct some typical call flows used on the GSM network.

Chapter 10 describes the methods used to support data services in a GSM system. We examine the issues of interoperability with wireline data services. We then describe circuit-switched data services in GSM. Next we examine the Short Message Service (SMS) used to provide two-way paging capabilities in GSM. Finally, we examine the GSM Packet Radio Service (GPRS).

In chapter 11 we concentrate on the security in GSM. We discuss the primary mechanisms used in the GSM system to achieve security—cryptographic security algorithms, SIM cards, and authentication procedures.

In chapter 12 we study three modulation methods—Minimum Shift Keying (MSK), Gaussian Minimum Shift Keying (GMSK), and π/4 Differential Quadrature Phase Shift Keying (π/4-DQPSK). These modulation methods are used in GSM, DECT, and PWT. Many of the figures in chapter 12 were generated using Matlab 5, and useful information about the modulators can be obtained by studying the Matlab programs. Therefore we include Matlab source files used to generate the figures on the attached disk. A student version of Matlab is available from Prentice-Hall.

In chapter 13 we discuss propagation and multipath characteristics of a radio wave. The concepts of delay spread, which causes channel dispersion and intersymbol interference, are also presented.

In chapter 14 we first present the teletraffic models required in cellular/PCS network planning and design. We then provide an overview of the methods used for subscriber location management in the GSM system. We include a design example of the GSM system to illustrate a systematic procedure for determining equipment needs in the system.

In chapter 15 we present the traditional approaches to network management (NM). We then briefly introduce TMN concepts. We provide management requirements for wireless networks and focus on the platform-centered NM approaches by presenting two widely used network management approaches.

Companions to cellular systems are wireless local loop phones and extended cordless phones. Chapter 16 describes DECT (and its U.S. equivalent PWT) that is used to provide voice and data services for wireless local loop and cordless phones as adjuncts to GSM systems.

In chapter 17 we provide a brief overview of SS7 for the benefits of those of you who are not familiar with SS7.

In chapter 18 we provide definitions of the terms often used in teletraffic engineering. Several numerical examples are provided to illustrate applications for calculating MSC traffic. We present the results of Erlang B traffic engineering for 1–100 servers in appendix A. We also provide a Windows 95 and Windows NT-4 program for calculating the offered load for any number of servers from 1–130 with an arbitrary selected blocking probability.

In chapter 19 we focus on the three TDMA-based PCS/cellular systems (DAMPS-1900, DCS-1900/GSM 1900, and PDC) that have been deployed or are being deployed in North America and Japan. We compare these systems.

In chapter 20 we examine the future of wireless communications and focus on the activities under way in ETSI for the next generation of wireless systems under the banner of IMT-2000.

We would like to thank the many people who helped us prepare the material in this book. Bernard Goodwin provided his encouragement in motivating us to write the book. Professor Theodore Rappaport of Virginia Tech took us under the banner of his new series and also reviewed the manuscript. Our coworkers at Bellcore and Lucent have answered our many questions on GSM. Reed Fisher provided us with his insight on the future of wideband CDMA; Don Zelmer helped in obtaining information on the latest ETSI activities and reviewed the manuscript. Mike Loushine and Zygmond Turski provided reviews of the manuscript. Lois Wilkes read parts of the manuscript and improved our writing. We especially want to thank Peter Wilkes for preparing the list of abbreviations, and editing some of the files. We also thank Peter for his excellent work in writing the traffic engineering program for the supplied disk and the data in appendix A.

We have included a disk of software with our book. While appendix A has a set of traffic tables, you may want to determine the number of servers for an arbitrary blocking probability and offered load. The Microsoft Windows-based program WErlangB provides that capability. A setup program can be found on the disk under the traffic directory. Double-click on Setup and the program will be installed on your hard disk with an icon established on the Start menu. The program runs under Windows NT-4 and Window 95.

The matlab directory on the included floppy disk contains the Matlab files for generating the signal and spectrum plots in chapter 12. The programs were generated in Matlab 5 on a Windows platform. If we need to update our m-files, we will use The Mathworks, Inc., web page. For additional information about Matlab please contact

The MathWorks, Inc.
24 Prime Park Way
Natick, MA 01760-1500
Phone: (508) 647-7000
Fax: (508) 647-7001
E-mail: info@mathworks.com
WWW: http://www.mathworks.com

Vijay Garg
Joe Wilkes
August 1998

An Overview of Wireless Communications Systems

1.1 INTRODUCTION

During the last two decades, the world has experienced phenomenal changes in the telecommunications industry. Many of the communications that were formerly carried on wires are now provided by radio (wireless). Thus, wireless communication, which uncouples the telephone from the wires to the local telephone exchange, has exploded. During the early 1980s, six incompatible analog systems were operational in Western Europe. This resulted in mobile phones designed for one system that could not be used with another system, making it impossible to roam between countries of Europe. With the growth of the European Common Market, roaming between European countries became important.

In 1982, the Conference Europeenne des Postes et Telecommunications (CEPT) set up a committee known as Groupe Special Mobile (GSM), later known as the Global System for Mobile Communications, under the auspices of its Committee on Harmonization. Its purpose was to define a mobile system that could be introduced across Europe in the 1990s. CEPT allocated two new radio frequency bands at 900 MHz.

The GSM initiative gave the European mobile communications industry a home market of about 300 million subscribers, while at the same time provided it with a significant technical challenge. The early years of the GSM were devoted mainly to the selection of the radio techniques for air interface. In 1986 field trials of different candidate systems proposed for the GSM air interface were conducted in Paris. The committee established a rank-ordered set of criteria to assess these candidates. Some of the criteria to be met by the candidate system included

☞   Spectral efficiency
☞   Subjective voice quality
☞   Cost of mobile
☞   Hand portable feasibility
☞   Cost of base station (BS)
☞   Ability to support the new services
☞   Coexistence with existing systems

The GSM interfaces, protocols, and protocol stacks are aligned with the Open System Interconnect (OSI) principles. The open GSM architecture provides maximum independence between network elements such as the BS Controller (BSC), the Mobile Switching Center (MSC), and the Home Location Register (HLR). This approach simplifies design, testing, and implementation of the system. It additionally favors an evolutionary growth path, since network element independence implies that modification to one network element can be made with minimum or no impact on the others. A system operator therefore has a choice of using network elements from different manufacturers.

GSM 900 has been adopted in many countries, including most of Europe, North Africa, the Middle East, many East Asian countries, and Australia. In most cases, roaming agreements exist to make it possible for subscribers to travel to different parts of the world and enjoy continuity of their telecommunications services with a single number and a single bill. The adaptation of GSM at 1800 megahertz (MHz) (DCS 1800) is also spreading outside Europe to East Asia and some South American countries. PCS 1900, a derivative of GSM for North America, is planned to cover a substantial area of the United States. Initially all these systems will also enjoy a form of roaming, referred to as Subscriber Identity Module (SIM) roaming. A subscriber from any of these systems can access telecommunication services by using a SIM card in a handset suitable for the network in the visited system. Ultimately, when the subscriber has a multiband phone, then one phone could be used worldwide. This globalization is making GSM and its derivatives the leading contenders to offer digital cellular and Personal Communications Services (PCS) worldwide. When a three-band handset (900, 1800, and 1900 MHz) is available, true worldwide roaming will be possible.

In this chapter we describe first-generation analog cellular systems, second-generation digital cellular systems and emerging digital cellular communications, and Personal Communications Networks (PCNs) being envisioned by operators and manufacturers. The chapter discusses the recent history of GSM technology that is being used to synthesize one version of PCN. Although the primary focus of this chapter is on the Pan-European GSM system and its worldwide derivatives technologies, wireless technologies used in the other parts of the world are also discussed briefly.

1.2 GSM MoU

In 1987, a Memorandum of Understanding (MoU) was signed by the European governments to support the implementation of GSM networks in Europe, and the GSM MoU Association was established. Five years later, an addendum was signed that extended membership to non-European operators, and an associated membership was created for DCS 1800 operators. Presently, the nonprofit GSM MoU Association has three types of members/signatories which include telecommunications regulators, GSM 900 license holders, and DCS 1800 license holders. By spring 1998, networks in 109 countries were in operation, and the original 15 GSM MoU signatories had grown to 251 (administrations, GSM, DCS, PCS operators).

The objective of the GSM MoU Association is "the promotion and evolution of the GSM 900/DCS 1800 and PCS 1900 system and GSM platform." The association develops the operational and commercial procedures necessary for international roaming. In addition, the specifications for exchange of billing data of roamers are elaborated. The association organizes the exchange of experience and know-how. It has a cooperation agreement with the European Telecommunications Standards Institute (ETSI) to ensure the efficient production of future standards.

The present GSM committee structure is shown in Figure 1.1.

1.3 GSM IN NORTH AMERICA

In the past few years, several North American companies have become GSM MoU signatories, including American Personal Communications/Sprint Spectrum, BellSouth Mobility DCS, Omnipoint Communications Inc., Pacific Bell Mobile Services, Powertel Inc., Microcell Telecommunications Inc., and Western Wireless Corporation. These companies participate in the meetings of the MoU Association and ETSI Technical Committee (TC) Special Mobile Group (SMG). During the September 1996 plenary meeting of the GSM MoU, the North American interest group decided that PCS 1900 should be called "GSM North America (GSM-NA)."

By the summer of 1998, there were about 2 million customers in 1500 U.S. markets that used GSM digital services. GSM networks were operational in 41 U.S. states. Standardized roaming agreements between U.S. service pro-

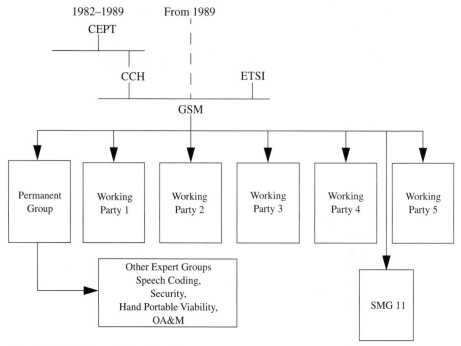

Fig. 1.1 GSM Committee Structure

viders were created and operational roaming was possible. An SS7 gateway will be installed for international roaming.

GSM-NA supports the recommendations of the SMG speech strategy experts group to focus on a multirate codec to avoid unnecessary proliferation of voice coders. The aim is to have a commercially available multirate (AMR) codec by the year 2000.

1.4 GSM MARKETS

Most of the European city areas are covered by GSM. By the fall of 1997, rural coverage often exceeded 90 percent. It is estimated that, worldwide, around 80 million subscribers are connected to GSM, and new subscribers are signing up for GSM services at a rate of about 10,000 per day. GSM is a market success based on a standardization success. Specifying a mobile telecommunications platform was completed in Europe earlier and in a more comprehensive manner than in other regions of the world. The availability of a type approval procedure, agreed between relevant parties, has also been an important factor. Based on the information available the market share of different technologies is as shown in Table 1.1.

Table 1.1 Worldwide Market Shares for Digital Cellular Mobile Systems (1997–1998)

Standards	Technologies	Users (million)
GSM (GSM900, DCS1800, PCS1900)	Advanced TDMA	80 (1998)
Public Digital Cellular (PDC) (Japan)	Basic TDMA	9 (1997)
IS 54 (800, 1900 MHz)	Basic TDMA	9.2 (1997)
IS 95 (800, 1900 MHz)	Narrowband CDMA	8 (1998)

1.5 OSI MODEL

In recent years, the International Standards Organization (ISO) has developed a reference model for data communications. The OSI reference model is used by many computer systems for computer-to-computer communications. In the model, seven layers are defined to segment different aspects and needs for communications from each other so that the communications can be conducted in an orderly fashion. In this section, we will describe the OSI model in sufficient detail so that the protocols and messages in subsequent chapters can be understood.

Each of the seven layers in the model communicates with its peer layer at the distant end and with the local layers immediately above and below it. The protocols at each layer define how peer-to-peer communications take place by defining message sets and state diagrams. For example, the layer 5 software in one computer communicates with the corresponding layer 5 software in another computer. The software in the two computers might be implemented in two different languages with two different operating systems and two completely different host computers. Thus, the layer 5 software that operates on one computer will not necessarily operate on another computer. But as long as both layer 5 software packages agree on how they will meet the OSI specifications, they will be able to communicate with each other. The model permits computer systems from widely different manufacturers to communicate with each other. For GSM phones, the messaging is done within the OSI model and allows phones from any manufacturer to communicate with networks from any other manufacturer.

Figure 1.2 shows the OSI reference model for communications between computer systems. In the figure, two protocol stacks are shown. One stack is for signaling, and the other stack is for voice or data communications. With the reference model, each layer communicates with the layer immediately above and below it and with its peer layer at the other computer. With a properly designed system, the software at one layer can be replaced without affecting the other layers. Similarly, the software/hardware at the physical layer can be replaced without affecting the layers above it. In the following sections, we discuss each layer in more detail.

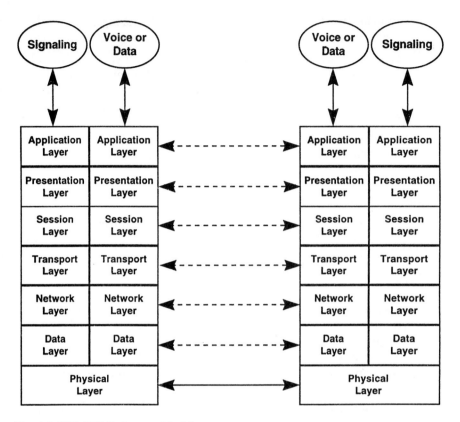

Fig. 1.2 ISO OSI Reference Model

The seven layers of the OSI model are

Layer 1—Physical layer. This layer describes the voltages or waveform for a bit (1 and 0), the time duration of a bit, the pin connections and type of connector for baseband systems or the frequencies used for radio systems, the handshaking to start and stop a connection, and whether the connection is one way or two way.

Layer 2—Link layer (Data Link). This layer converts bits into frames of data. Methods must be determined for obtaining bit sync and frame sync of the frames, preventing the data from causing the false transmission of frame sync, and retransmitting data when errors occur or when frames are lost or duplicated. Buffers must be designed to cope with fast and slow transmitters and receivers.

Layer 3—Network layer. This layer passes packets of information between two different end points. This layer is often designed to let layer 4 see an error-free channel. Unfortunately this is not usually true. Thus, layer 4 must often also cope with errors. At this layer billing and routing information for packets must be done. Buffering must be done at this layer to prevent too many packets from causing network congestion.

Layer 4—Transport layer. The transport layer is the last layer to do error correction. All higher layers assume the layer below provides a perfect connection. Thus, the goal of this layer is to provide an error-free channel for the higher layers. If layer 3 is free of errors, this job is easy. If layer 3 has errors, the transport protocol must allow for retransmission of data, error detection, and correction. This layer will set up and tear down calls to another host (addressing information is needed). It will also multiplex data from multiple processes in the host. Like other layers, it too must buffer data.

Layer 5—Session layer. This layer allows user or presentation layer processes to communicate. Events that occur during the layer are log-on messages and log-on IDs and passwords. Exchange of communications parameters occurs here, such as baud rate and full/half duplex. Grouping of messages and automatic requests for a new connection can also occur here.

Layer 6—Presentation layer. This layer copes with things like protocol conversion, terminal type, encryption, definition of primitives, message compression, and file format conversion. Layer 6 software is sometimes combined with layer 7 software. Sometimes the work done by the application layer is minimal, and a null application layer is implemented.

Layer 7—Application layer. Here the definition of what goes on is up to the end user.

1.6 First-Generation Analog Cellular Systems

In the United States, the Federal Communications Commission (FCC) has allocated a 50-MHz spectrum in the bands 824–849 MHz and 869–894 MHz to cellular mobile radio. In a given geographical licensing region, two service providers are assigned. Each of the service providers controls 25 MHz of spectrum. The A and B bands are allocated to nonwireline and wireline service providers, respectively. The total spectrum is divided into 832 radio frequency (RF) channels, each 30 kilohertz (kHz) wide. Out of 832 RF channels, 416 channels are assigned to the A-band service provider and the remaining 416 to the B-band service provider. Each service provider uses 21 channels for control and signaling, leaving 395 channels to carry users' traffic. FM of 8 kHz deviation is used for speech, and the signaling channels use binary Frequency Shift Keying (FSK) with a 10 kilobit per second (kbps) rate. The Advanced Mobile Telephone System (AMPS) has been available to the public since 1983. AMPS is also used in Canada, Central and South America, Australia, and certain countries of Asia.

In Europe, several first-generation systems similar to AMPS have been deployed. These include the Total Access Communications System (TACS) in the United Kingdom, Italy, Spain, Austria, and Ireland; the Nordic Mobile Telephone (NMT) in many Scandinavian countries; C-450 in Germany and Portugal; RadioCom 2000 in France; and the Radio Telephone Mobile System (RTMS) in Italy. Those systems use FM for speech, FSK for signaling, and channel spacing of 25 kHz (TACS, NMT-450, RTMS); 10 kHz (C-450); and 12.5 kHz (NMT-900, RadioCom 2000). Handover decisions are made on the basis of the power received at the BSs surrounding the MS. C-450 is an exception and uses round-trip delay measurements for handover decisions.

In Japan, a total of 56-MHz spectrum was allocated for the analog cellular system (860–885/915–940 MHz and 843–846/898–901 MHz). The Nippon Telephone and Telegram (NTT) analog system began its operation in Tokyo in 1979. The frequency was 925–940 MHz (mobile station [MS] to BS) paired with 870–885 MHz (BS to MS). The channel spacing was 25 kHz, giving a total of 600 duplex channels. The control channel signaling rate was 0.6 kbps. In 1988 a high capacity system was introduced in the same band, with an increased control channel signaling of 2.4 kbps and reduced channel spacing of 12.5 kHz. The number of RF channels was further increased by frequency interleaving (channel center frequencies 6.25 kHz apart), resulting in an overall increase in the number of channels to 2400. In addition, an associated control channel below the voice band with a rate of 100 bps was introduced to allow signaling without interrupting voice communications. To improve transmission quality and frequency reuse pattern, diversity reception was introduced into both the BS and MS. The mobile terminal of the high-capacity system is dual mode and can access the initial (25 kHz) system. In 1987 cellular radio was deregulated in Japan. Two new operators were introduced—IDO and DDI. IDO began operation of the NTT high-capacity system in December 1988 in the bands 860–863.5/915–918.5 MHz. The DDI cellular group provides coverage outside the metropolitan areas and uses the JTACS/NTACS system (based on the European TACS system) in the bands 860–870/915–925 MHz and 843–846/898–901 MHz. Following the DDI group, IDO has also introduced NTACS in the bands 843–846/898–901 MHz and 863.5–867/918.5–922 MHz. IDO and DDI have joined together to form a partnership to provide nationwide service by introducing roaming capabilities between the two systems.

1.7 SECOND-GENERATION DIGITAL CELLULAR SYSTEMS

The development of low-rate digital speech codecs and the continuous increase in the device density of integrated circuits have made completely digital second-generation cellular systems viable. The performance of a cellular radio system depends primarily upon cochannel interference, and a given voice quality can be achieved at a much higher level of cochannel interference when digital transmission is used. Digital cellular systems use either Time-Division

Multiple Access (TDMA) or Code-Division Multiple Access (CDMA) as alternatives to Frequency-Division Multiple Access (FDMA). With TDMA, each radio channel is partitioned into multiple time slots, and each user is assigned a specific frequency/time slot combination. Thus, only a single MS in a given cell uses a given frequency at any particular time. With CDMA, a frequency channel is used simultaneously by multiple mobiles in a given cell, and the signals are distinguished by spreading them with different codes. One obvious advantage of both TDMA and CDMA is the sharing of the radio hardware in the BS among multiple users.

A second factor that has encouraged the use of digital transmission for GSM is that the telecommunications industry worldwide has been migrating to digital methods; thus, digital transmission is perceived to be a modern approach. After considerable debate over the most suitable transmission mode (FDMA, TDMA, and CDMA), in 1987 TDMA was adopted for GSM. In 1989 the responsibility for generating specifications for GSM was moved from CEPT to the newly formed ETSI. The specifications for GSM Phase 1 were completed by 1990. They are divided into 12 sets of recommendations and cover different aspects of the GSM system. GSM Phase 1 is a self-contained version of the standard that supports only a subset of the services that were originally thought to be in GSM. To reflect the worldwide deployment of GSM, the name was changed to Global System for Mobile Communications.

GSM Phase 2 is a full-fledged version of the standard and differs from Phase 1 primarily in a number of additional supplementary services. However, upon close examination one finds that the signaling protocol, the Mobile Application Part (MAP), and the protocol between MS and the infrastructure have been modified in many areas. GSM Phase 2+ activities are organized as a set of independent tasks, so that each of them could be introduced with little or no impact on the others (e.g., the SIM tool kit). So far, more than 80 tasks have been identified by the SMG committee. They cover aspects from radio transmission to call management. The real challenge of GSM Phase 2+ is to introduce important changes while maintaining upward compatibility.

GSM uses eight channels per carrier with a gross data rate of 22.8 kbps (a net rate of 13 kbps) in the full rate channel and a frame of 4.6 milliseconds (ms) duration. Each user transmits in every eighth time slot (of duration 0.575 ms) and receives in a corresponding time slot. With this approach, the BS requires only one transceiver for eight channels. Also, transmit/receive time slot staggering allows a relaxation of duplex filter requirements for the MS. The intermittent activity of the mobile transceiver also provides the opportunity for the MS to measure the strength of the signals from the neighboring BSs. These measurements are reported to the serving BS and are used for handover decisions.

A complex TDMA frame structure is used in GSM to accommodate the needs of the radio channel and the various signaling/control requirements. Each time slot transmits 114 useful bits of information in each frame. In the center of the time slot, 26 bits are assigned as a training sequence for the equalizer in the receiver. The equalizer creates a model of the radio channel and uses the model to counteract the effects of multipath time dispersion. At the end of each time slot, a guard time of 8.25 bits is provided to allow for uncertainties in the arrival time of TDMA time slots at the BS from MSs at varying distances.

Control information is mapped onto time slot 0 within the frame, and a 51-frame multiframe is developed to further multiplex various control information channels. Frequency correction and synchronization data are delivered with the time slot 0 structure at periodic intervals. Traffic channels in the GSM architecture may be organized as full rate or half rate (16 per TDMA frame). It is possible to mix both full- and half-rate channels within a frame, although a given time slot can be used only in either full- or half-rate mode at any particular time.

The data rate over the radio channel is 270 kbps. Gaussian Minimum Shift Keying (GMSK) modulation is used with a bandwidth (B) times bit period (T) equal to 0.3 (BT = 0.3) and channel spacing of 200 kHz. See chapter 12 for more information on GMSK. Frequency hopping is an optional network capability in GSM. Hopping occurs at the TDMA frame rate of about 217 hops/ s with the hop sequence being communicated to the MS during call setup and during handover. Frequency hopping counteracts multipath fading in addition to that already achieved with channel coding, interleaving, and antenna diversity. Also, frequency hopping provides a better statistical distribution of interference; thus its use enables very efficient frequency reuse within the cellular environment.

In the United States, the Telecommunications Industry Association (TIA) adopted the IS-54 TDMA standard to meet the growing need for increased cellular capacity in high-density areas. IS-54 retains the 30-kHz channel spacing of AMPS to facilitate evolution from analog to digital systems. Each frequency channel provides a raw RF bit rate of 48.6 kbps. This is achieved by using $\pi/4$ Differential Quadrature Phase Shift Keying ($\pi/4$-DQPSK) modulation at a 24.3-kilosymbols/s channel rate. The channel is divided into six time slots, two of which are assigned to each user. A 7.95-kbps vector sum excited linear prediction (VSELP) speech codec is used to digitize the voice. Thus, each 30-kHz frequency pair serves three users simultaneously. IS-54 provides three times the capacity of AMPS. The IS-54 standard provides for an adaptive equalizer to mitigate the intersymbol interference caused by large delay spreads, but due to the relatively low channel rate of 24.3 kilosymbols/s, the equalizer will be unnecessary in many situations.

Since IS-54 systems must operate in the same spectrum used by the existing systems, it provides for both analog (AMPS) and digital operation. This is necessary to accommodate roaming subscribers, given a large embedded base of AMPS equipment. Initially, the IS-54 standard used the AMPS control channel with 10-kbps Manchester-encoded FSK. IS-136, the new version of IS-54, includes a digital control channel (DCCH) that uses the 48.6-kbps modem. With an increased signaling rate, the DCCH offers capabilities

such as point-to-point short messaging, broadcast messaging, group addressing, and private user groups. IS-54 equipment has already been deployed and
is operational in a majority of the top ten cellular markets in the United
States. IS-136 equipment is being deployed for PCS in several major cities.

The IS-95 standard is based on CDMA. The basic user channel rate is 9.6
kbps which is spread to a channel chip rate of 1.2288 million chips per second
(Mcps), giving a total spreading factor of 128. The spreading processes used on
the forward (BS-to-MS) and reverse (MS-to-BS) links are different. On the forward link, the user data stream is encoded using a rate 1/2 convolutional code,
interleaved, and spread by one of the 64 orthogonal Walsh functions. Each MS
in a given cell is assigned a different Walsh function to provide (under ideal
conditions) perfect separation among signals from different users. To reduce
interference between mobiles that use the same Walsh function in different
cells and to provide the desired widespread spectral characteristics, all signals
in a particular cell are scrambled using a pseudorandom sequence of length
2^{15} chips. Orthogonality among users within a cell is preserved because their
signals are scrambled identically. A pilot channel (code) is provided on the forward link for channel estimation. The pilot channel is transmitted at higher
power than the user channels.

On the reverse link, a different spreading strategy is used because each
received signal arrives at the BS via a different propagation channel. The user
data stream is first convolutionally encoded at a rate of 1/3. After interleaving,
each block of six encoded symbols is mapped to one of the 64 orthogonal Walsh
functions. A final fourfold spreading, giving a rate of 1.2288 Mcps, is obtained
by spreading the resulting 307.2-kilochip per second (kcps) stream by user-
specific and base-specific codes of periods $2^{42} - 1$ and 2^{15} chips, respectively.
The rate 1/3 coding and mapping onto Walsh functions results in a greater tolerance for interference than would be realized from traditional spreading
using a repetition code. The reverse link tightly controls the mobile's transmit
power to avoid the "near-far" problem. The near-far problem is caused from
the different fading, shadowing, and path loss situations experienced by the
different signals arriving at the same BS.

RAKE receivers are used to resolve and combine multipath components
to reduce fading amplitude at both the BS and MS. The receiver architecture
also is used to provide BS diversity during "soft handover," whereby an MS
making the transition between cells maintains links with both BSs during the
transition. The mobile's receiver combines the signals from two or more BSs in
the same manner as it would combine signals associated with different multipath components. The variable rate speech codec, power control, reduced fade
margin, and forward error correction all contribute to the reduction of the
required RF transmit power. IS-95 is a dual-mode standard designed for the
existing North American cellular bands. IS-95 terminals can operate either in
CDMA mode or the AMPS mode.

In Japan in 1989 a development study of digital cellular systems with a
common air interface was initiated under the auspices of the Ministry of Posts

and Telecommunications (MPT). The new digital system, called Personal Digital Cellular (PDC), was established in 1991. The PDC system is also based on TDMA, with three time slots multiplexed onto each carrier, similar to IS-54. The channel spacing is 25 kHz with interleaving to facilitate migration from analog to digital. The RF signaling rate is 42 kbps and the modulation is $\pi/4$-DQPSK. A key feature of PDC is the MS-assisted handover, which facilitates the use of small cells for efficient frequency usage. The full-rate VSELP speech codec operates at 6.7 kbps (11.2 kbps with error correction). A 5.6-kbps Code-Excited Linear Prediction (CELP) half-rate codec has also been standardized and will soon be introduced. A total of 80-MHz spectrum is allocated to PDC. The frequency bands are 810–826 MHz paired with 940–956 MHz, and 1429–1453 MHz paired with 1477–1501 MHz. With antenna diversity, the required signal-to-interference (S/I) ratio is reduced, giving a reuse factor of 4. Group 3 fax (2.4 kbps), as well as 4.8-kbps modem transmission with MNP class 4, are supported using an adaptor to provide the required transmission quality.

Tables 1.2 and 1.3 provide a summary of the first- and second-generation cellular systems.

Table 1.2 First-Generation Analog Cellular Systems

Standard	Region	Frequency (MHz)	Channel Spacing (kHz)	No. of Channels	Modulation	Data Rate (kbps)
AMPS	USA	824–849 869–894	30	832	FM	10
TACS	Europe	890–915 935–960	25	1000	FM	8
ETACS	UK	872–905 917–950	25	1240	FM	8
NMT 450	Europe	453–457.5 463–467.5	25	180	FM	1.2
NMT 900	Europe	890–915 935–960	12.5	1999	FM	1.2
C-450	Germany Portugal	450–455.74 460–465.74	10	573	FM	5.28
RTMS	Italy	450–455 460–465	25	200	FM	—
Radiocom 2000	France	414.8–418 424.8–428	12.5	256	FM	—
NTT	Japan	870–885 925–940	25	600	FM	0.3
JTACS/ NTACS	Japan	860–870 915–925	25	400	FM	8.0

Table 1.3 Second-Generation Cellular and Cordless Systems

System	IS-54	GSM	IS-95	CT-2	CT-3 DCT-90	DECT
Country	USA	Europe	USA	Europe, Asia	Sweden	Europe
Access Technology	TDMA/ FDMA	TDMA/ FDMA	CDMA/ FDMA (DS)	FDMA	TDMA/ FDMA	TDMA/ FDMA
Primary Use	cellular	cellular	cellular	cordless	cordless	cordless/ cellular
Frequency Band BS(MHz) MS (MHz)	869–894 824–849	935–960 890–915	869–894 824–849	864–868	862–866	1800–1900
Duplexing	FDD*	FDD	FDD	TDD†	TDD	TDD
RF Channel Spacing (kHz)	30	200	1250	100	1000	1728
Modulation	$\pi/4$ DQPSK	GMSK	BPSK/ QPSK	GFSK	GFSK	GFSK
Handset Power, Maximum/ Average in megawatts (mW)	600/200	1000/125	600	10/5	80/5	250/10
Frequency Assignment	Fixed	Fixed	Fixed	Dynamic	Dynamic	Dynamic
Power Control MS BS	Y Y	Y Y	Y Y	N N	N N	N N
Speech Coding	VSELP	RPE-LTP	QCELP	ADPCM	ADPCM	ADPCM
Speech rate (kbps)	7.95	13	8 (variable rate)	32	32	32
Speech Channel per RF Channel	3	8	13-40	1	8	12
Channel Bit Rate (kbps)	48.6	270.833	1228.8	72	640	1152
Channel Coding	1/2 rate convolutional	1/2 rate convolutional	1/2 rate forward, 1/3 rate reverse, CRC	None	CRC	CRC
Frame Duration (ms)	40	4.615	20	2	16	10

*Frequency Division Duplex.

†Time-Division Duplex.

1.8 THIRD-GENERATION SYSTEMS

The focus of third-generation mobile systems is on economical networks and radio protocols to deliver seamless services for use across many networks (wireless and wireline). In Europe, three related network platforms are currently the subject of intensive research. These are Universal Mobile Telecommunications Systems (UMTS, now known as IMT-2000), Mobile Broadband Systems (MBS), and Wireless Local Area Network (WLAN).

One major distinction of IMT-2000, relative to second-generation systems, is the hierarchical cell structure. Second-generation systems use a one-layer cell structure and employ frequency reuse within adjacent cells. Thus, each single cell manages its own radio zone and radio circuit control within the mobile network, including traffic management and handover procedures. The traffic supported in each cell is fixed because of frequency limitations and little flexibility of radio transmission, which is mainly optimized for voice and low data rate. Increasing traffic leads to costly cellular reconfiguration such as cell splitting and cell sectorization. Hierarchical cell structures can be designed to support a wide range of multimedia broadband services within the various cell layers by using advanced transmission and protocol technologies that span several cells.

The multilayer cell structure in UMTS aims to overcome these problems by overlaying picocells and microcells over the wide coverage area macrocell structure. Global/satellite cells can also be used to provide area coverage where macrocell constellations are not economical to deploy or support long-distance traffic.

In picocells, with low user mobility and smaller delay spread, high bit rates and high traffic density can be supported with low complexity, whereas in larger macrocells, only low bit rates and traffic load can be supported because of the higher user mobility and higher delay spread. The users will expect common services across wireline and wireless networks. Freedom of location and means of access will be facilitated by smart cards to allow customers to register on different terminals with varying capabilities (speech, multimedia, data, short messaging).

The choice of the radio link protocol to support a multiple access scheme is a critical issue and must consider spectral efficiency, the ever increasing market demand for mobile communications, and the scarce radio spectrum resource. A comparative assessment of several different schemes has been carried in the framework of the Research in Advanced Communications Equipments (RACE) program. Two solutions have emerged, one using a wideband CDMA approach for outdoor high-mobility communications and one using a hybrid CDMA/TDMA/FDMA approach for indoor and low-mobility communications.

1.9 SUMMARY

This chapter included the development of the GSM system in Europe. We presented functions of the GSM MoU Association and traced its growth from 15 to 208 signatories, in addition to discussing the worldwide GSM market and comparing it with other digital systems. We had a brief discussion of the OSI seven-layer model to familiarize those of you who are not exposed to the OSI concept in order to facilitate the understanding of subsequent chapters in this book. The chapter covered the first-generation analog and second-generation digital cellular systems, along with derivatives of the GSM both in Europe and North America. We concluded with third-generation digital cellular/PCS systems which are being explored to replace second-generation and second-generation-plus systems.

1.10 REFERENCES

1. Garg, V. K., and Wilkes, J. E., *Wireless and Personal Communications Systems*, Prentice Hall, Upper Saddle River, NJ, 1996.

2. Balston, D. M., and Macario, R. C. V., *Cellular Radio Systems*, Artech House, Norwood, MA, 1993.

3. Balston, D. M., "The Pan-European Cellular Technology," IEE Conference Publication, 1988.

4. Marley, N., "GSM and PCN Systems and Equipment," JRC Conference, Harrogate, 1991.

5. Mouly, M., and Pautet, M. B., *The GSM System for Mobile Communications*, Mouly and Pautet, Palaiseau, France, 1992.

6. Dasilva, J. S., Ikonomou, D., and Erben, H., "European R&D Programs on Third-Generation Mobile Communications Systems," *IEEE Personal Communications* 4 (1), February 1997, pp. 46–52.

7. "The European Path Towards UMTS," *IEEE Personal Communications*, special issue, February 1995.

8. Rapeli, J., "UMTS: Targets, System Concepts, and Standardization in a Global Framework," *IEEE Personal Communications*, February 1995.

Standards for Wireless Communications Systems

2.1 INTRODUCTION

GSM was originally designed to operate only in the bands around 900 MHz. In early 1989 the UK Department of Trade and Industry began an initiative that finally led to the assignment of 150 MHz in the 1.8-gigahertz (GHz) band for PCN in Europe. GSM was selected as the standard for PCN. This system is called Digital Cellular System 1800 (DCS 1800). Its definition meant translating the GSM specifications to a new band and modifying some parts for accommodating overlays of micro- and macrocells. Cellular and PCN are certainly the most prominent applications, but GSM has been extended to include "group calls" and "push to talk" for private mobile radio (PMR) applications.

International standards bodies are currently engaged in defining the third-generation mobile systems, which are expected to start services around year 2002. In Europe the ETSI is defining UMTS. Parallel standardization activities on a worldwide level are undertaken within the International Telecommunications Union (ITU) under the banner of International Mobile Telecommunications in the year 2000 (IMT-2000).

Section 2.2 presents the European cordless systems. Section 2.3 discusses the GSM standard developed by ETSI. In section 2.4 we present the

Universal Personal Telecommunication (UPT) based on the use of a Personal Telecommunication Number (PTN). Section 2.5 and 2.6 are devoted to IMT-2000 and UMTS standards. In section 2.7 we discuss the North American standards IS-54 and IS-95. Section 2.8 includes the Japanese standards and presents the PDC and Personal Handy phone System (PHS).

2.2 CORDLESS SYSTEMS

Cordless systems offer spatially limited terminal mobility. In 1985 the CEPT initiated the standardization of second-generation digital cordless systems (Cordless Telephone 2 [CT-2]). The first-generation analog CT-1 is a simple system with one BS connecting one mobile device to the fixed public switching telephone network (PSTN) on a fixed frequency. In residential areas, CT-1 phones offer wireless access to the PSTN with restricted mobility and low speech quality.

The telepoint service based on CT-2 was introduced in the United Kingdom in 1988. It allowed cordless access to PSTN in a certain area around a fixed BS. This service offers limited mobility (moving to a dedicated area and using one's own phone). In parallel, CEPT in 1988 introduced a new European system in a different frequency range. This resulted in the Digital Enhanced Cordless Telecommunication (DECT) standard in 1992. This system focuses on applications other than telepoint by providing wireless Private Automatic Branch Exchange (PABX) service with high speech quality and local loop replacement, as well as cordless data services for wireless Local Area Network (LAN) applications with large data throughputs.

The DECT standard uses forty 100-kHz channels in the 900-MHz band in an FDMA/Time-Division Duplex (TDD) mode. DECT systems also work at 1880–1900 MHz to allow a high traffic density and use 10 carrier frequencies which are subdivided into frames with 12 time slots. The radio interface is based on TDMA/TDD/Multiple Carrier (MC) media access, realized as a real-time dynamic channel allocation. A terminal has access to all frequency/time slot combinations. When a channel is required, the channel with the least interference is assigned to the handset. DECT has a typical cell radius in buildings of 20–100 meters (m) and outdoors of up to 300 m. It also allows handover between BSs. Transmissions in CT-2 are limited to one 32-kbps channel, whereas DECT explicitly supports data application by enabling the aggregation of up to twelve 32-kbps time slots for one connection. A seamless handover from one BS to another is based on channel measurements and initiated by the mobile device. The geographical extension of a DECT network is usually restricted to a limited area due to the expensive infrastructure required to cover a wider area. Within the covered region, DECT allows full subscriber mobility.

Most DECT networks are administered by one radio controller, which reduces the management complexity. All subscriber locations are kept in one central database. The physical limitations of such a cordless network therefore significantly simplifies location tracking. If a more complex network topology with multiple radio controllers is necessary to provide cordless mobility to a wider area, standards based on intelligent networking (IN) may be used to manage station mobility.

The DECT MSs periodically perform channel measurements, and channel usage is controlled based on these measurements. One 32-kbps channel is available for each phone call. Adaptive Differential Pulse Code Modulation (ADPCM) at 32 kbps is used for speech coding. While this rate is one half of the usual 64-kbps PCM used in the wireline network, it causes no perceptible degradation in speech quality.

DECT offers many services for business use, public access, residential use, and local loop replacement. It has also been designed to satisfy the demand for data communications, but the system is not widely used for this application.

The DECT standard defines OSI-compliant protocols for interworking with Integrated Services of Digital Network (ISDN) and GSM. For DECT/ISDN interworking, two reference configurations have been defined: with ISDN terminating either in the DECT fixed subsystem or in the DECT intermediate system and the ISDN S interface realized in the portable device.

The DECT/GSM profile describes the necessary air interface functions to connect DECT to a GSM network. In this case, the DECT system may benefit from the wide area mobility of GSM, and GSM features may be accessed with high-quality DECT offers. Other solutions are under development for a device with dual interfaces (one for DECT and the other for GSM). Thus, offering local mobility with high quality and wide area mobility with lower quality based on the GSM network can be achieved.

There are few problems with a DECT network regarding the compatibility with other networks. The DECT standard specifies only the lower layers of communication system. Its architecture is closely related to the lower layers of the OSI reference model; higher-layer functionality is not required. Once a link has been established, DECT's data link protocol is in charge of maintaining it, even if the physical resources have to be changed during the dialogue (e.g., handover). Routing calls to their destinations is provided by the DECT network. All functionality of data link layer and network layer is required only to establish and maintain an association between controller and mobile device; this is DECT's signaling functionality. For data transmission (e.g., a connection of a mobile terminal to an OSI network using a DECT system), the interface of the medium access layer is used. If, for example, link-level data integrity is required (e.g., CRC for transmitted data), this is achieved on top of the DECT protocol stack. In this case, a protocol like a High-Level Data Link Control (HDLC) could be used to achieve OSI conformance. To guarantee secure wireless access, DECT needs additional functionality in the network layer. For more information on DECT see chapter 16.

2.3 GSM

GSM Phase 1 (1991–1994) is a self-contained version of the standard which supports only a subset of services considered to be in GSM. Phase 1 is the basis for the first commercially operated GSM system. Phase 2 (1994–1995) is the full-fledged version of the standard. Although Phase 2 basically differs from Phase 1 in number of supplementary services offered, a close examination suggests that major signaling protocols such as application protocols between Mobile Switching Center (MSC) and databases (i.e., MAP) and the protocols between the MS and the infrastructure have been modified in several areas. The reason behind these changes is to achieve upward compatibility between GSM Phase 1 and Phase 2.

The Phase 2+ (1995 on) activities are organized as a set of independent tasks, so that each of them could be introduced with little or no impact on the others. To date more than 80 tasks are included in the SMG committee, covering aspects from radio transmission to call management. The challenge of Phase 2+ is to introduce important changes gradually while maintaining upward compatibility. The Phase 2+ program is devoted to

☞ **Speech.** The improvement of spectral efficiency aims at the search of a speech encoding/decoding scheme that provides a good compromise between the cost of the system and the quality of the service to the users. The GSM system can accept different speech encoding algorithms. Mechanisms exist in the signaling protocol for necessary negotiations between MS and the infrastructure. The half-rate speech codec (5.6 kbps) has been standardized. The quality of speech in the GSM half-rate speech codec is reasonably comparable to the GSM full-rate codec in most of the conditions except it shows some weaknesses in case of a communication between two GSM users. A task in Phase 2+ is aimed at mitigating this shortcoming by transporting GSM-encoded speech from end to end and using the Tandem-Free Operation (TFO). Other evolutionary ideas for speech coding have been introduced in a new full-rate (16-kbps) codec, the Enhanced Full-Rate Codec (EFRC), to achieve wireline-quality telephony. The U.S. (PCS-1900) version of GSM also supports the enhanced full-rate speech codec.

☞ **Group Call and Related Services.** In 1992 GSM was selected for future standardized means of communications between trains and ground-based staff. Several features are required to fulfill the requirements for this new service. The feature to support high speed is already identified as a task in SMG. Other new services are associated with PMR, including the support of group calls (i.e., calls involving several users) on a push-to-talk basis and sharing radio channels for efficiency. In a group call, a single speaker is heard by all the participants of the group in the service area. The service area includes several cells. A given group call is supported in each cell by a single traffic channel, comparable to the one used in a point-to-point speech call. In the downlink (BS to MS) direction, the speech of the speaker is broadcast in each cell on a group channel. The right to talk in the uplink (MS to BS) direction is managed by exchanging signaling messages. The infrastructure guarantees that only a single speaker is allowed in the uplink direction. The proposed technical solution uses elementary mechanisms already present in the system and supported by MSs.

☞ **General Packet Radio Service (GPRS).** This service is a packet-switched data service with a capacity around the raw data rate of 12 kbps on a traffic channel. It will support broadcast, multicast, and point-to-point transmissions. In contrast to the other data services (except SMS), GPRS will not permanently allocate separate resources on the radio interface or the GSM fixed network, respectively. This reduces the waste of bandwidth when bursty communications are supported over a continuous circuit.

☞ **Short Message Service (SMS).** SMS offers a connectionless packet-oriented data service for small traffic requirements. This service allows the point-to-point transport of short messages to and from MSs, as well as a cell broadcast service for the transmission of the same information to many MSs. SS7 is used for the transport of short messages. SMS capacity is limited to 640 b/s with a maximum message size of 160 bytes. SMS was not designed as a general-purpose packet-switched data service. Due to the signaling overhead and capacity of the channels of the radio interface used for short message transmission, SMS was intended to offer an extended paging service. Recently, SMS has been used as an additional signaling means to indicate to the user other waiting events such as voice mail messages, e-mail messages, and fax. Several Phase 2+ work items are aimed at making SMS more versatile.

2.4 UPT

UPT is based on the use of a PTN to establish a dynamic association between a user and a terminal. The terminal can be either wireline or mobile terminal, leading to PCS in the latter case. A user may initiate an association on any network terminal by entering his or her PTN and Personal Identification Number (PIN). The *in call registration service* enables receiving of incoming calls at the terminal. Similarly, the *outgoing UPT call service* can be used to make phone calls from any terminal charged to the UPT account. UPT is currently being standardized within ITU-T and is closely related to the emerging technology of IN. The major aspect of INs is the separation between service control and call control, supporting for example the address transformations required to implement UPT services.

2.5 IMT-2000

IMT-2000 standardization work is taking place in ITU-T Study Group 11. IMT-2000 is intended to provide telecommunication services to mobile and stationary users via a wireless link, covering wide ranges of user sectors (e.g., public, private, business, residential) and accommodating different types of user equipment (e.g., personal pocket terminals, vehicle-mounted terminals, special mobile terminals, standard PSTN/ISDN terminal equipment connected to MS). One of the service objectives of IMT-2000 is to provide multimedia services. Therefore, the requirements for network functions in IMT-2000 must include the support for multimedia services.

Although the primary focus of IMT-2000 is on public access, provisioning of IMT-2000 services in private networks is also being considered (e.g., the interworking of a mobile PBX or LAN on a ship or train with the public networks or use of personal pocket stations as extensions to a PBX). Public radio access to a PBX is also being envisioned for hotels and hospitals. IMT-2000 radio interfaces will be applied to fixed services in all types of environments including urban, rural, and remote. The aim of IMT-2000 is to allow for small and simple start-up systems to grow in capacity and functionality as needed. It will also be possible to use an IMT-2000 radio connection for a residential cordless telephone or as a replacement for the local loop. Satellite access is considered as an integral part of IMT-2000.

A number of different radio environments are involved, from indoor picocells, to large outdoor terrestrial cells, to satellite coverage. A major focus of the ITU standard on IMT-2000 is to have maximum commonality between various radio interfaces to simplify the task of developing multimode mobile terminals to cover more than one radio environment. A unified support to the various radio interfaces by the backbone network is important. In IMT-2000, functions dependent on radio access technology are identified and separated from functions that are not dependent on radio access technology. Therefore, an attempt is made to define the network independently from the radio access technologies.

IMT-2000 may be implemented as a stand-alone network with gateways and interworking units to support networks, in particular PSTN, ISDN, and B-ISDN. This is comparable to the current implementations of public land mobile networks, and it is also a solution where the fixed network and the radio network are operated by different service providers.

IMT-2000 may also be integrated with fixed networks. In this case functionality required to support a radio network, such as location registration, paging, and handover, will become an integral part of the fixed network. Such integration will be feasible and will require development of IN, ISDN, and B-ISDN switches.

2.6 UMTS

UMTS standardization is taking place in ETSI's Special Mobile Group 5 (SMG5). It began in 1994 and aims to provide complete standards in 1997–98. ETSI's work on UMTS consists of three phases. Phase 1 was completed in 1995. It involved the definition of UMTS framework, services, radio interface(s), network platform, management, satellite systems, and so on [3–6].These requirements are intended to guide the definition of related work (completed in 1996) and finally will lead to detailed specifications in Phase 3 (1996–98).

UMTS will provide a wide range of services to mobile and wireline users. Many of these services and environments in which they will be used are already provided by various existing systems such as wireline, cordless, cellular, and satellite. UMTS will provide an integrated system in which users have access to the desired service via uniform service access procedures irrespective of the environment they find themselves in. Thus, UMTS aims for the integration of both services and networks.

The integration of services implies maximum commonality of teleservices and supplementary services between wireline and mobile users; it also requires a common user interface. UMTS services shall, as far as possible, be identical to those offered to users of wireline terminals, with a comparable quality of service. Therefore, UMTS envisions radio bearer service providing up to 144 kbps (with future extension to 2 Megabytes per second [Mbps]) for various multimedia and multiparty services. Therefore, the types of radio access would range from a constant/variable-bit-rate connection, connection-oriented packet access up to connectionless packet access. Service quality should at least be comparable to present wireline services.

The integration of networks includes the common use of infrastructure, control functions, and protocols [7,8]. In general, each mobile communications system will consist of a **radio access network** and a **wireline network** which will comprise a control network and a switching network. Three components must be investigated when looking at integration and evolution issues:

☞ The UMTS core (switching) network should provide the switching and transmission functions (i.e., call and bearer/connection control). A candidate UMTS should aim for integration with B-ISDN as well as backward compatibility with pre-B-ISDN networks.

☞ The UMTS (service) control network should provide additional services and control logic required to support users (mobility management and so on). The control network will be based on IN principles since IN allows the flexible integration of new services. Consequently, the support for personal, terminal, and service mobility will be provided as IN service capabilities.

☞ The UMTS radio access system will comprise all radio-related aspects, including air interface, BSs, and other equipment enabling interworking between the MS and the core network. Thus, the radio access system will implement UMTS radio-related functions. It will also contain the basic UMTS interface (air interface) and two other important wireline network interfaces: one toward the core network (i.e., B-ISDN user-network interface) and one toward the control network (i.e., IN Application protocol, INAP interface). UMTS will aim for a highly flexible and efficient radio interface in the 1885–2025 MHz and 2110–2200 MHz frequency spectrum allocated during the World Administration Radio Conference (WARC) in 1992. It should operate in a variety of environments—such as home/business indoor, rural, urban, and public outdoor, aeronautical, and satellite based—with seamless and global radio coverage. Thus, B-ISDN and IN standards will have major impacts on UMTS development.

In contrast to GSM, UMTS services will be realized by IN concepts. Mobile service control points will be added in the network. Although Capability Set One (CS-1) of IN primarily contains call-related services, the introduction of IN will provide the means for a variety of different service profiles and flexible service offerings to end users. Therefore, the third-generation mobile network will rely heavily on distributed processing and distributed databases in contrast to the centralized database approach of the mobility management in second-generation systems.

All mobile systems, including UMTS, share the problem of connecting to other data networks when the error rate on the radio link is high. Thus additional protocol layers (e.g., GSM's Radio Link Protocol [RLP]) are needed. The addition of RLP or its equivalent does not have a correspondence in the wireline data network.

Since interoperability is one of the major goals of UMTS, the system is likely to have a variety of logical radio interfaces. Therefore, an open architecture is desirable. The subscriber profile handling, including quality of service (QoS) management, will be an essential component in the next-generation networks, but it is still not adequately covered by current signaling concepts in the IN environment. Thus new concepts like the reference model for Open Distributed Processing (ODP) will be considered to overcome this gap. The integration of ODP, TMN, and in particular X.700 services and protocols into a new Telecommunication Information Networking Architecture (TINA) will also be considered. The different views on service switching and service management need to converge before the idea of true open systems interconnection can succeed in the world of global telecommunications networks.

2.7 NORTH AMERICAN STANDARDS

2.7.1 TDMA-based Digital Cellular System—IS-54

In North America, where a common analog air interface was available and where roaming anywhere in Canada, the United States, or Mexico was possible, there was no need to replace the existing analog systems. Therefore, the Cellular Telecommunication Industry Association (CTIA) requested the TIA to specify a system that could be retrofitted into the existing AMPS system. The high cost of the cell site was the major driving force. Thus the important factor in the IS-54 (North American TDMA system) was to maximize the number of voice channels that can be supported by a cell site within the available cellular spectrum. IS-54 uses three TDMA 8-kbps-encoded speech channels into each 30-kHz AMPS channel. The IS-54 has been extended to PCS operation at 1.8 GHz and is referred to as upbanded IS-136.

IS-54 uses a linear modulation technique, $\pi/4$-DQPSK, to provide a better bandwidth efficiency. The transmission rate is 48.6 kbps with a channel spacing of 30 kHz. This provides bandwidth efficiency of 1.62 bps/Hz, a 20-percent improvement over GSM. The IS-54 speech coder is a type of code book excited linear predictive coding called VSELP. The source rate is 7.95 kbps and the transmission rate is 13 kbps. Recently, a new enhanced vocoder Algebraic Code Book Excited Linear Prediction (ACELP) was introduced to replace the VSELP.

2.7.2 CDMA-based Digital Cellular System—IS-95

IS-95, based on CDMA, was standardized in the United States. It is aimed at dual-mode operation with the existing analog cellular system. IS-95 uses a wideband RF channel with a channel width of 1.25 MHz. Each RF channel is shared by many users with different codes. IS-95 provides soft handover capability to improve voice quality and a RAKE receiver to reduce the impact of multipath fading. Other factors that affect the channel capacity include use of a variable-rate-8 kbps vocoder, voice activity, and power control in the forward and reverse channels. J-STD-008 is the PCS version of IS-95. J-STD-008 and IS-95 have been combined into IS-95 B.

2.8 JAPANESE STANDARDS

2.8.1 PDC

The development of the PDC system started in 1989, and it was standardized in 1991. Initially, the PDC system was intended to increase the system capacity for a cellular voice communication system; it has since developed into a personal communications system that supports both voice and data communications. There are about 10 million subscribers on the PDC system.

The PDC system was designed to improve system capacity, voice quality, and privacy over the conventional analog cellular systems. The PDC system uses $\pi/4$-DQPSK modulation and a low-bit-rate voice codec. In the first stage, full-rate VSELP codec with a bit rate of 11.2 kbps was introduced to achieve similar capacity and better voice quality than the existing analog systems. A half-rate codec with 5.6 kbps was introduced in 1995 to further improve the system capacity.

The PDC system uses three-channel TDMA. Two frequency bands are used—the 800-MHz band with 130 MHz of duplex separation and the 1.5-GHz band with 48 MHz of duplex separation. The 800 MHz band is used for the cellular system. The 1.5-GHz band will be used for the PCS system. The carrier spacing in PDC is 25 kHz. In the 800-MHz band, the uplink transmission (from the MS to the BS) frequency is 940–956 MHz. The downlink transmission (from the BS to the MS) frequency is 810 to 826. Thus, there are 640 carriers and 3 full-rate channels per carrier. A total of 1920 channels is available. The number of channels is doubled with the half-rate speech codec. The PDC system supports these services:

- ☞ Voice (full-rate and half-rate)
- ☞ Supplementary services (call waiting, voice mail, three-way calling, call forwarding, and so on)
- ☞ Data service (up to 9.6 kbps)
- ☞ Packet-switched wireless data

2.8.2 PHS

PHS is a digital microcellular system that is intended to support personal multimedia communication services. PHS became a standard in 1993, and its service was launched in 1995 in Japan. For PHS, 12 MHz from the 1.9-GHz band is exclusively allocated for a maximum of three PHS operators in each regional block. PHS operators are requested to make an effort to provide services in the area where not more than 50 percent of the population of the licensed regional block resides, within five years after they initiate services. Also, public telecommunications operators are requested to interconnect PHS operators to their networks with fair conditions to promote sound development of PHS business in competitive circumstances.

The cell radius of PHS is limited to 100–500 m to achieve high system capacity, to reduce transmission power for saving battery life, and to reduce the cost of the cell site. PHS is basically a digital cordless system, in which a PHS terminal can be used as a handset of a cordless phone in the business office or home. This operation is called a **private mode**. In this case, a parent phone at home or a PBX in the office is connected to the PSTN. Outside the home or business office, the terminal accesses the public cell site. This operation is referred to as a **public mode**. To use a PHS terminal in a public mode, the terminal has to have another 10-digit subscriber number. In this case, each cell site is connected to the ISDN network. Also, the PHS has a mode that enables direct communication between terminals. This is known as the **transceiver mode.** This mode of operation is available when the terminals are located close together. This is one of the unique design features of the PHS.

Each user has a choice to select one of the three modes based on the operational environment.

PHS has 77 carriers with a bandwidth of 300 kHz in the frequency range of 1895–1918.1 MHz (23.1 MHz of system bandwidth). Carriers 1–37 (37 carriers) are used for private mode of operation. Of these carriers, 1–10 are also used for the transceiver mode of operation. Carriers 38–77 (40 carriers) are used for the public mode of operation. Among these, the high-power cell sites (20 mW) can use only carriers 38–53 for the traffic channels, and low-power cell sites can use carriers 38–69 for traffic channels.

The features of PHS are

☞ High-quality voice communication and enhanced security

☞ SIM card

☞ High system capacity due to microcellular and dynamic channel allocation (DCA) technologies

☞ Long continuous call operating time (5 hours) and standby time (17 days) due to small cell radius, intermittent reception during standby mode, and other battery-saving technologies.

☞ Multimedia service capability using bearer service function with bit rate up to 64 kbps

☞ Simple spectrum management due to DCA

☞ Common terminal between cordless phone, public microcellular phone and transceiver.

2.9 SUMMARY

In this chapter, we presented standards for wireless communications used in the different parts of the world. In Europe, CT-1 was introduced as the first-generation cordless system which was followed by digital cordless system CT-2. Along with CT-2, DECT was standardized by ETSI, whereas in Japan PHS was adopted as the cordless as well as a microcellular system. In Europe, digital cellular systems based on the GSM standard replaced the incompatible first-generation analog cellular systems. Recently, GSM has been extended to DCS 1800 for PCN in Europe. Second-generation digital cellular systems in North America are based on IS-54 or IS-95 standards. Recently, both IS-54 and IS-95 have been extended for PCS at 1.9-GHz frequency. North American cellular and PCS systems are dual-mode dual-frequency systems that support both analog as well as digital technologies. The PDC is the Japanese second-generation digital cellular system; it is designed to operate at 900-MHz and 1800-MHz frequencies. Third-generation systems being standardized in Europe include ETSI's UMTS and ITU's IMT-2000. The third-generation system in North America includes an extension of CDMA and is called wideband CDMAone.

2.10 REFERENCES

1. Department of Trade and Industry, "Second Generation Cordless Telephone (CT-2) Common Air Interface Specifications," London, May 1989.
2. Code RES—3, "Digital European Cordless Telephone (DECT) System—Common Interface Specifications," 1989.
3. GSM Standard Committee, "Physical Layer on Radio Path: GSM System," GSM Recommendations, 05.02, Volume G, July 1988.
4. Project 2215, EIA/TIA IS-54, "Cellular System Dual Mode Mobile Station-Base Station Compatibility Standard," Washington, DC, December 1989.
5. EIA/TIA IS-95, "Mobile Station-Base Station Compatibility Standard for Dual Mode Wideband Spread Spectrum Cellular System," 1993.
6. ITU-R, "Future Public Land Mobile Telecommunication Systems (FPLMTS)," Report 1153, 1990.
7. Recommendations ITU-R, "Requirements for Radio Interface(s) for FPLMTS," M.1034, 1994.
8. RCR, "Personal Digital Cellular (PDC) Telecommunications System," RCR Standard-27, April 1991.
9. RCR, "Personal Handy Phone System (PHS)," RCR Standard-28, May 1993.

Access Technologies

3.1 INTRODUCTION

In this chapter, we discuss narrowband channelized and wideband nonchannelized systems for wireless communications and focus on access technologies including FDMA, TDMA, and CDMA. We examine these access technologies from a capacity, performance, and spectral efficiency viewpoint. The GSM system uses FDMA, TDMA, and frequency hopping to provide access to the system by users. We examine all of the access systems and calculate the efficiencies of the systems. As networks have evolved, the demand for higher capacities has encouraged researchers and system designers to examine access methods that are even more spectrally efficient than TDMA. Therefore, we also examine the CDMA system. Work in standards bodies around the world indicates that the next generation of digital systems will evolve to wideband CDMA systems to achieve high efficiencies and high access data rates.

3.2 NARROWBAND CHANNELIZED SYSTEMS

Traditional architectures for analog and digital wireless systems are channelized. In a channelized system, the total spectrum is divided into a large number of relatively narrow radio channels that are defined by carrier frequency. Each radio channel consists of a pair of frequencies. The frequency used for transmission from the BS to the MS is called the **forward channel**, and the frequency used for transmission from the MS to the BS is called the **reverse channel**. A user is assigned both frequencies for the duration of the call. The forward and reverse channels are assigned widely separated frequencies to keep cochannel interference between transmission and reception to a minimum.

A narrowband channelized system demands precise control of output frequencies for an individual transmitter. In this system, the transmission by a given MS is confined within a specified narrow bandwidth to avoid interference with adjacent channels. The tightness of bandwidth limitations plays a dominant role in the evaluation and selection of modulation technique. It also influences the design of transmitter and receiver elements, particularly the filters, which can greatly affect the cost of an MS.

We examined narrowband analog systems in detail in our previous book [4]. In this chapter we focus primarily on the digital systems in use worldwide.

A critical issue with regulators and operators around the world is how efficiently the radio spectrum is being used. Regulatory bodies want to encourage competition for the cellular services. Since spectrum must be allocated for each operator in a country or region of a country, a more efficient system can use less bandwidth. Thus, for a given availability of bandwidth, more competitors can be licensed. For a particular operator, a more efficient technology can support more users within the assigned spectrum and thus (hopefully) increase profits.

When we examine efficiencies of various technologies, we find that each system has made different trade-offs in determining the optimum method for access. Some of the parameters that are used in the trade-off include bandwidth per user, guard bands between channels, frequency reuse between different cells in the system, the signal-to-noise ratio (SNR) and S/I ratio, the methods of channel and speech coding, and the implementation complexity of the system.

3.2.1 Narrowband Digital Channelized Systems

First-generation analog cellular systems showed signs of capacity saturation in major urban areas, even with a modest total user population. A major capacity increase was needed to meet future demand. Several digital techniques were deployed to solve the capacity problem of the analog cellular systems. There are two basic digital strategies whereby a fixed spectrum resource can be allocated to different users. These are

☞ Using different frequencies—FDMA
☞ Using different time slots—TDMA

3.2.1.1 FDMA   In FDMA, signals from various users are assigned different frequencies, just as in an analog system. Frequency guard bands are maintained between adjacent signal spectra to minimize crosstalk between channels (see Figure 3.1).

The advantages and disadvantages of FDMA are

☞ **Advantages**

1. Capacity increase can be obtained by reducing the information bit rate and using efficient digital codes.

2. Technological advances required for implementation are simple. A system can be configured so that improvements in terms of speech codec bit rate reduction could be readily incorporated.

☞ **Disadvantages**

1. Since the system architecture, based on FDMA, does not differ significantly from the analog system, the improvement available in capacity depends on operation at a reduced S/I ratio. But the narrowband digital approach gives only limited advantages in this regard so that modest capacity improvements could be expected from a given spectrum allocation.

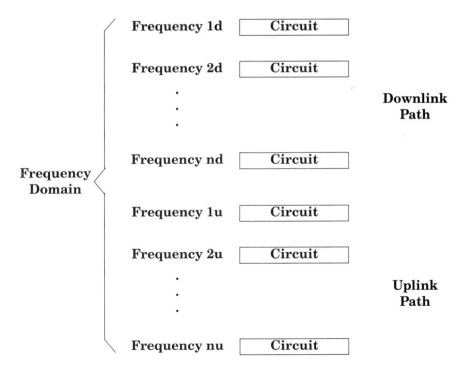

Fig. 3.1 FDMA/FDD Channel Architecture

2. The maximum bit rate per channel is fixed and low, inhibiting the flexibility in bit-rate capability that is needed for computer file transfers.

Traditional systems use FDD in which the transmitter and receiver operate simultaneously on different frequencies. Separation is provided between the forward or downlink and reverse or uplink channels to keep the transmitter from interfering with or desensing the receiver. Other precautions are also needed to prevent desensing, such as the use of two antennas or alternatively one antenna with a duplexer (a special arrangement of RF filters protecting the receiver from the signal power of the transmitter). A duplexer adds weight, size, and cost to a radio transceiver and can limit the minimum size of a subscriber unit.

3.2.1.2 TDMA   In a TDMA system, data from each user is conveyed in time intervals called slots (see Figure 3.2). Several slots make up a frame. Each slot is made up of a preamble plus information bits addressed to various stations as shown in Figure 3.3. The functions of the preamble are to provide identification and incidental information and to allow synchronization of the slot at

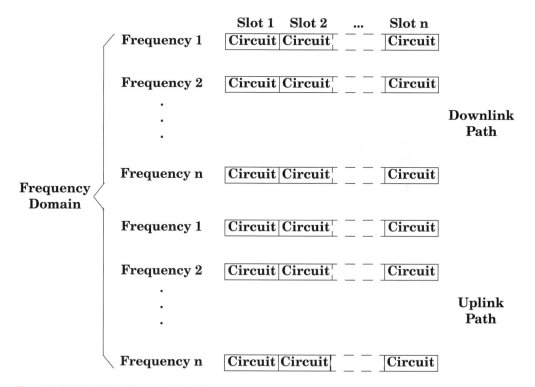

Fig. 3.2 TDMA/FDD Channel Architecture

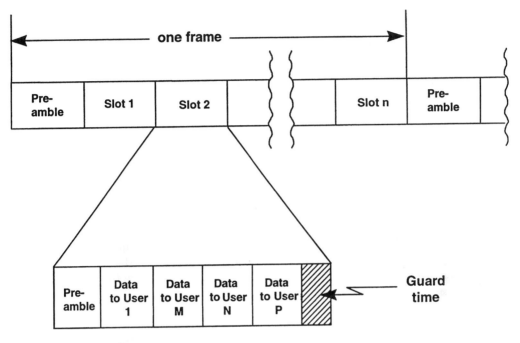

Fig. 3.3 TDMA Frame

the intended receiver. Guard times are used between each user's transmission to minimize crosstalk between channels.

Most TDMA systems time divide a frame into multiple slots used for different transmissions. This approach, called Time-Division Multiplex (TDM), uses several fixed-rate bit streams time-division multiplexed onto a TDMA bit stream. The data is transmitted via a radio carrier from a BS to several active MSs in the downlink. In the reverse direction (uplink), transmission from MSs to BSs is time sequenced and synchronized on a common frequency for TDMA.

A TDMA system using multiple slots can support a wide range of user bit rates by selecting the lowest multiplexing rate or a multiple of it. This enables supporting a variety of voice coding techniques at different bit rates with different voice qualities. Data communications customers could make the same kinds of decisions, choosing and paying for digital data rate as required. This would allow customers to request and pay for bandwidth on demand.

The advantages and disadvantages of TDMA are

☞ **Advantages**

    1. TDMA permits a flexible bit rate, not only for multiples of basic single channel rate but also submultiples for low-bit-rate broadcast-type traffic.

2. TDMA offers the opportunity for frame-by-frame monitoring of signal strength and bit error rates to enable either MSs or BSs to initiate and execute handovers.

3. TDMA transmits each signal with sufficient guard time between time slots to accommodate

   ✗ Source time inaccuracies because of clock instability

   ✗ Predetermined delay spread

   ✗ The transmission time delay because of propagation distance

   ✗ The "tails" of signal pulses in TDMA because of transient responses

☞ **Disadvantages**

1. TDMA requires a substantial amount of signal processing for matched filtering and correlation detection for synchronizing with a time slot.

A TDMA system can be designed to use one frequency band by using TDD. In TDD, a bidirectional flow of information is achieved using a simplex-type scheme by automatically alternating in time the direction of transmission on a single frequency. At best TDD can only provide a quasisimultaneous bidirectional flow, since one direction must be off while the other is using the frequency. However, with a high enough transmission rate on the channel, the off time is not noticeable during conversations and, with a digital speech system, the only effect is a very short delay.

The amount of spectrum required for both FDD and TDD is similar. TDD's strength lies in the ease of finding a single band of unassigned frequencies compared to finding two bands separated by the required bandwidth.

With TDD systems, the transmit time slot and the receiver time slot of the subscriber unit occur at different times. With the use of a simple RF switch in the subscriber unit, the antenna can be connected to the transmitter when a transmit burst is required (thus disconnecting the receiver from the antenna) and to the receiver for the incoming signal. The RF switch thus performs the function of the duplexer, but is less complex, smaller in size, and less costly. TDD uses a burst-mode scheme like TDMA and therefore also does not require a duplexer.

Since the bandwidth of the transmitter and receiver in a TDD system is twice that of a transmitter and receiver in an FDD system, RF filters in all the transmitters and receivers for TDD systems must be designed to cover twice the bandwidth of FDD system filters.

Depending on the data rate used and the number of slots per frame, a TDMA system can use the entire bandwidth of a system or can use FDD. The resultant multiplexing is a mixture of frequency division and time division. Thus, the entire frequency band is divided into a number of duplex channels (spaced about 350–400 kHz apart). These channels are deployed in a frequency-reuse pattern, where radio-port frequencies are assigned using an autonomous adaptive frequency assignment algorithm. Each channel is configured in a TDM mode for downlink (BS to MS) direction and a TDMA mode for the uplink (MS to BS) direction.

3.3 SPECTRAL EFFICIENCY

An efficient use of the frequency spectrum is the most desirable feature of a mobile communications system. To realize an efficient use of a spectrum, a number of techniques have been proposed or are already implemented in mobile communications systems. Some of these techniques employed to improve spectral efficiency include reduction of the channel bandwidth, information compression, variable bit rate control, and improved channel assignment algorithms. Spectral efficiency of a mobile communications system also depends on the choice of a multiple access scheme. A precise measure of spectral efficiency enables one to estimate the capacity of a mobile communications system and allows the setup of a minimum standard as a reference of measure.

The overall efficiency of a mobile communications system can be estimated by knowing the modulation and the multiple access spectral efficiencies separately.

3.3.1 Spectral Efficiency of Modulation

Spectral efficiency with respect to modulation is defined as:

$$\eta_m = \frac{\text{(Total Number of Channels Available in the System)}}{\text{(Bandwidth)(Total Coverage Area)}} \tag{3.1a}$$

$$\eta_m = \frac{\dfrac{B_w}{B_c} \times \dfrac{N_c}{N}}{B_w \times N_c \times A_c} \tag{3.1b}$$

$$\eta_m = \frac{1}{B_c \times N \times A_c} \tag{3.1c}$$

where:
η_m = modulation efficiency (channels/MHz/km^2),
B_w = bandwidth of the system (MHz),
B_c = channel spacing (MHz),
N_c = total number of cells in the covered area,
N = frequency reuse factor of system, and
A_c = area covered by a cell (km^2).

Eq. (3.1c) shows that the spectral efficiency of modulation does not depend on the bandwidth of the system. It depends only on the channel spacing, the total coverage area, and the frequency reuse factor. By reducing the channel spacing, the spectral efficiency of modulation for the system can be increased provided the total coverage area remains unchanged. If a radio plan or modulation scheme can be designed to reduce N, then more channels are available in a cell and efficiency is improved.

Another definition of spectral efficiency of modulation is Erlangs per MHz per km^2.

$$\eta_m = \frac{(\text{Total Traffic Carried by the System})}{(\text{Bandwidth})(\text{Total Coverage})} \tag{3.2a}$$

$$\eta_m = \frac{\text{Total Traffic Carried by } \left(\dfrac{B_w/B_c}{N}\right) \text{Channels}}{B_w A_c} \tag{3.2b}$$

By introducing the trunking efficiency factor, η_t, in Eq. (3.2b), the total traffic carried through the system is given as:

$$\eta_m = \frac{\eta_t \left(\dfrac{B_w/B_c}{N}\right)}{B_w A_c} \tag{3.2c}$$

$$\eta_m = \frac{\eta_t}{B_c N A_c} \tag{3.2d}$$

where:

η_t is a function of the blocking probability and $\dfrac{B_w}{B_c}$.

Several observations can be made from Eq. (3.2d).

1. The voice quality depends on the frequency reuse factor N, which is a function of the S/I ratio of the modulation scheme used in the mobile communications system.

2. The relationship between system bandwidth B_w and the amount of traffic carried by the system is nonlinear; in other words, for a given percentage increase in B_w, the increase in the traffic carried by the system is more than the increase in B_w.

3. From the average traffic per user (Erlang/user) during the busy hour and Erlang per km^2 per MHz, the capacity of the system in terms of users per km^2 per MHz can be obtained.

4. Spectral efficiency depends on the blocking probability.

E X A M P L E   3 – 1

Problem Statement

In the GSM 900 digital channelized cellular system, the one-way bandwidth of the system is 12.5 MHz. The channel spacing is 200 kHz. Eight users share each channel and three channels per cell are used (or reserved) for control channels. Calculate the spectral efficiency (for a dense metropolitan area with small cells) using the following parameters. Assume

- Omnidirectional cells
- Area of a cell = 8 km^2
- Total coverage area = 4000 km^2
- Average number of calls per user during the busy hour = 1.2
- Average holding time of a call = 100 s
- Call blocking probability = 2%
- Frequency reuse factor = 4

Solution

Number of 200-kHz channels = $\dfrac{12.5 \times 1000}{200}$ = 62

Number of traffic channels = 62×8 = 496

Number of signaling channels per cell = 3

Number of traffic channels per cell = $\dfrac{496}{4} - 3$ = 121

Number of cells = $\dfrac{4000}{8}$ = 500

With 2% blocking for the omnidirectional case, the total traffic carried by 121 channels (using Erlang B formula) = 108.4 Erlangs/cell = 13.55 Erlangs/ km^2

Number of calls per hour = $\dfrac{108.4 \times 3600}{100}$ = 3902.5 , Calls/hour/cell = $\dfrac{3902.5}{8}$ = 487.81 calls/hour/km^2

Number of users/cell = $\dfrac{3902.5}{1.2}$ = 3252, Users/hour/channel = $\dfrac{3252}{121}$ = 26.88

$\eta_m = \dfrac{108.4 \times 500}{4000 \times 12.5}$ = 1.08 Erlangs/MHz/km^2

3.3.2 Multiple Access Spectral Efficiency

In FDMA, users share the radio spectrum in the frequency domain; the multiple access spectral efficiency is reduced because of guard bands between channels and also because of signaling channels. In TDMA, the efficiency is reduced because of guard time and synchronization sequence.

Multiple access spectral efficiency is defined as the ratio of the total time-frequency domain dedicated for voice transmission to the total time-frequency domain available to the system. Thus, the multiple access spectral efficiency is a dimensionless number with an upper limit of unity.

3.3.2.1 FDMA Spectral Efficiency   For FDMA, multiple access spectral efficiency is given as:

$$\eta_a = \frac{B_c N_T}{B_w} \leq 1 \tag{3.3}$$

where:
η_a = multiple access spectral efficiency, and
N_T = total number of voice channels in the covered area.

3.3.2.2 TDMA Spectral Efficiency   For the wideband TDMA, multiple access spectral efficiency is given as:

$$\eta_a = \frac{\tau M_t}{T_f} \tag{3.4}$$

where:
τ = duration of a time slot,
T_f = frame duration, and
M_t = number of time slots per frame.

In Eq. (3.4) it is assumed that the total available bandwidth is shared by all users. For the narrowband TDMA schemes, the total band is divided into a number of subbands, each using the TDMA technique. For the narrowband TDMA system, frequency domain efficiency is not unity as the individual user channel does not use the whole frequency band available to the system. The efficiency of the narrowband TDMA system is given as:

$$\eta_a = \left(\frac{(\tau M_t)}{T_f}\right)\left(\frac{(B_u N_u)}{B_w}\right) \tag{3.5}$$

where:
B_u = bandwidth of an individual user during his or her time slot and
N_u = number of users sharing the same time slot in the system, but having access to different frequency subbands.

3.3.2.3 Overall Spectral Efficiency of FDMA and TDMA System   The overall spectral efficiency η of a mobile communications system can be obtained by considering both the modulation and multiple access efficiencies.

$$\eta = \eta_m \eta_a$$

E X A M P L E   3 – 2

Problem Statement
In the North American narrowband TDMA cellular system, the one-way bandwidth of the system is 12.5 MHz. The channel spacing is 30 kHz, and there are 395 total voice channels in the system. The frame duration is 40 ms, with 6 time slots per frame. The system has an individual user data rate of 16.2 kbps in which the speech with error protection has a rate of 13 kbps. Calculate the efficiency of the TDMA system.

Solution

The time slot duration, $\tau = \left(\dfrac{13}{16.2}\right)\left(\dfrac{40}{6}\right) = 5.35$ ms

$T_f = 40$ ms, $M_t = 6$, $N_u = 395$, $B_u = 30$ kHz, and $B_w = 12.5$ MHz.

$\eta_a = \dfrac{5.35 \times 6}{40} \times \dfrac{30 \times 395}{12500} = 0.76$

The overhead portion of the frame = 1.0–0.76 = 24%

3.3.2.4 Capacity and Frame Efficiency of a TDMA System
Capacity
The capacity of a TDMA system is given by:

$$N_u = \frac{\eta_b \mu}{v_f} \times \frac{B_w}{RN} \tag{3.6}$$

where:
N_u = number of channels (mobile users) per cell,
η_b = bandwidth efficiency factor,
μ = bit efficiency (= 2 for QPSK, = 1.354 for GMSK as used in GSM),
v_f = voice activity factor (equal to one for TDMA),
B_w = one-way bandwidth of the system,
R = Information bit rate plus overhead, and
N = Frequency reuse factor.

$$\text{Spectral Efficiency } \eta = \frac{N_u \times R}{B_w} \text{ bit/sec/Hz} \tag{3.7}$$

E X A M P L E   3 – 3

Problem Statement
Calculate the capacity and spectral efficiency of a TDMA system using the following parameters: bandwidth efficiency factor $\eta_b = 0.9$, bit efficiency (with QPSK) $\mu = 2$, voice activity factor $v_f = 1.0$, one-way system bandwidth $B_w = 12.5$ MHz, information bit rate $R = 16.2$ kbps, and frequency reuse factor $N = 19$.

$N_u = \dfrac{0.9 \times 2}{1.0} \times \dfrac{12.5 \times 10^6}{16.2 \times 10^3 \times 19}$

$N = 73.1$ (say 73 mobile users per cell)

Spectral Efficiency $\eta = \dfrac{73 \times 16.2}{12.5 \times 1000} = 0.094$ bit/sec/Hz

Efficiency of a TDMA Frame
The number of overhead bits per frame (see Figure 3.3) is:

$$b_o = N_r b_r + N_t b_p + (N_t + N_r) b_g \tag{3.8}$$

where:
N_r = Number of reference bursts per frame,
N_t = Number of traffic bursts (slots) per frame,
b_r = Number of overhead bits per reference burst,
b_p = Number of overhead bits per preamble per slot, and
b_g = Number of equivalent bits in each guard time interval.

The total number of bits per frame is:

$$b_T = T_f \times R_{rf} \tag{3.9a}$$

where:
T_f = Frame duration and
R_{rf} = Bit rate of the radio frequency channel.

$$\text{Frame efficiency } \eta = (1 - b_o/b_T) \times 100\% \tag{3.9b}$$

It is desirable to keep the efficiency as high as possible.

The number of bits per data channel (user) per frame $b_c = RT_f$, where R = bit rate of each channel.

No. of channels/frame $N_{CF} = \dfrac{(\text{Total data bits})/(frame)}{(\text{Bits per channel})/(frame)}$

$$N_{CF} = \frac{\eta R_{rf} T_f}{R T_f} \tag{3.10a}$$

$$N_{CF} = \frac{\eta R_{rf}}{R} \tag{3.10b}$$

E X A M P L E   3 – 4

Problem Statement

Consider a GSM TDMA system with the following parameters:

N_r = 2,

N_t = 24 frames of 120 ms each with 8 time slots per frame,

b_r = 148 bits in each of 8 time slots,

b_p = 34 bits in each of 8 time slots,

b_g = 8.25 bits in each of 8 time slots,

T_f = 120 ms,

R_{rf} = 270.8333333 kbps, and

R = 22.8 kbps.

Calculate the frame efficiency and the number of channels per frame.

Solution

b_o = $2 \times 8 \times 148 + 24 \times 8 \times 34 + 8 \times 8.25 = 10{,}612$ bits per frame

$$b_T = 120 \times 10^{-3} \times 270.8333333 \times 10^3 = 32,500 \text{ bits per frame}$$

$$\eta = \left(1 - \frac{10612}{32500}\right) \times 100 = 67.35\%$$

$$\text{Number of channels/frame} = \frac{0.6735 \times 270.8333333}{22.8} = 8$$

The last calculation, with an answer of 8 channels, confirms that our calculation of efficiency is correct.

3.4 WIDEBAND SYSTEMS

In wideband systems, the entire system bandwidth is made available to each user and is many times larger than the bandwidth required to transmit information. Such systems are known as Spread Spectrum (SS) systems. There are two fundamental types of spread spectrum: Direct Sequence Spread Spectrum (DSSS) (see Figure 3.4) and Frequency Hopping Spread Spectrum (FHSS).

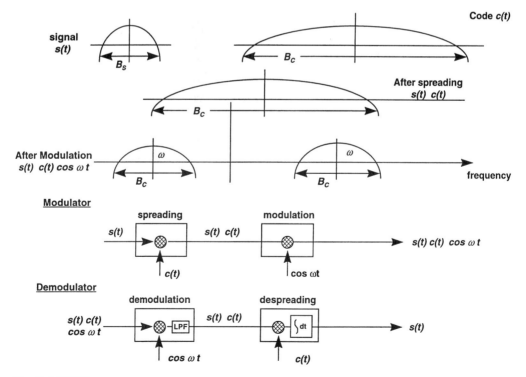

Fig. 3.4 DSSS

One advantage of DSSS systems is that the transmission bandwidth exceeds the coherence bandwidth. The received signal, after despreading, resolves into multiple signals with different time delays. A RAKE receiver can recover the multiple time-delayed signals and combine them into one signal, providing an inherent time diversity receiver with a lower frequency of deep fades. Thus, the DSSS systems provide an inherent robustness against mobile channel degradations. Another potential benefit of DSSS systems is the greater resistance to interference effects in a frequency reuse situation.

Frequency Hopping (FH) is the periodic changing of the frequency or the frequency set associated with transmission. If the modulation is multiple FSK, two or more frequencies are in the set that change at each hop. For other modulations, a single center or carrier frequency is changed at each hop.

An FH signal may be considered as a sequence of modulated pulses with pseudorandom carrier frequencies. The set of possible carrier frequencies is called the **hop set**. Hopping occurs over a frequency band that includes a number of frequency channels. The bandwidth of a frequency channel is called the **instantaneous bandwidth** (B_I). The bandwidth of the frequency band over which the hopping occurs is called the **total hopping bandwidth** (B_H). The time duration between hops is called the **hop duration** or **hopping period** (T_H).

FH may be classified as fast or slow. Fast FH occurs if there is frequency hop for each transmitted symbol. Thus, fast FH implies that the hopping rate equals or exceeds the information symbol rate. Slow FH occurs if two or more symbols are transmitted in the time interval between frequency hops.

FH allows communicators to hop out of frequency channels with interference or to hop out of fades. To exploit this capability, error-correcting codes, appropriate interleaving, and disjointed frequency channels are nearly always used.

A frequency synthesizer is used for an FH system to convert a stable reference frequency into the various frequencies of hop set.

FH communicators do not often operate in isolation. Instead, they are usually elements of a network of FH systems that cause mutual multiple access interference. This network is called a Frequency Hopping Multiple Access (FHMA) network.

If the hoppers of an FHMA network all use the same M frequency channels, but coordinate their frequency transitions and their hopping sequence, then the multiple access interference for a lightly loaded system can be greatly reduced over a nonhopped system. For the number of hopped signals (M_h) less than the number of channels (N_c), a coordinated hopping pattern can eliminate interference. As the number of hopped signals increases beyond N_c, then the interference will increase in proportion to the ratio of the number of signals to the number of channels. In the absence of fading or multipath interference, since there is no interference suppression system in frequency hopping, for high channel loadings the performance of an FH system is no better than a nonhopped system. FH systems are best for light channel loadings in the presence of conventional nonhopped systems. When fading or multipath interference is present, the FH system has better error performance than a non-hopped system. If the transmitter hops to a channel in a fade, the errors are limited in duration since the system will shortly hop to a new frequency where the fade may not be as deep.

Network coordination is simpler to implement for FH systems than for DS-CDMA systems because the timing alignments must be within a fraction of a hop duration, rather than a fraction of a sequence chip. Because of operational complications of coordination, asynchronous FHMA networks are usually preferable.

FH systems work best when a limited number of signals are sent in the presence of nonhopped signals where mutual interference can be avoided. In general, FH systems reject interference by trying to avoid it, whereas Direct Sequence (DS) systems reject interference by spreading it. The interleaving and error-correcting codes that are effective with FH systems are effective with DS systems. Error-correcting codes are more essential for FH systems than for DS systems because partial-band interference is a more pervasive threat than high-power pulsed interference.

The two major problems with FH systems with increasing hopping rates are that the cost of a frequency synthesizer increases and its reliability decreases, and that synchronization becomes more difficult.

3.5 COMPARISONS OF **FDMA**, **TDMA**, AND **DS-CDMA**

The primary advantage of DS-CDMA is its ability to tolerate a fair amount of interfering signals compared to FDMA and TDMA, which typically cannot tolerate any such interference (Figure 3.5). As a result of the interference tolerance of CDMA, the problems of frequency band assignment and adjacent cell interference are greatly simplified. Also, flexibility in system design and deployment are significantly improved since interference to others is not a problem. On the other hand, FDMA and TDMA radios must be carefully assigned a frequency or time slot to assure that there is no interference with other similar radios. Therefore, sophisticated filtering and guard band protection is needed with FDMA and TDMA technologies.

Capacity improvements with DS-CDMA also result from voice activity patterns during two-way conversation (i.e., times when a party is not talking) that cannot be exploited cost effectively in FDMA or TDMA systems. DS-CDMA radios can, therefore, accommodate more mobile users than FDMA/TDMA radios on the same bandwidth.

With DS-CDMA, adjacent microcells share the same frequencies, whereas with FDMA/TDMA it is not feasible for adjacent microcells to share the same frequencies because of interference.

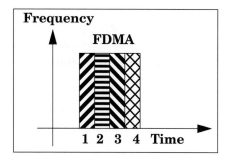

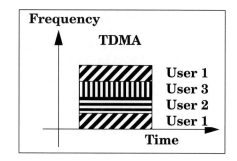

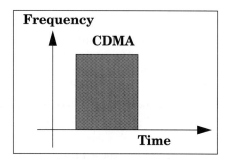

Fig. 3.5 Comparison of Multiple Access Methods

Further capacity gains can also result from antenna technology advancement by using directional antennas that allow the microcell area to be divided into sectors.

Table 3.1 provides a summary of access technologies for wireless systems.

Table 3.1 Access Technologies for Wireless Systems

System	Access Technology	Mode of Operation	Frame Rate (kbps)
North American IS-54 (Dual Mode)	TDMA/FDD FDMA/FDD	Digital/ Analog FM	48.6 —
North American IS-95 (Dual Mode)	DS-CDMA/FDD FDMA/FDD	Digital/ Analog FM	1228.8 —
GSM	TDMA/FDD	Digital	270.833
CT-2 Cordless	FDMA/TDD	Digital	72.0
DECT Cordless	TDMA/TDD	Digital	1152.0

3.6 SUMMARY

This chapter described the access technologies used for wireless communications. FDMA, TDMA, and CDMA technologies were discussed and their advantages and disadvantages were listed. We used examples to illustrate calculations for determination of capacity of the TDMA system. Brief descriptions of the FDD, TDD, TDM/TDMA, and TDM/TDMA/FDD approaches were also presented.

3.7 PROBLEMS

1. In a proposed TDMA cellular system, the one-way bandwidth of the system is 40 MHz. The channel spacing is 30 kHz, and there are 1333 total voice channels in the system. The frame duration is 40 ms divided equally between 6 time slots. The system has an individual user data rate of 16.2 kbps in which the speech with error protection has a rate of 13 kbps. Calculate the efficiency of the TDMA system. What is the efficiency of the system with 20, 60, 80, and 100 MHz?.

2. In Problem 1, what is the optimum bandwidth for a TDMA system? What are the practical limits that prevent one from using the optimum bandwidth?

3. Recompute the capacity of the GSM system in Example 2-1 when a sectorized system is used. With sectorization, there are 12 channel sets of 39 channels each with 3 sets assigned at each cell, 1 for each sector.

4. In the IS-54 (TDMA/FDD), the frame duration is 40 ms. The frame contains 6 time slots. The transmit bit rate is 48.6 kbps. Each time slot carries 260 bits of user information. The total number of 30 kHz voice channels available is 395 and the total system bandwidth is 12.5 MHz. Calculate the access efficiency of the system.

5. Calculate the capacity and spectral efficiency (η) of the IS-54 system using the following parameters: $\eta_b = 0.96$, $\mu = 1.62$ (i.e., modulation efficiency with $\pi/4$- DQPSK), voice activity factor $v_f = 1.0$, information bit rate = 19.5 kbps, frequency reuse factor = 7, and system bandwidth = 12.5 MHz.

3.8 REFERENCES

1. Bellamy, J. C., *Digital Telephony*, 2d Ed., John Wiley & Sons, Inc., New York, 1990.

2. Bell Communications Research, "Generic Framework Criteria for Version 1.0 Wireless Access Communication System (WACS)," FA-NWT-001318, Piscataway, NJ, June 1992.

3. Calhoun, G., *Digital Cellular Radio*, Artech House, Norwood, MA, 1988.

4. Garg, V. K., and Wilkes, J. E., *Wireless and Personal Communications*, Prentice Hall, 1996.

5. Gilhousen, S., et al., "Increased Capacity Using CDMA for Mobile Satellite Communication," *IEEE Journal on Selected Areas in Communications* 8 (4), May 1990.

6. Jacobs, I. M., et al., "Comparison of CDMA and FDMA for the MobileStar System," Proc. Mobile Satellite Conf., Pasadena, CA, May 3–5, 1988, pp. 283–90.

7. Lee, W. C. Y., "Spectrum Efficiency in Cellular," *IEEE Transactions on Vehicular Technology* 38, May 1989, pp. 69–75.

8. Lee, W. C. Y., "Overview of Cellular CDMA," *IEEE Transactions on Vehicular Technology* 40 (2), May 1991, pp. 291–302.

9. Lee, W. C. Y., *Mobile Communications Design Fundamentals*, 2d ed., John Wiley & Sons, Inc., New York, 1993.

10. Mehrotra, A., *Cellular Radio—Analog & Digital Systems*, Artech House, Norwood, MA, 1994.

11. Parsons, J. D., and Gardiner, J. G., *Mobile Communication Systems*, Halsted Press, New York, 1989.

12. Torrieri, J., *Principles of Secure Communication Systems*, 2d Ed., Artech House, Norwood, MA, 1992.

13. Viterbi, A. J., "When Not to Spread Spectrum-A Sequel," *IEEE Communications Magazine* 23, April 1985, pp. 12–17.

14. Ziemer, E., and Peterson, R. L., *Introduction to Digital Communications*, Macmillan Publishing Company, New York, 1992.

Cellular Communications Fundamentals

4.1 INTRODUCTION

In this chapter, we present the fundamentals of cellular communications. We also develop a relationship between the reuse ratio (q) and the reuse factor or cluster size (N) for hexagonal cell geometry, as well as study cochannel interference for omnidirectional and sectorized cells. The chapter also covers cell splitting procedures used in cellular communications.

4.2 CELLULAR SYSTEMS

Most commercial radio and television systems are designed to cover as much area as possible. These systems typically operate at maximum power and with the highest antennas allowed by the FCC. The frequency used by the transmitter cannot be reused again until there is enough geographical separation so that one station does not interfere significantly with another station assigned to that frequency. There may even be a large region between two transmitters using the same frequency where neither signal is received.

The cellular system takes the opposite approach. It seeks to make an efficient use of available channels by using low-power transmitters to allow frequency reuse at much smaller distances (see Figure 4.1). Maximizing the

47

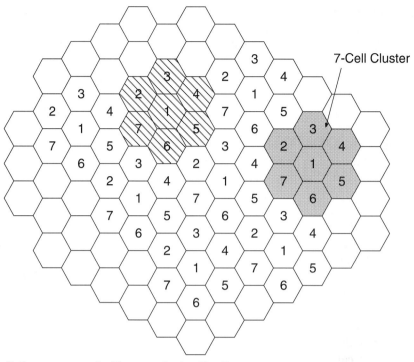

Cell arrangement with reuse factor N = 7

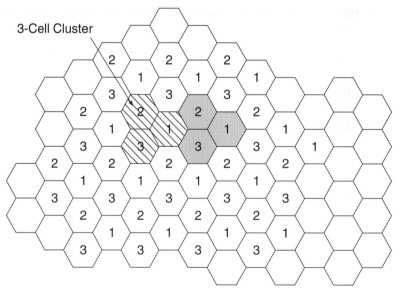

Cell arrangement with reuse factor N = 3

Fig. 4.1 Frequency Reuse

number of times each channel may be reused in a given geographic area is the key to an efficient cellular system design.

Cellular systems are designed to operate with groups of low-power radios spread out over the geographical service area. Each group of radios serve MSs presently located near them. The area served by each group of radios is called a **cell**. Each cell has an appropriate number of low-power radios for communications within itself. The power transmitted is chosen to be large enough to communicate with MSs located near the edges of its cell. The radius of each cell may be chosen to be perhaps 28 km (about 16 miles) in a start-up system with relatively few subscribers, down to less than 2 km (about 1 mile) for a mature system requiring considerable frequency reuse.

As the traffic grows, new cells and channels are added to the system. If an irregular cell pattern is selected, it would lead to an inefficient use of the spectrum due to its inability to reuse frequencies because of cochannel interference. In addition, it would also result in an uneconomical deployment of equipment, requiring relocation from one cell site to another. Therefore, a great deal of engineering effort would be required to readjust the transmission, switching, and control resources every time the system goes through its development phase. The use of a regular cell pattern in a cellular system design eliminates all these difficulties.

In reality, cell coverage is an irregularly shaped circle. The exact coverage of the cell will depend on the terrain and many other factors. For design convenience and as a first-order approximation, we assume that the coverage areas are regular polygons. For example, for omnidirectional antennas with constant signal power, each cell site coverage area would be circular. To achieve full coverage without dead spots, a series of regular polygons are required for cell sites. Any regular polygon such as an equilateral triangle, a square, or a hexagon can be used for cell design. The hexagon is used for two reasons—a hexagonal layout requires fewer cells and, therefore, fewer transmitter sites, and a hexagonal cell layout is less expensive compared to square and triangular cells. In practice, after the polygons are drawn on a map of the coverage area, radial lines are drawn and the SNR calculated for various directions using the propagation models, which we will discuss in chapter 13, or using computer programs. For the remainder of this chapter, we will assume regular polygons for the coverage areas even though in practice that is only an approximation.

4.3 GEOMETRY OF A HEXAGONAL CELL

We use the u-v axes to calculate the distance D between points C_1 and C_2 (refer to Figure 4.2). C_1 and C_2 are the centers of the hexagonal cells with coordinates (u_1, v_1) and (u_2, v_2).

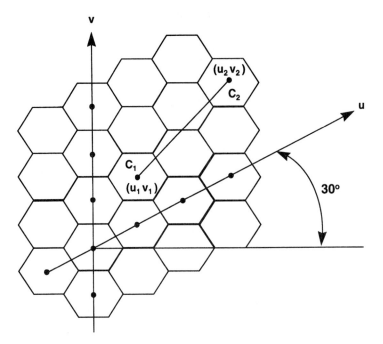

Fig. 4.2 Coordinate System

$$D = \{(u_2 - u_1)^2(\cos 30°)^2 + [(v_2 - v_1) + (u_2 - u_1)\sin 30°]^2\}^{1/2}$$

$$D = \{(u_2 - u_1)^2 + (v_2 - v_1)^2 + (v_2 - v_1)(u_2 - u_1)\}^{1/2} \tag{4.1}$$

If we assume $(u_1, v_1) = (0, 0)$, or the origin of the coordinate system is the center of a hexagonal cell, and restrict (u_2, v_2) to be integer value (i, j) then Eq. (4-1) can be written as:

$$D = [i^2 + j^2 + ij]^{1/2} \tag{4.2}$$

The normalized distance between two adjacent cells is unity ($i = 1, j = 0$) or ($i = 0, j = 1$). The actual center-to-center distance between two adjacent hexagonal cells is $2R\cos 30°$ or $\sqrt{3}R$ where R is the center-to-vertex distance.

We assume the size of all the cells is roughly the same. As long as the cell size is fixed, and each cell transmits the same power, cochannel interference will be independent of the transmitted power of each cell. The cochannel interference is a function of q where $q = D/R$. Furthermore, D is a function of N_I and S/I in which N_I is number of cochannel-interfering cells in the first tier (refer to Figure 4.3) and S/I is the received signal-to-interference (S/I) ratio at the desired mobile receiver.

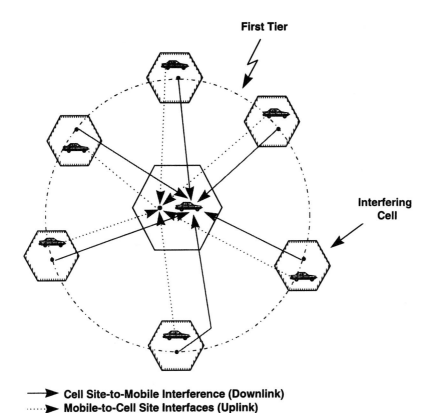

Fig. 4.3 Cochannel Interference with Omnidirectional Cell Site

From Figure 4.4, we see the radius of the large cell D (cochannel separation) is given as:

$$D^2 = 3R^2(i^2 + j^2 + ij) \qquad (4.3)$$

Since the area of a hexagon is proportional to the square of distance between center and vertex, the area of the large hexagon is:

$$A_{large} = k[3R^2(i^2 + j^2 + ij)] \qquad (4.4)$$

where:
k is a constant.

Similarly the area of the small hexagon is given as:

$$A_{small} = k(R^2) \qquad (4.5)$$

Comparing Eq. (4.4) and Eq. (4.5) and using Eq. (4.3), we can write

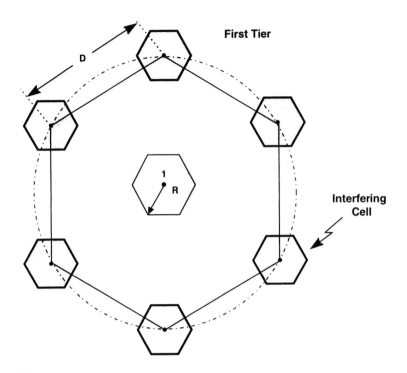

Fig. 4.4 Six Effective Interfering Cells of Cell 1

$$\frac{A_{large}}{A_{small}} = 3(i^2 + j^2 + ij) = \frac{D^2}{R^2} \tag{4.6}$$

From symmetry, we can see the large hexagon encloses the center cluster of N cells plus one-third the number of the cells associated with six other peripheral hexagons. Thus, the total number of cells enclosed is equal to $N + 6(1/3\ N) = 3N$. Since the area is proportional to the number of cells, $A_{large} = 3N$, and $A_{small} = 1$.

$$\frac{A_{large}}{A_{small}} = 3N \tag{4.7}$$

Substituting Eq. (4.7) into Eq. (4.6) we get:

$$3N = 3(i^2 + j^2 + ij) \tag{4.8}$$

$$\therefore \frac{D^2}{R^2} = 3N \tag{4.9}$$

$$\frac{D}{R} = q = \sqrt{3N} \tag{4.9a}$$

where:
q = reuse ratio (refer to Figure 4.5).

Table 4.1 lists the values of q for different values of N.

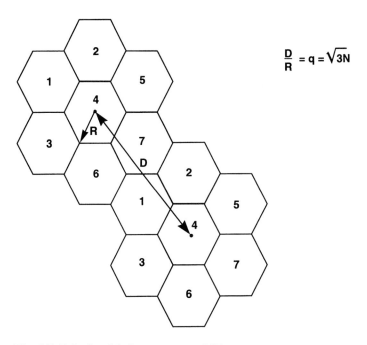

$$\frac{D}{R} = q = \sqrt{3N}$$

Fig. 4.5 Relationship between q and N

Table 4.1 Cochannel Reuse Ratio vs. Frequency Reuse Pattern

i	j	N	$q = D/R$
1	0	1	1.73
1	1	3	3.00
2	0	4	3.46
2	1	7	4.58
3	0	9	5.20
2	2	12	6.0
3	1	13	6.24
4	0	16	6.93
3	2	19	7.55
4	1	21	7.94
4	2	28	9.17

Eq. (4.9a) is important because it affects traffic-carrying capacity of a cellular system and the cochannel interference. By reducing q the number of cells per cluster is reduced. If total RF channels are constant, then the number of channels per cell is increased, thereby increasing the system traffic capacity. On the other hand, cochannel interference is increased with small q. The reverse is true when q is increased—an increase in q reduces cochannel interference and also the traffic capacity of the cellular system.

Table 4.1 shows that a 2-cell, 5-cell, etc., reuse pattern does not exist. However, the basic assumption in Eq. (4.8) is that all six first-tier interferers are located at the same distance from the desired cell. Asymmetrical reuse arrangements, where interferers are located at various distances, do allow for 2-cell, 5-cell, etc., reuse. Figure 4.6 shows how 5-cell reuse with 90-degree

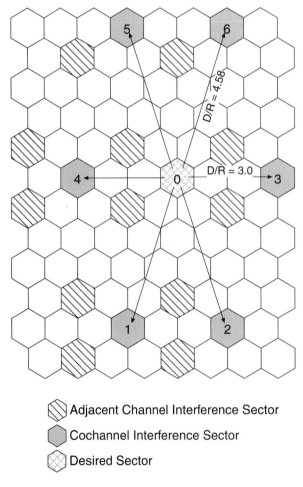

Fig. 4.6 Five-Cell, Four-Sector Reuse Pattern

sectors can be used. In that arrangement four interferer sources are located at D/R = 4.58 and two interferer sources are at D/R = 3.0.

4.4 COCHANNEL INTERFERENCE RATIO

The S/I ratio at the desired mobile receiver is given as:

$$\frac{S}{I} = \frac{S}{\sum\limits_{k=1}^{N_I} (I_k)} \qquad (4.10)$$

In a fully equipped hexagonal-shaped cellular system, there are always six cochannel-interfering cells in the first tier (i.e., N_I = 6, Figure 4.3). Most of the cochannel interference results from the first tier. Contribution from second and higher tiers amounts to less than 1 percent of the total interference and, therefore, it is ignored. Cochannel interference can be experienced both at the cell site and at the MSs in the center cell. In a small cell system, interference will be the dominating factor and thermal noise can be neglected. Thus the S/I ratio can be given as:

$$\frac{S}{I} = \frac{1}{\sum\limits_{k=1}^{6} \left(\dfrac{D_k}{R}\right)^{-\gamma}} \qquad (4.11)$$

where:
$2 \le \gamma \le 5$ is the propagation path-loss slope, and
γ depends upon the terrain environment (refer to chapter 13).

If we assume D_k is the same for the six interfering cells for simplification, or $D = D_k$, then Eq. (4.11) becomes:

$$\frac{S}{I} = \frac{1}{6(q)^{-\gamma}} = \frac{q^{\gamma}}{6} \qquad (4.12)$$

$$\therefore q = \left[6\left(\frac{S}{I}\right)\right]^{\frac{1}{\gamma}} \qquad (4.13)$$

For analog systems using FM, normal cellular practice is to specify the S/I ratio to be 18 decibels (dB) or higher based on subjective tests. An S/I ratio of 18 dB is the measured value for the accepted voice quality from present-day cellular mobile receivers.

Using an S/I ratio equal to 18 dB (i.e., 63.1) and γ = 4 in Eq. (4.13), then

$$q = [6 \times 63.1]^{0.25} = 4.41 \qquad (4.14)$$

Substituting q from Eq. (4.14) into Eq. (4.9a) yields

$$N = \frac{(4.41)^2}{3} = 6.49 \approx 7 \tag{4.15}$$

Eq. (4.15) indicates that a 7-cell reuse pattern is needed for an S/I ratio of 18 dB. Based on $q = D/R$, we can select D by choosing the cell radius R.

E X A M P L E   4 – 1

Problem Statement

We consider a cellular system with 395 total allocated voice channel frequencies. If the traffic is uniform with an average call holding time of 120 seconds and the call blocking during the system busy hour is 2%, calculate:

1. The number of calls per cell site per hour

2. Mean S/I ratio for cell reuse factors equal to 4, 7, and 12.

Assume omnidirectional antennas with six interferers in the first tier and a slope for the path loss of 40 dB/decade ($\gamma = 4$).

Solution

For a reuse factor N = 4, the number of voice channels per cell site = $\frac{395}{4} \approx 99$ and $q = \sqrt{3 \times 4} = 3.5$. Using the Erlang-B traffic table for 99 channels with 2% blocking, we find a traffic load of 87 Erlangs. The offered load will be $(1 - 0.02) \times 87 = 85.26$ Erlangs

$$\therefore \frac{\text{No. of Calls per Cell Site per hour} \times 120}{3600} = 85.26$$

$$\therefore \text{No. of Calls per Cell Site per hour} = 85.26 \times 30 = 2558$$

Using Eq. (4.12) we can calculate the mean S/I ratio as $\frac{S}{I} = \frac{(3.5)^4}{6} = 25 = 14\ \text{dB}$.

The results for N = 7 and N = 12 are given in Table 4.2.

Table 4.2 Cell Reuse Factor vs. Mean S/I Ratio & Call Capacity

N	q	Voice Channels per Cell	Calls per Cell per Hour	Mean S/I dB
4	3.5	99	2558	14.0
7	4.6	56	1376	18.7
12	6.0	33	739	23.3

It is evident from the results that, by increasing the reuse factor from N = 4 to N = 12, the mean S/I ratio is increased from 14 dB to 23.3 dB (i.e., 66.4-

percent improvement). However, the call capacity of the cell site is reduced from 2558 to 739 calls per hour (i.e., a 72-percent reduction).

EXAMPLE 4 – 2

Problem Statement

Consider a GSM system with a one-way spectrum of 12.5 MHz and channel spacing of 200 kHz. There are three control channels per cell, and the reuse factor is 4. Assuming an omnidirectional antenna with six interferers in the first tier and a slope for path loss of 40 dB/decade ($\gamma = 4$), calculate the number of calls per cell site per hour with 2% blocking during system busy hour and an average call holding time of 120 seconds.

Solution

$$\text{No. of Voice Channels per Cell Site} = \frac{12.5 \times 10^{6} \times 8}{200 \times 10^{3} \times 4} - 3 = 122$$

Using the Erlang-B traffic table for 122 channels with 2% blocking, we find a traffic load of 110 Erlangs. The offered load will be $(1 - 0.02 \times 110) = 107.8$ Erlangs:

$$\therefore \frac{\text{No. of Calls per Cell Site per hour} \times 120}{3600} = 107.8$$

$$\therefore \text{No. of Calls per Cell Site per hour} = 107.8 \times 30 = 3234$$

$$q = \sqrt{3 \times 4} = 3.5$$

Using Eq. (4.12) we can calculate the mean S/I ratio as $\frac{S}{I} = \frac{(3.5)^{4}}{6} = 25 = 14$ dB.

4.5 CELLULAR SYSTEM DESIGN IN WORST CASE WITH AN OMNIDIRECTIONAL ANTENNA

In the previous section we showed that the value of $q \approx 4.6$ is adequate for the normal interference case with a 7-cell reuse pattern. We reexamine the 7-cell reuse pattern and consider the worst case in which the MS is located at the cell boundary (refer to Figure 4.7) where it receives the weakest signal from its own cell but is subjected to strong interference from all the interfering cells in the first tier. The distances from the six interfering cells are shown in Figure 4.7.

The S/I ratio can be expressed as:

$$\frac{S}{I} = \frac{R^{-\gamma}}{2(D-R)^{-\gamma} + 2D^{-\gamma} + 2(D+R)^{-\gamma}} \tag{4.16}$$

Using $\gamma = 4$ and $D/R = q$, we rewrite Eq. (4.16) as:

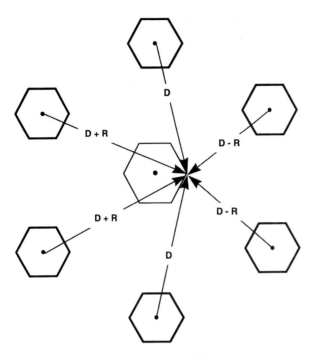

Fig. 4.7 Worst Case for Cochannel Interference

$$\frac{S}{I} = \frac{1}{2(q-1)^{-4} + 2q^{-4} + 2(q+1)^{-4}} \tag{4.17}$$

where:
$q = 4.6$ for a normal 7-cell reuse pattern.

Substituting $q = 4.6$ in Eq. (4.17), we get S/I = 54.3 or 17.3 dB. This S/I ratio is lower than 18 dB. For a conservative estimate, if we use the shortest distance (D-R) then

$$\frac{S}{I} = \frac{1}{6(q-1)^{-4}} = \frac{1}{6(3.6)^{-4}} = 28 \text{ or } 14.47 \text{ dB} \tag{4.18}$$

In the real situation, because of imperfect cell site locations and the rolling nature of the terrain configuration, the S/I ratio is often less than 17.3 dB. It could be 14 dB or lower. Such a condition can easily occur in a heavy traffic. Therefore, the cellular system should be designed around the S/I ratio of the worst case. If we consider the worst case for a 7-cell reuse pattern, we conclude that a cochannel interference reduction factor of $q = 4.6$ is not enough in an omnidirectional cell system. In an omnidirectional cell system, N = 9 (q = 5.2) or N = 12 (q = 6.0) cell reuse pattern would be a better choice. These cell reuse patterns would provide the S/I ratio of 19.78 dB and 22.54 dB, respectively.

4.6 COCHANNEL INTERFERENCE REDUCTION WITH USE OF DIRECTIONAL ANTENNAS

In the case of increased call traffic, the frequency spectrum should be used efficiently. We should avoid increasing the number of cells N in a frequency reuse pattern. As N increases, the number of frequency channels assigned to a cell becomes smaller, thereby decreasing the efficiency of the frequency reuse pattern.

Instead of increasing N in a set of cells, we use a directional antenna arrangement to reduce the cochannel interference. In this scheme, each cell is divided into 3 or 6 sectors and uses 3 or 6 directional antennas at the BS (refer to Figures 4.8 and 4.9). Each sector is assigned a set of channels (frequencies). The cochannel interference decreases as we'll soon see.

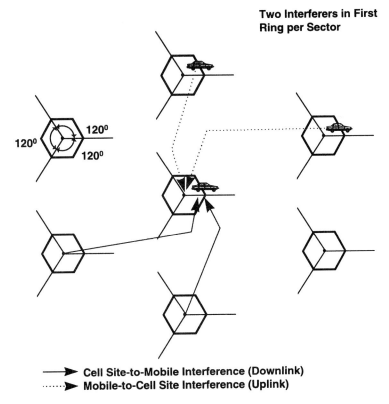

Two Interferers in First Ring per Sector

——▶ Cell Site-to-Mobile Interference (Downlink)
······▶ Mobile-to-Cell Site Interference (Uplink)

Fig. 4.8 Cochannel Interference with 120° Sectorized Cell Sites

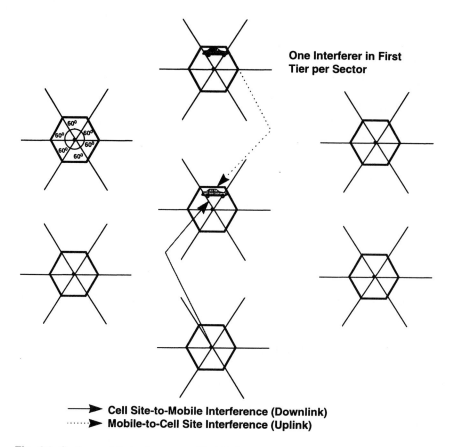

**One Interferer in First
Tier per Sector**

⟶ **Cell Site-to-Mobile Interference (Downlink)**
·····▶ **Mobile-to-Cell Site Interference (Uplink)**

Fig. 4.9 Cochannel Interference with 60° Sectorized Cell Sites

4.7 DIRECTIONAL ANTENNAS IN 7-CELL REUSE PATTERN

4.7.1 Three-Sector Case

The three-sector case for the 7-cell reuse pattern is shown in Figure 4.8. We consider the worst case where the MS is at position M in Figure 4.10a. In this situation, the MS receives the weakest signal from its own cell and fairly strong interference from two interfering cells 1 and 2. Because of the use of directional antennas, the number of interfering cells is reduced from six to two. At point M, the distances between the MS and the two interfering antennas are D and $(D + 0.7\,R)$, respectively. The S/I ratio in the worst case is given as:

$$\frac{S}{I} = \frac{R^{-4}}{D^{-4} + (D + 0.7R)^{-4}} \tag{4.19}$$

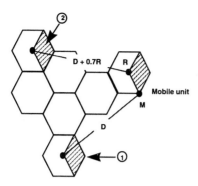

Fig. 4.10a Worst Case in 120° Sectorized Cell Sites

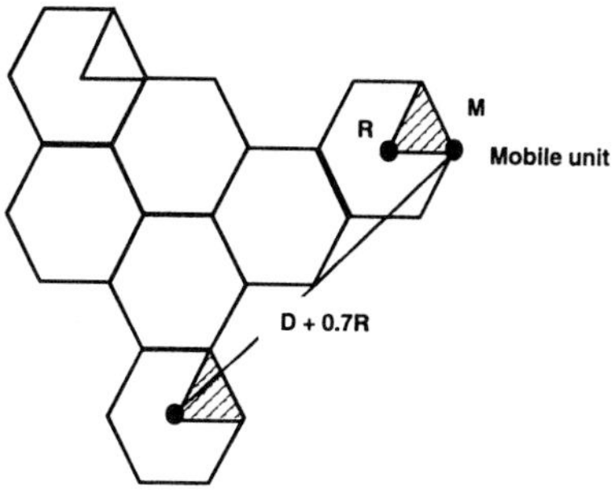

Fig. 4.10b Worst Case in 60° Sectorized Cell Sites

$$\frac{S}{I} = \frac{1}{q^{-4} + (q + 0.7)^{-4}} \qquad (4.19a)$$

Using q = 4.6 in Eq. (4.19a), we get S/I = 285 or 24.5 dB. The S/I for a MS served by a cell site with a 120° directional antenna exceeds 18 dB in the worst case. It is evident from Eq. (4.19a) that the use of directional antennas is helpful in reducing the cochannel interference. In real situations, under heavy traffic, the S/I could be 6 dB weaker than in Eq. (4.19a) due to irregular terrain configurations and imperfect site locations. The resulting 18.5 dB S/I is still adequate.

4.7.2 Six-Sector Case

In this case the cell is divided into six sectors by using six 60° beamwidth directional antennas as shown in Figure 4.9. In this case, only one interference can occur. The worst case S/I ratio will be (refer to Figure 4.10[b]):

$$\frac{S}{I} = \frac{R^{-4}}{(D + 0.7R)^{-4}} = (q + 0.7)^4 \qquad (4.20)$$

For q = 4.6, Eq. (4.20) gives S/I = 789 or 29 dB. This indicates a further reduction of cochannel interference. Using the argument as used for the three-sector case and subtracting 6 dB from 29 dB, the remaining 23 dB is still more than adequate. For heavy traffic, the 60° sector configuration can be used to reduce cochannel interference. However, with the six-sector configuration, the trunking efficiency is decreased.

E X A M P L E   4 – 3

Problem Statement

Figure 4.11 shows the carried load in Erlangs for a 7-cell cellular system located in a busy metropolitan area. There are 395 total available channels in the system.

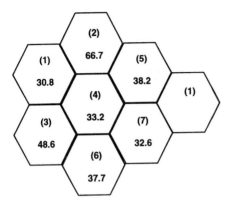

Fig. 4.11 Traffic in Erlangs for a 7-Cell Cellular System

Assuming each subscriber in the system generates 0.03 Erlangs of traffic with an average call holding time of 120 seconds and the system covers an area of 1200 square miles with cells designed for a Grade-of-Service (GoS) of 2%, compute (see Table 4.3):

- Number of channels required in each cell
- Number of subscribers served by the system
- Average number of subscribers per channel
- Number of calls supported by the system
- Subscriber density per square mile
- Call density per square mile
- Cell radius in miles
- Channel reuse factor

Table 4.3 Required Number of Channels, Subscribers, and Calls per Cell

Cell Number	Traffic (Erlangs)	No. of Subscribers per Cell	No. of Calls per Cell	No. of Channels Required
1	30.8	1026.7	924	40
2	66.7	2223.3	2001	78
3	48.6	1620.0	1458	59
4	33.2	1106.7	996	43
5	38.2	1273.3	1146	48
6	37.8	1260.0	1134	48
7	32.6	1086.7	978	42
Total	287.9	9597.0	8637	358

Solution

$$\frac{\text{No. of Calls per hour per Subscriber} \times 120}{3600} = 0.03$$

$$\text{No. of Calls per hour per Subscriber} = 0.9$$

- Average Number of Subscribers/Channel = $\frac{9597}{358} = 26.8$

- Erlangs per mile^2 = $\frac{287.9}{1200} = 0.24$

- Subscriber Density = $\frac{9597}{1200} = 8.0$ Subscribers per mile^2

- Call Density = $\frac{8637}{1200} = 7.2$ Calls per mile^2

- Area of each cell = $\frac{1200}{7} = 171.4$ miles^2

- $A_{\text{hex}} = 2.6\ R^2 = 171.4$
- Radius of the cell, R = 8.12 miles

Since there are a total of 395 voice channels, the channel reuse factor is $\frac{358}{395} = 0.906$

E X A M P L E   4 – 4

Problem Statement

Compare the spectral efficiency of the digital system with respect to the present analog system using the following data:

- Total number of channels = 416
- Number of control channels = 21
- Number of voice channels = 395
- Channel band width = 30 kHz
- Reuse factor N = 7
- Total available band width in each direction = 12.5 MHz
- Total coverage area = 10,000 km^2
- Required S/I ratio for analog system = 18 dB
- Required S/I ratio for digital system = 16 dB
- Call blocking = 2%

Solution

Analog System

- Number of voice channels per cell site = $\frac{395}{7} = 56$

- Area of the hexagonal cell = $A = 2.6R^2$

- The offered traffic load (using Erlang-B traffic tables) = 45.9 Erlangs per cell site
- The carried traffic load = $(1.0 - 0.02) \times 45.9 = 44.98$ Erlangs per cell site

$$\text{Spectral Efficiency} = \frac{44.98 \times \left(\dfrac{10000}{2.6R^2}\right)}{12.5 \times 10000} = \frac{1.384}{R^2} \text{ Erlangs/\{km}^2\}/\text{MHz}$$

$$q = \left(6\frac{S}{I}\right)^{\frac{1}{4}}$$

$$q^2 = \left(6\frac{S}{I}\right)^{\frac{1}{2}}$$

$$\therefore \frac{q_{digital}^2}{q_{analog}^2} = \sqrt{\frac{39.8}{63.1}} = 0.7943$$

Digital System
- Number of channels per 30 kHz = 3
- Number of voice channels per cell site = 56 x 3 = 168
- Offered traffic load = 154.5 Erlangs per cell site
- Carried traffic load = 151.4 Erlangs per cell site

$$\text{Spectral Efficiency} = \frac{151.4}{12.5 \times 2.6R^2 \times 0.7943} = \frac{5.8648}{R^2} \text{ Erlangs/\{km}^2\}/\text{MHz}$$

Thus, the relative spectral efficiency of a digital system with respect to the present analog system is $\dfrac{5.8648}{1.384} = 4.24$.

E X A M P L E   4 – 5

Problem Statement

We consider a cellular system with 395 total allocated voice channels of 30 kHz each. The total available bandwidth in each direction is 12.5 MHz. The traffic is uniform with average call holding time of 120 seconds, and call blocking during the system busy hour is 2%. Calculate:

1. Calls per cell site per hour
2. Mean S/I ratio
3. Spectral efficiency in Erlangs/km^2/MHz

For a cell reuse factor N equal to 4, 7, and 12, respectively, and for omnidirectional, 120° and 60° systems, calculate the call capacity.

Plot spectral efficiency vs. cell radius for N = 7 and comment on the results. Assume that there are 10 MSs/km² with each MS generating traffic of 0.02 Erlangs. The slope of path loss is γ = 40 dB per decade.

Solution

We consider only the first-tier interferers and neglect the effects of cochannel interference from the second and other higher tiers. The mean S/I ratio can then be given as:

$$\text{Mean}\frac{S}{I} = \gamma\log(\sqrt{3N}) - 10\log m$$

where:

γ = slope of path loss (dB/decade),
m = number of interferers in the first tier,
m = 6 for omnidirectional system,
m = 2 for 120° sectorized system, and
m = 1 for 60° sectorized system.

The traffic carried per cell site = $V \times t \times A_c = V \times t \times 2.6R^2$

where:

V = number of MSs per km²,
t = traffic in Erlangs per MS, and
A_c = area of hexagonal cell (i.e., $A_c = 2.6\ R^2$).

The traffic carried per cell site = $10 \times 0.02 \times 2.6R^2 = 0.52R^2$.

$$\text{Spectral Efficiency} = \frac{\text{Traffic carried per cell} \times N_c}{B_w \times A}$$

where:

N_c = number of cells in the system (i.e., $\dfrac{A}{2.6R^2}$), and

A = area of the system.

$$\therefore \text{Spectral Efficiency} = \frac{\text{Traffic carried per cell}}{2.6R^2 \times B_w}$$

We will demonstrate the procedure for calculating the results in one row in Tables 4.4 and 4.5; the remaining calculations can be made easily.

120° Sectorized Cell Site

- N = 7

- Number of voice channels per sector = $\dfrac{395}{7 \times 3} \approx 19$

- Offered traffic load per sector from Erlang-B tables = 12.3 Erlangs

- Offered traffic load per cell site = $3 \times 12.3 = 36.9$ Erlangs

- Carried traffic load per cell site = $(1 - 0.02) \times 36.9 = 36.2$ Erlangs

$$\therefore \frac{\text{No. of calls per cell site per hour} \times 120}{3600} = 36.2$$

Thus, the number of calls per cell site per hour = 1086

The spectral efficiency for a cell radius = 2 km will be:

$$\text{Spectral Efficiency} = \frac{36.2}{2.6 \times 12.5 \times 2^2} = 0.278$$

$$\therefore R = \sqrt{\frac{36.2}{0.52}} = 8.3 \text{ km, and}$$

$$\text{mean } \frac{S}{I} = 40\log\sqrt{21} - 10\log 2 = 26.44 - 30.1 = 23.43 \text{ dB}$$

Table 4.4 Omni vs. Sectorized Cellular System Performance

System	N	Channels per Sector	Offered Load (E) per Cell	Carried Load (E) per Cell	Calls per Cell per Hour	Cell Radius (km)	Mean S/I (dB)
Omni	4	99	87.0	85.3	2559	12.8	13.8
	7	56	45.9	45.0	1350	9.3	18.7
	12	33	24.6	24.1	723	6.8	23.3
120° Sector	4	33	73.8	72.3	2169	11.8	18.6
	7	19	36.9	36.2	1086	8.3	23.4
	12	11	17.5	17.2	516	5.8	28.1
60° Sector	4	17	64.2	62.9	1887	11.0	21.6
	7	9	26.0	25.5	765	7.0	26.4
	12	6	13.7	13.4	402	5.1	31.1

Table 4.5 Spectral Efficiency in Erlangs/km²/MHz vs. Cell Radius (km) [see Figure 4.12]

System	N	q = D/R	Cell Radius (km)				
			2	4	6	8	10
Omni	4	3.5	0.656	0.164	0.073	0.041	0.026
	7	4.6	0.346	0.087	0.038	0.022	0.014
	12	6.0	0.185	0.046	0.021	0.012	0.007
120° Sector	4	3.5	0.556	0.139	0.062	0.035	0.022
	7	4.6	0.278	0.070	0.031	0.017	0.011

Table 4.5 Spectral Efficiency in Erlangs/km²/MHz vs. Cell Radius (km) [see Figure 4.12] (Continued)

| System | N | q = D/R | \multicolumn{5}{c}{Cell Radius (km)} |
			2	4	6	8	10
	12	6	0.132	0.033	0.015	0.008	0.005
60° Sector	4	3.5	0.484	0.121	0.054	0.030	0.019
	7	4.6	0.196	0.049	0.022	0.012	0.008
	12	6.0	0.103	0.026	0.012	0.006	0.004

From the results in Tables 4.4 and 4.5, we draw the following conclusions:

1. Sectorization reduces cochannel interference and improves the mean S/I ratio for a given cell reuse factor. However, it reduces trunking efficiency since the channel resource is distributed more thinly among the various sectors. As a result, spectrum efficiency of a sectorized system is reduced if the reuse factor is kept constant.

2. Since a sectorized cellular system has fewer cochannel interferers, it is possible to reduce the reuse factor, hence increasing the spectrum efficiency of the overall system.

3. An omnidirectional cellular system requires a cluster size of 7, while a 120° sectorized system requires a reuse factor of 4; 60° sectorized system requires a reuse factor of 3 for desired mean S/I ratio of $\approx$ 18 dB.

4.8 CELL SPLITTING

As the traffic within a particular cell increases, the cell is split into smaller cells. This is done in such a way that cell areas, or the individual component coverage areas of the cellular system, are further divided to yield yet more cell areas (refer to Figure 4.13). The splitting of cell areas by adding new cells provides for increasing the amount of channel reuse and, hence, increasing subscriber serving capacity.

Decreasing cell radii imply that cell boundaries will be crossed more often. This will result in more handovers per call and a higher processing load per subscriber. Simple calculations show that a reduction in a cell radius by a factor of four will produce about a tenfold increase in the handover rate per subscriber. Since the call processing load tends to increase geometrically with the increase in the number of subscribers, with cell splitting the handover rate will increase exponentially. Therefore, it is essential to perform a cost-benefit study to compare the overall cost of cell splitting vs. other available alternatives to handle increased traffic load.

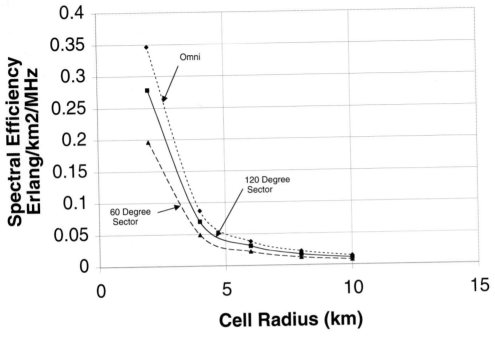

Fig. 4.12 Spectral Efficiency vs. Cell Radius

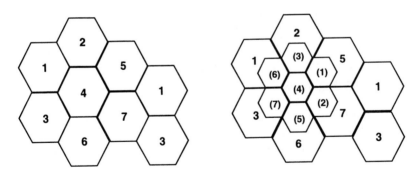

Growing by splitting cell 4 into cells of small size

Fig. 4.13 Cell Splitting

4.9 SUMMARY

In this chapter, we developed a relationship between the the reuse ratio (q) and the reuse factor (N) for a hexagonal cell geometry. We derived cochannel interference ratios for the omnidirectional and sectorized cell and presented several numerical examples to show that, for a given reuse factor, sectorization provides a higher S/I ratio, but reduces the spectral efficiency. However, it is possible to achieve a higher spectrum efficiency by reducing the reuse factor in a sectorized system without lowering the S/I ratio below the minimum requirement. The North American TDMA system often uses a reuse factor of N = 7, whereas the GSM system employs N = 4. We concluded the chapter by discussing the cell splitting procedures used in cellular communications.

4.10 PROBLEMS

1. Sometimes it is possible to delay cell splitting by allowing the blocking probability to increase for a short period of time. This will increase the carried load of the system at the expense of some customer dissatisfaction. If only one or two cells are experiencing high blocking, this may be a viable alternative to cell splitting. For example 4-1, calculate the increase in carried traffic load when the blocking probability is allowed to increase to 3%, 5%, and 10%. If the wireless system is growing at a rate of 20% per year, how much extra time is gained before cell splitting for each blocking rate compared to the normal 2% blocking rate? Is this a viable alternative? Explain your answer.

2. In section 4.4, the S/I ratio was calculated by neglecting the interference from cells other than the nearest ones on the same frequency. Calculate the amount of interference from the next ring of cells. Is it reasonable to neglect this interference?

3. In example 4-4, the offered and carried load of the digital system was calculated for 168 digital voice channels per cell site. If the number of channels per cell site is lower, the trunking efficiency (from the Erlang B tables) is lower and the system becomes more expensive per user. If a new spectrum was opened up that allowed less than 168 digital channels per cell site but allowed a cost per channel of 25% less than the current system, how many channels would be needed to have the same cost per user as the competing systems with larger channel capacity.

4. Consider a GSM system with a one-way spectrum of 12.5 MHz and channel spacing of 200 kHz. There are three control channels per cell and the reuse factor is 4. Assuming an omnidirectional antenna with six interferers in the first tier and a slope for path loss of 45 dB/decade ($\gamma = 4.5$), calculate the number of calls per cell site per hour with 2% blocking during system busy hour and an average call holding time of 120 seconds. What is the S/I ratio?

5. Repeat question 4 with 3-sector and 6-sector antennas. Discuss your results with an omnidirectional antenna and provide comments.

6. Compare the spectral efficiency of the GSM system with respect to the TACS using the following data:

- GSM channel spacing: 200 kHz
- TACS channel spacing: 25 kHz
- The required S/I for GSM: 12 dB
- The required S/I for TACS: 16 dB
- The total available one-way spectrum: 25 MHz
- The number of control channels for GSM per cell = 3
- The number of control channels for TACS per cell = 6
- The reuse factor for TACS: 7
- The reuse factor for GSM: 4
- Total coverage area = 10,000 km^2

4.11 REFERENCES

1. AT&T Technical Education Center, "Cellular System Design & Performance Engineering I," CC1400, version 1.12, 1993.

2. Chen, G. K., "Effects of Sectorization on the Spectrum Efficiency of Cellular Radio Systems," *IEEE Transactions on Vehicular Technology* 41 (3), August 1992, pp. 217–25.

3. Dersch, U., and Braun, W., "A Physical Mobile Radio Channel Model," Proceedings of IEEE Vehicular Technology Conference, May 1991, pp. 289–94.

4. French, R. C., "The Effects of Fading & Shadowing on Channel Reuse in Mobile Radio," *IEEE Transactions on Vehicular Technology* 28, August 1979.

5. Lee, W. C. Y., *Mobile Cellular Telecommunications System*, McGraw-Hil, New York, 1989.

6. Lee, W. C. Y., "Spectrum Efficiency in Cellular," *IEEE Transactions on Vehicular Technology* 38, May 1989, pp. 69–75.

7. Lee, W. C. Y., "Elements of Cellular Mobile Radio System," *IEEE Transactions on Vehicular Technology* 35, May 1986, pp. 48–56.

8. Lee, W. C. Y., "Spectrum Efficiency and Digital Cellular," presented at 38th IEEE Vehicular Technology Conference, Philadelphia, PA, June 1988.

9. MacDonald, V. H., "The Cellular Concept," *Bell System Technical Journal* 58 (1), January 1979, pp. 15–41.

10. Mehrotra, A., *Cellular Radio Analog & Digital System*, Artech House, Boston, 1994.

11. Whitehead, J. F., "Cellular System Design: An Emerging Engineering Discipline," *IEEE Communications Magazine* 24 (2), February 1986, pp. 8–15.

12. Young, W. R., "Advanced Mobile Phone Service: Introduction, Background, and Objectives," *Bell System Technical Journal* 58 (1), January 1979, pp. 1–14.

GSM Architecture and Interfaces

5.1 INTRODUCTION

In this chapter we present an overview of the GSM as described in ETSI's recommendations. The chapter discusses GSM frequency bands, the GSM Public Land Mobile Network (PLMN) and its objectives and services, GSM architecture and GSM subsystem entities, interfaces, and protocols between GSM entities. We address the mapping between GSM protocols and OSI layers and present the architecture of the North American PCS-1900.

5.2 GSM FREQUENCY BANDS

The GSM system is a frequency- and time-division system; each physical channel is characterized by a carrier frequency and a time slot number. GSM system frequencies include two bands at 900 MHz and 1800 MHz commonly referred to as the GSM-900 and DCS-1800 systems. For the primary band in the GSM-900 system, 124 radio carriers have been defined and assigned in two sub-bands of 25 MHz each in the 890–915 MHz and 935–960 MHz ranges, with channel widths of 200 kHz. Each carrier is divided into frames of 8 time slots (for full rate), with a frame duration of about 4.6 ms. For DCS-1800, there are two sub-bands of 75 MHz in the 1710–1785 MHz and 1805–1880 MHz ranges.

5.3 GSM PLMN

ETSI originally defined GSM as a European digital cellular telephony standard. GSM interfaces defined by ETSI lay the groundwork for a multivendor network approach to digital mobile communication. Figure 5.1 shows a GSM PLMN.

GSM offers users good voice quality, call privacy, and network security. SIM cards provide the security mechanism for GSM. SIM cards are like credit cards and identify the user to the GSM network. They can be used with any GSM handset, providing phone access, ensuring delivery of appropriate services to that user and automatically billing the subscriber's network usage back to the home network.

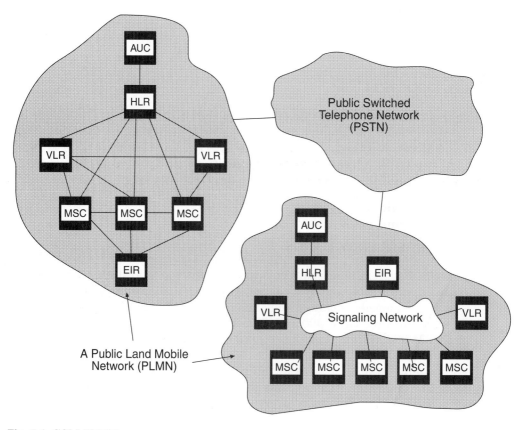

Fig. 5.1 GSM PLMN

Roaming arrangements between most GSM networks in Europe allow subscribers to have access to the same services no matter where they travel. The real gem of GSM is its MAP and its flexibility. This coupled with the SIM tool kit will allow service providers far more flexibility in the future than anything currently offered even in IS-41.

A major importance of GSM is its potential for delivering enhanced services requiring multimedia communication: voice, image, and data. Several mobile service providers offer free voice mailboxes and phone answering services to subscribers.

The key to delivering enhanced services is SS7, a robust set of protocol layers designed to provide fast, efficient, reliable transfer and delivery of signaling information across the signaling network and to support both the switched voice and nonvoice applications. With SS7 on the enhanced services platform and integrated mailbox parameters, subscribers can be notified about the number of stored messages in their mailboxes, time and source of last messages, message urgency, and whether the messages are voice or fax. Future applications such as fax store-and-forward, and audiotex can also use the platform's voice and data handling capabilities.

5.4 OBJECTIVES OF A GSM PLMN

A GSM PLMN cannot establish calls autonomously other than local calls between mobile subscribers. In most cases, the GSM PLMN depends upon the existing wireline networks to route the calls. Most of the time the service provided to a subscriber is a combination of the access service by a GSM PLMN and the service by some existing wireline network. Thus, the general objectives of a GSM PLMN network with respect to services to a subscriber are

☞ To provide the subscriber a wide range of services and facilities, both voice and nonvoice, that are compatible with those offered by existing networks (e.g., PSTN, ISDN)

☞ To introduce a mobile radio system that is compatible with ISDN

☞ To provide certain services and facilities exclusive to mobile situations

☞ To give access to the GSM network for a mobile subscriber in a country that operates the GSM system

☞ To provide facilities for automatic roaming, locating, and updating of mobile subscribers

☞ To provide service to a wide range of MSs, including vehicle-mounted stations, portable stations, and handheld stations

☞ To provide for efficient use of the frequency spectrum

☞ To allow for a low-cost infrastructure and terminal and to keep cost of service low

5.5 GSM PLMN SERVICES

A telecommunication service supported by the GSM PLMN is defined as a group of communication capabilities that the service provider offers to the subscribers. The basic telecommunication services provided by the GSM PLMN are divided into three main groups (for additional details on GSM services, refer to chapter 9):

☞ **Bearer services.** These services give the subscriber the capacity required to transmit appropriate signals between certain access points (i.e., user-network interfaces).

☞ **Teleservices.** These services provide the subscriber with necessary capabilities including terminal equipment functions to communicate with other subscribers.

☞ **Supplementary services.** These services modify or supplement basic telecommunications services and are offered together or in association with basic telecommunications services.

The GSM system offers the opportunity for a subscriber to roam freely through countries where a GSM PLMN is operational. Agreements are required between the various service providers to guarantee access to services offered to subscribers.

5.6 GSM SUBSYSTEMS

A series of functions are required to support the services and facilities in the GSM PLMN. The basic subsystems of the GSM architecture are (Figure 5.2) the Base Station Subsystem (BSS), Network and Switching Subsystem (NSS), and Operational Subsystem (OSS).

The BSS provides and manages transmission paths between the MSs and the NSS. This includes management of the radio interface between MSs and the rest of the GSM system. The NSS has the responsibility of managing communications and connecting MSs to the relevant networks or other MSs. The NSS is not in direct contact with the MSs. Neither is the BSS in direct contact with external networks. The MS, BSS, and NSS form the operational part of the GSM system. The OSS provides means for a service provider to control and manage the GSM system. In the GSM, interaction between the subsystems can be grouped in two main parts:

☞ **Operational.** External networks to/from NSS to/from BSS to/from MS to/from subscriber

☞ **Control.** OSS to/from service provider

The operational part provides transmission paths and establishes them. The control part interacts with the traffic-handling activity of the operational part by monitoring and modifying it to maintain or improve its functions.

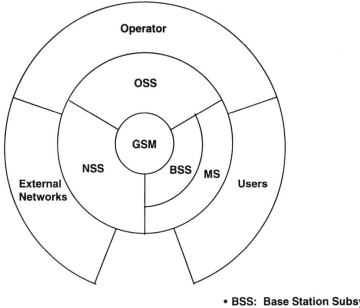

- **BSS: Base Station Subsystem**
- **NSS: Network and Switching Subsystem**
- **OSS: Operational Subsystem**
- **MS: Mobile Station**

Fig. 5.2 GSM Subsystems

5.6.1 GSM Subsystem Entities

Figure 5.3 shows the functional entities of the GSM and their logical interconnection. We will briefly describe these functional entities here.

5.6.1.1 MS The MS consists of the physical equipment used by the subscriber to access a PLMN for offered telecommunication services. Functionally, the MS includes a Mobile Termination (MT) and, depending on the services it can support, various Terminal Equipment (TE), and combinations of TE and Terminal Adaptor (TA) functions (the TA acts as a gateway between the TE and the MT) (see Figure 5.4). Various types of MS, such as the vehicle-mounted station, portable station, or handheld station, are used.

The MSs come in five power classes which define the maximum RF power level that the unit can transmit. Tables 5.1 and 5.2 provide the details of maximum RF power for various classes in GSM and DCS-1800. Vehicular and portable units can be either class I or class II, whereas handheld units can be class III, IV, and V. The typical classes are II and V. Table 5.3 provides the details of maximum RF power for GSM and DCS-1800 micro-BSs.

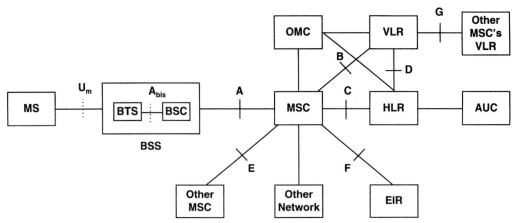

- **MS:**   **Mobile Station**
- **BSS:** **Base Station Subsystem**
- **BTS:** **Base Transceiver Station**
- **BSC:** **Base Station Controller**
- **MSC:** **Mobile Service Switching Center**
- **OMC:** **Operations and Maintenance Center**
- **HLR:** **Home Location Register**
- **VLR:** **Visitor Location Register**
- **EIR:** **Equipment Identity Register**
- **AUC:** **Authentication Center**

Fig. 5.3 GSM Reference Model

Table 5.1 Maximum RF Power for MS in GSM

Class	MS Max. RF Power (watts)
I	20 (not currently implemented)
II	8
III	5
IV	2
V	0.8

Table 5.2 Power Level in DCS-1800

Power Class	Max. MS RF Power watts (dBm)	Max. BS RF Power watts (dBm)
1	1 (30)	20 (43)
2	0.25 (24)	10 (40)
3		5 (37)
4		2.5 (34)

Table 5.3 Power Levels for Micro-BS in GSM and DCS-1800

Power Class	Max. RF Power of GSM Micro-BS, watts (dBm)	Max. RF Power of DCS-1800 Micro-BS, watts (dBm)
M1	0.25 (24)	1.6 (32)
M2	0.08 (19)	0.5 (27)
M3	0.03 (14)	0.16 (22)

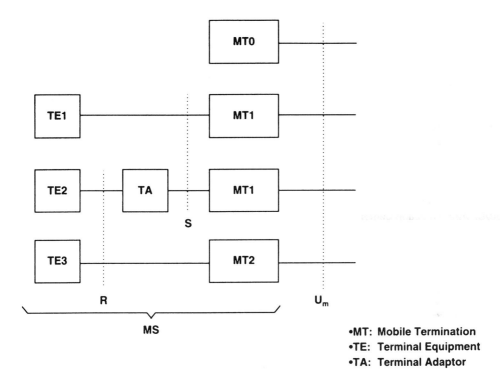

•MT: Mobile Termination
•TE: Terminal Equipment
•TA: Terminal Adaptor

Fig. 5.4 Types of MSs

Basically, an MS can be divided into two parts. The first part contains the hardware and software to support radio and human interface functions. The second part contains terminal/user-specific data in the form of a smart card, which can effectively be considered a sort of logical terminal. The SIM card plugs into the first part of the MS and remains in for the duration of use. Without the SIM card, the MS is not associated with any user and cannot make or receive calls (except possibly an emergency call if the network allows). The SIM card is issued by the mobile service provider after subscription, while the first part of the MS would be available at retail shops to buy or rent. This type of SIM card mobility is analogous to terminal mobility, but provides a personal-mobility-like service within the GSM mobile network (refer to chapter 11 for more details).

An MS has a number of identities including the International Mobile Equipment Identity (IMEI), the International Mobile Subscriber Identity (IMSI), and the ISDN number. The IMSI is stored in the SIM. The SIM card contains all the subscriber-related information stored on the user's side of the radio interface.

☞ **IMSI.** The IMSI is assigned to an MS at subscription time. It uniquely identifies a given MS. The IMSI will be transmitted over the radio interface only if necessary. The IMSI contains 15 digits and includes

  ✗ Mobile Country Code (MCC)—3 digits (home country)

  ✗ Mobile Network Code (MNC)—2 digits (home GSM PLMN)

  ✗ Mobile Subscriber Identification (MSIN)

  ✗ National Mobile Subscriber Identity (NMSI)

☞ **Temporary Mobile Subscriber Identity (TMSI).** The TMSI is assigned to an MS by the VLR. The TMSI uniquely identifies an MS within the area controlled by a given VLR. The maximum number of bits that can be used for the TMSI is 32.

☞ **IMEI.** The IMEI uniquely identifies the MS equipment. It is assigned by the equipment manufacturer. The IMEI contains 15 digits and carries

  ✗ The Type Approval Code (TAC)—6 digits

  ✗ The Final Assembly Code (FAC)—2 digits

  ✗ The serial number (SN)— 6 digits

  ✗ A Spare (SP)—1 digit

☞ **SIM.** The SIM carries the following information (see chapter 11 for more details):

  ✗ IMSI

  ✗ Authentication Key (K_i)

  ✗ Subscriber information

  ✗ Access control class

  ✗ Cipher Key (K_c)[*]

  ✗ TMSI[*]

  ✗ Additional GSM services[*]

  ✗ Location Area Identity (LAI)[*]

  ✗ Forbidden PLMN

---

*Updated by the network.

5.6.1.2 BSS The BSS is the physical equipment that provides radio coverage to prescribed geographical areas, known as the cells. It contains equipment required to communicate with the MS. Functionally, a BSS consists of a control function carried out by the BSC and a transmitting function performed by the BTS. The BTS is the radio transmission equipment and covers each cell. A BSS can serve several cells because it can have multiple BTSs.

The BTS contains the Transcoder Rate Adapter Unit (TRAU). In TRAU, the GSM-specific speech encoding and decoding is carried out, as well as the rate adaptation function for data. In certain situations the TRAU is located at the MSC to gain an advantage of more compressed transmission between the BTS and the MSC.

5.6.1.3 NSS The NSS includes the main switching functions of GSM, databases required for the subscribers, and mobility management. Its main role is to manage the communications between GSM and other network users. Within the NSS, the switching functions are performed by the MSC. Subscriber information relevant to provisioning of services is kept in the HLR. The other database in the NSS is the VLR.

The MSC performs the necessary switching functions required for the MSs located in an associated geographical area, called an MSC area (see Figure 5.5).

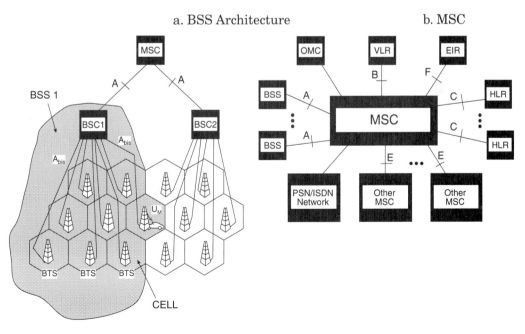

Fig. 5.5 MSC Area in GSM

The MSC monitors the mobility of its subscribers and manages necessary resources required to handle and update the location registration procedures and to carry out the handover functions. The MSC is involved in the interworking functions to communicate with other networks such as PSTN and ISDN. The interworking functions of the MSC depend upon the type of the network to which it is connected and the type of service to be performed. The call routing and control and echo control functions are also performed by the MSC.

The HLR is the functional unit used for management of mobile subscribers. The number of HLRs in a PLMN varies with the characteristics of the PLMN. Two types of information are stored in the HLR: subscriber information and part of the mobile information to allow incoming calls to be routed to the MSC for the particular MS. Any administrative action by the service provider on subscriber data is performed in the HLR. The HLR stores IMSI, MS ISDN number, VLR address, and subscriber data (e.g., supplementary services).

The VLR is linked to one or more MSCs. The VLR is the functional unit that dynamically stores subscriber information when the subscriber is located in the area covered by the VLR. When a roaming MS enters an MSC area, the MSC informs the associated VLR about the MS; the MS goes through a registration procedure. The registration procedure for the MS includes these activities:

☞ The VLR recognizes that the MS is from another PLMN.

☞ If roaming is allowed, the VLR finds the MS's HLR in its home PLMN.

☞ The VLR constructs a Global Title (GT) from the IMSI to allow signaling from the VLR to the MS's HLR via the PSTN/ISDN networks.

☞ The VLR generates a Mobile Subscriber Roaming Number (MSRN) that is used to route incoming calls to the MS.

☞ The MSRN is sent to the MS's HLR.

The information in the VLR includes MSRN, TMSI, the location area in which the MS has been registered, data related to supplementary service, MS ISDN number, IMSI, HLR address or GT, and local MS identity, if used.

The NSS contains more than MSCs, HLRs, and VLRs. In order to deliver an incoming call to a GSM user, the call is first routed to a gateway switch, referred to as the Gateway Mobile Service Switching Center (GMSC). The GMSC is responsible for collecting the location information and routing the call to the MSC through which the subscriber can obtain service at that instant (i.e., the visited MSC). The GMSC first finds the right HLR from the directory number of the GSM subscriber and interrogates it. The GMSC has an interface with external networks for which it provides gateway function, as well as with the SS7 signaling network for interworking with other NSS entities.

5.6.1.4 Operation and Maintenance Subsystem (OMSS) The OMSS is responsible for handling system security based on validation of identities of various telecommunications entities. These functions are performed in the Authentication Center (AuC) and EIR.

The AuC is accessed by the HLR to determine whether an MS will be granted service.

The EIR provides MS information used by the MSC. The EIR maintains a list of legitimate, fraudulent, or faulty MSs.

The OMSS is also in charge of remote operation and maintenance functions of the PLMN. These functions are monitored and controlled in the OMSS. The OMSS may have one or more Network Management Centers (NMCs) to centralize PLMN control.

The Operational and Maintenance Center (OMC) is the functional entity through which the service provider monitors and controls the system. The OMC provides a single point for the maintenance personnel to maintain the entire system. One OMC can serve multiple MSCs.

5.7 GSM INTERFACES

5.7.1 The Radio Interface (MS to BTS)

The U_m radio interface (between MS and base transceiver stations [BTS]) is the most important in any mobile radio system, in that it addresses the demanding characteristics of the radio environment. The physical layer interfaces to the data link layer and radio resource management sublayer in the MS and BS and to other functional units in the MS and network subsystem (which includes the BSS and MSC) for supporting traffic channels. The physical interface comprises a set of physical channels accessible through FDMA and TDMA.

Each physical channel supports a number of logical channels used for user traffic and signaling. The physical layer (or layer 1) supports the functions required for the transmission of bit streams on the air interface. Layer 1 also provides access capabilities to upper layers. The physical layer is described in the GSM Recommendation 05 series (part of the ETSI documentation for GSM). At the physical level, most signaling messages carried on the radio path are in 23-octet blocks. The data link layer functions are multiplexing, error detection and correction, flow control, and segmentation to allow for long messages on the upper layers.

The radio interface uses the Link Access Protocol on Dm channel (LAPDm). This protocol is based on the principles of the ISDN Link Access Protocol on the D channel (LAPD) protocol. Layer 2 is described in GSM Recommendations 04.05 and 04.06. The following logical channel types are supported (see chapter 7 for the details of logical channel types):

☞ Speech traffic channels (TCH)
  ✗ Full-rate TCH (TCH/F)
  ✗ Half-rate TCH (TCH/H)
☞ Broadcast channels (BCCH)
  ✗ Frequency correction channel (FCCH)
  ✗ Synchronization channel (SCH)
  ✗ Broadcast control channel (BCCH)
☞ Common control channels (CCCH)
  ✗ Paging channel (PCH)
  ✗ Random access channel (RACH)
  ✗ Access grant channel (AGCH)
☞ Cell broadcast channel (CBCH)
  ✗ Cell broadcast channel (CBCH) (the CBCH uses the same physical channel as the DCCH)
☞ Dedicated control channels (DCCH)
  ✗ Slow associated control channel (SACCH)
  ✗ Stand-alone dedicated control channel (SDCCH)
  ✗ Fast associated control channel (FACCH)

The radio resource layer manages the dialog between the MS and BSS concerning the management of the radio connection, including connection establishment, control, release, and changes (e.g., during handover). The mobility management layer deals with supporting functions of location update, authentication, and encryption management in a mobile environment. In the connection management layer, the call control entity controls end-to-end call establishment and management, and the supplementary service entity supports the management of supplementary services. Both protocols are similar to those used in the fixed wireline network (for more details refer to chapter 9). The SMS protocol of this layer supports the high-level functions related to the transfer and management of short message services.

5.7.2 A_bis Interface (BTS to BSC)

The interconnection between the BTS and the BSC is through a standard interface, A_{bis} (most A_{bis} interfaces are vendor specific). The primary functions carried over this interface are traffic channel transmission, terrestrial channel management, and radio channel management. This interface supports two types of communications links: traffic channels at 64 kbps carrying speech or user data for a full- or half-rate radio traffic channel and signaling channels at 16 kbps carrying information for BSC-BTS and BSC-MSC signaling. The BSC handles the LAPD channel signaling for every BTS carrier. The first three layers are based on the following OSI/ITU-T recommendations:

☞ Physical layer: ITU-T Recommendation G.703 and GSM Recommendation 0-8.54

☞ Data link layer: GSM Recommendation 08.56 (LAPD)

☞ Network layer: GSM Recommendation 08.58

There are two types of messages handled by the traffic management procedure part of the signaling interface—**transparent** and **nontransparent**. Transparent messages are between the MS and BSC-MSC and do not require analysis by the BTS. Nontransparent messages do require BTS analysis.

5.7.3 A Interface (BSC to MSC)

The A interface allows interconnection between the BSS radio base subsystem and the MSC. The physical layer of the A interface is a 2-Mbps standard Consultative Committee on Telephone and Telegraph (CCITT) digital connection. The signaling transport uses Message Transfer Part (MTP) and Signaling Connection Control Part (SCCP) of SS7 (see chapter 17 for details). Error-free transport is handled by a subset of the MTP, and logical connection is handled by a subset of the SCCP. The application parts are divided between the BSS application part (BSSAP) and BSS operation and maintenance application part (BSSOMAP). The BSSAP is further divided into Direct Transfer Application Part (DTAP) and BSS management application part (BSSMAP). The DTAP is used to transfer layer 3 messages between the MS and the MSC without BSC involvement. The BSSMAP is responsible for all aspects of radio resource handling at the BSS. The BSSOMAP supports all the operation and maintenance communications of BSS (refer to chapter 15 for more details).

5.7.4 Interfaces between Other GSM Entities

Information transfer between GSM PLMN entities uses the MAP. The MAP contains a mobile application and several Application Service Elements (ASEs). It uses the service of the Transaction Capabilities Application Part (TCAP) of SS7. It employs the SCCP to offer the necessary signaling functions required to provide services such as setting mobile facilities for voice and non-voice application in a mobile network. The major procedures supported by MAP are

☞ Location registration and cancellation

☞ Handover procedures

☞ Handling supplementary services

☞ Retrieval of subscriber parameters during call setup

☞ Authentication procedures.

Figure 5.3 shows the various interfaces between the GSM entities. In Figure 5.6, protocols used between the GSM entities are given.

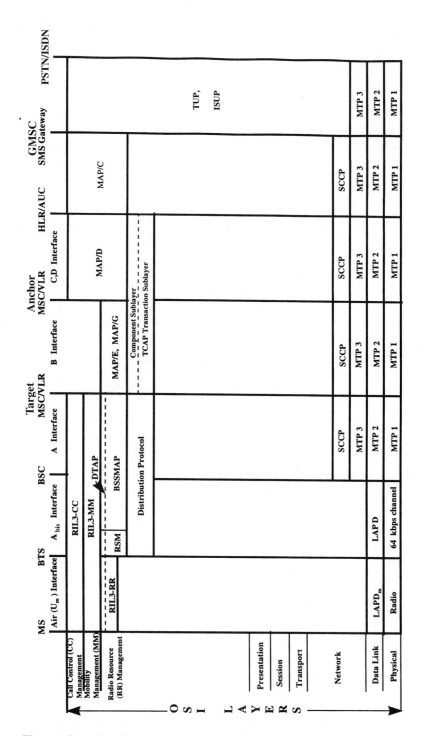

Fig. 5.6 Signaling Protocols between GSM Entities

5.8 Mapping of GSM Layers onto OSI Layers

When an MS is switched on somewhere, it first has to determine whether it has access to a PLMN. It initiates a location update to inform its home PLMN about its current location in order to enable the routing of incoming calls to the subscriber. The location of an MS is stored in a central database, the HLR of the PLMN where the customer has purchased service. In addition to other user-specific information, the HLR maintains the routing number to an MSC. The MSC is primarily responsible for switching and mobility management (MM). Once connected to an MSC, a BSS communicates with the MS via the radio interface. Each MS, positioned in a cell of a BSS of an MSC, is registered with a specific database associated with the MSC and the VLR. If a call to a GSM subscriber is generated from an external network, the call is routed to a Gateway MSC (GMSC) first. This GMSC interrogates the HLR of the called subscriber to obtain the routing number of the visited MSC. The latter then initiates the transmission of a paging message within each of its associated cells. If the called subscriber answers, the BSS assigns a traffic channel to be used for the communication, and the link is fully established.

During a call, the MS is allowed to move from cell to cell in the whole GSM service area, and GSM maintains the communication links without interruption of the end-to-end connection. The handover procedure in GSM is mobile assisted and performed by the BSS. The MS periodically measures downlink signal quality and reports it to its serving BTS, as well as to all cells in its neighborhood that are prospective candidates for handover. Different handover types can be performed, changing either a channel in the serving cell (i.e., the serving BTS remains the same) or changing the cell inside the area controlled by a BSC; between two BSCs within a location area; or between two location areas (i.e., MSCs).

The GSM protocol architecture for signaling and mapping onto the corresponding OSI layers is shown in Figure 5.7. GSM uses out-of-band signaling through a separate signaling network.

As discussed in section 5.7, at the data link layer the radio interface of the MS uses LAPDm protocol. The higher-layer protocols of GSM are grouped into the third layer. GSM layer 3 includes functionality of higher OSI layers and OSI management, such as connection management, subscriber identification, and authentication.

At the interface between BSC and MSC, the lower layers are realized by MTP of SS7. It covers functionality of layer 1, layer 2, and part of layer 3 of the OSI reference model. The MTP itself is layered into three levels. The two lower levels are mapped directly onto the corresponding OSI layers, and level 3 covers the lower part of the OSI network layer. The missing functionality of the higher part of the network layer is provided by SCCP. The BSSAP serves primarily as a bridge between the radio resource (RR) management and the MSC, handling for instance the assignment and switching at call setup and

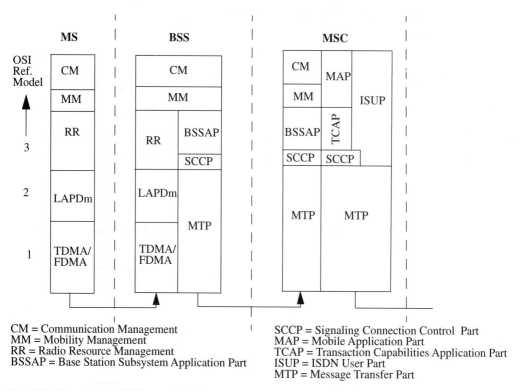

CM = Communication Management
MM = Mobility Management
RR = Radio Resource Management
BSSAP = Base Station Subsystem Application Part

SCCP = Signaling Connection Control Part
MAP = Mobile Application Part
TCAP = Transaction Capabilities Application Part
ISUP = ISDN User Part
MTP = Message Transfer Part

Fig. 5.7 Mapping of GSM onto OSI Layers

handover processing. It therefore provides the functionality typically provided by the transport layer, application layer, and network management of OSI.

The MSC is connected to the signaling network via SS7 and is responsible for exchange of all information required for call setup, maintenance, and management. TCAP contains functions to provide associations between two TCAP users as well as protocols and services to perform remote operations. It is closely related to the Remote Operation Service Element (ROSE) of the OSI application layer. Since TCAP directly uses the services of SCCP, the transport, session, and presentation layers are null layers. Hence, this part of the SS7 is a typical example of a system using a reduced protocol stack where functions of different OSI protocol layers are incorporated into the remaining layers. TCAP provides functionality of the OSI transport layer.

The call-related signaling between MSCs and external networks uses the ISDN User Part (ISUP), while all GSM-specific signaling between MSC and location registers is performed via the MAP. These protocols correspond to the OSI application layer, although their functionality is mainly used to maintain network-level connections. It can be noticed that the network complexity of telecommunication networks seems to yield protocols that combine functionality distributed across the higher layers and management part of the OSI protocol stack (for details of signaling protocols refer to chapter 17).

5.9 NORTH AMERICAN PCS-1900

Figure 5.8 shows the functional model that has been derived from the T1P1 reference model [1]. Several physical scenarios can be developed using the functional entities shown in Figure 5.8. Figure 5.9 shows the Functional Entity (FE) grouping in which the physical interface between the Radio System (RS) and the Switching System Platform (SSP) carries both the call control (CC) and mobility management messages.

- ☞ Radio Terminal Function (RTF) FE—it is the subscriber unit (SU). The only physical interface is to the Radio System (RS) using the air interface.
- ☞ Radio Control Function (RCF) FE and Radio Access Control Function (RACF) FE—these are included in the RS. Combining these FEs onto the same platform allows air-interface-specific functions (such as those that would impact handover) to be isolated from the other interfaces. OS

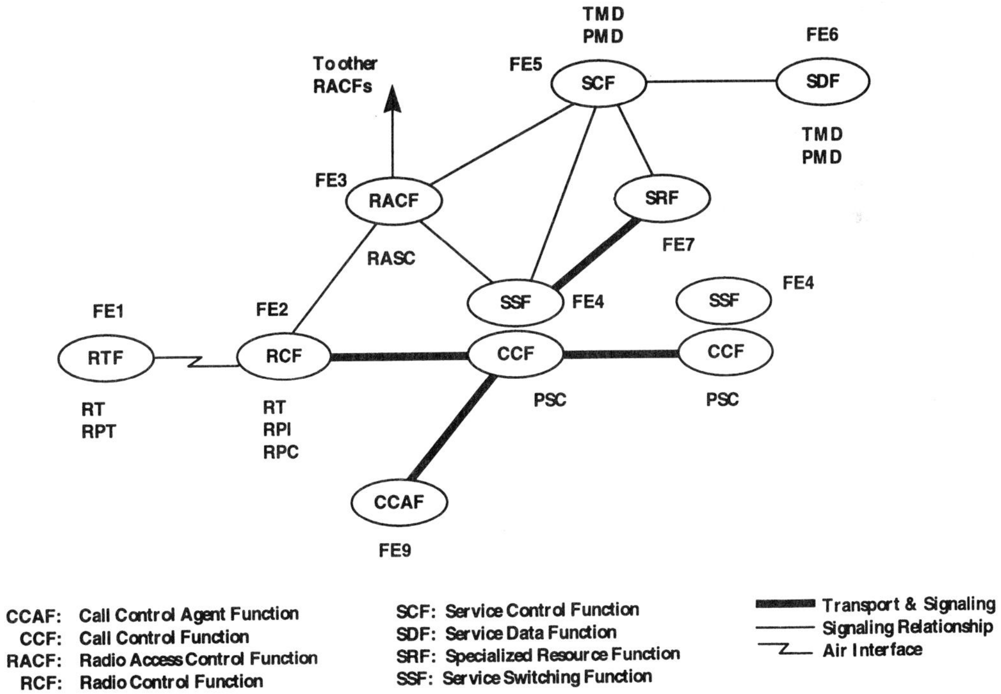
TMD
PMD
FE6
To other RACFs
FE5
SCF
SDF
TMD
PMD
FE3
RACF
SRF
FE7
RASC
FE4
FE1
FE2
SSF
FE4
SSF
RTF
RCF
CCF
CCF
RT
RPT
RT
RPI
RPC
PSC
PSC
CCAF
FE9
CCAF: Call Control Agent Function
CCF: Call Control Function
RACF: Radio Access Control Function
RCF: Radio Control Function
SCF: Service Control Function
SDF: Service Data Function
SRF: Specialized Resource Function
SSF: Service Switching Function
Transport & Signaling
Signaling Relationship
Air Interface

Fig. 5.8 Functional Model Derived from T1P1 Reference Model [1]

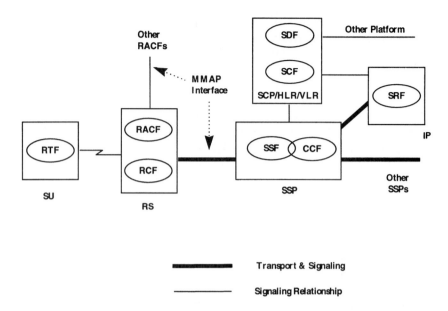

Fig. 5.9 Functional Entity Groupings

information, including performance data and accounting records, is generated, collected, and formatted on this platform. There is only one physical interface to the SSP to carry both the call control and mobility management signaling.

☞ SSF/CCF FE—it is contained in SSP and provides interfaces to operator services, E911, international calls, and network repair/maintenance centers. Physical interfaces for this collection include: to the RS, to the mobility management platform, to the information provider, and to other SSPs and external networks.

☞ Specialized Resource Function (SRF) FE and data interworking function—they are contained in the information provider. Physical interfaces for this collection include one to SSP and another to the mobility management platform.

☞ Individually the SSF/CCF FE and CCF FE represent interswitch and internetwork functional entity collections and physical interfaces.

As shown in Figure 5.9, only the interface to the RS is from the SSP. There is no direct physical path between the RS and the SCP/VLR. All operations to or from the RS pass through the SSP, whether or not the SSP terminates or ignores the operation.

The proposed North American PCS-1900 standard is an extension of the ETSI DCS-1800 that was initially developed for the frequency band of 1800

MHz. PCS-1900 consists of 200-kHz radio channels shared by 8 time slots, one per terminal. The PCS-1900 standard supports a frequency duplex arrangement for forward and reverse links. It uses a fixed rate Residual Pulse Excitation (RPE) based on a speech coder that operates at 13 kbps.

The North American types of handover are network initiated and Mobile Assisted Handover (MAHO). In the case of the network-initiated handover, both hard and soft handover are supported. The PCS-1900 standard defines support for MAHO and a form of network-initiated handover that applies only to hard handover. For PCS-1900 systems to function as an integral part of the North American PCS environment, handover needs will be supported between PCS-1900 and North American systems.

PCS-1900 supports voice privacy through the encryption capabilities. Encryption (voice privacy) is an air interface capability that can be controlled by the network operator rather than as a service that may not be controlled by the network operator but may also be offered as a service to the end user. The GSM encryption is only an air interface function and does not depend on the GSM MAP function.

The authentication algorithm in the PCS-1900 uses IMSI as one of its inputs. The terminal possesses a "key," which is the same "key" known by the home network. The network computes a signature that is specific for an end user. This signature is used to authenticate the end user through the duration of the service. This authentication scheme has its strength in the authentication algorithm. However, there is no mechanism to recognize clones.

To satisfy the PCS needs and requirements for ubiquity (accessibility) and seamless service, air interface transparency must exist. Transparency implies that an end user can have access to service regardless of the access method.

In the initial phase of PCS, multiple air interfaces may exist, and therefore dual-mode or dual-spectrum terminals may be used. The aim is to attain some level of interoperability with the existing North American networks. If interoperability does not exist between the PCS-1900 air interface and the analog AMPS 800 MHz air interface, the ubiquity of service is precluded. The PCS-1900 air interface may access the network that provides GSM services. The AMPS analog air interface may have access to IS-41 services.

5.10 SUMMARY

In this chapter, we presented an overview of the GSM system, which consists of four subsystems—MSS, BSS, NSS, and OSS. We also described functional entities in each of the subsystems and presented interfaces and protocols used between different functional entities of the GSM system. We included the mapping of the GSM protocols onto the OSI layers and provided the architecture of the PCS-1900 (a derivative of GSM) in North America.

5.11 REFERENCES

1. Garg, V. K., and Wilkes, J. E., *Wireless and Personal Communication Systems*, Prentice Hall, 1996.

2. ETSI, GSM Specification Series 01.02–1.06, "GSM Overview, Glossary, Abbreviations, Service Phases."

3. ETSI, GSM Specification Series 02.01–2.88, "GSM Services and Features."

4. ETSI, GSM Specification Series 03.01–3.88, "GSM PLMN Functions, Architecture, Numbering and Addressing Procedures."

5. ETSI, GSM Specification Series 04.01–4.88, "MS-BSS Interface."

6. ETSI, GSM Specification Series 06.01–6.32, "Radio Link."

7. ETSI, GSM Specification Series 07.01–7.03, "Terminal Adaptation."

8. ETSI, GSM Specification Series 08.01–8.60, "BSS-MSC Interface, BSC-BTS Interface."

9. ETSI, GSM Specification Series 09.01–9.11, "Network Interworking, MAP."

10. Mouly, M., and Pautet, M., *The GSM System for Mobile Communications*, Mouly and Pautet, Palaiseau, France, 1992.

CHAPTER 6

Radio Link Features in GSM Systems

6.1 INTRODUCTION

The GSM system uses a number of interference-reducing mechanisms. These include

☞ Adaptive Power Control (APC)
☞ Discontinuous Transmission (DTX)
☞ Slow Frequency Hopping (SFH)

The purpose of power control is to adjust the power of the radio transmitter and adapt to the needs of an actual radio link between the BTS and the MS. If the received power and quality are high enough, the transmitted power is reduced stepwise to a minimum necessary value.

Discontinuous transmission means that the transmitter is powered on only if the subscriber is actively talking. In the speech gaps nothing is transmitted except a minor portion of signaling traffic. The percentage time when a subscriber actively speaks is given by speech activity factor v. Measurements indicate that v can take values between 40 percent and 60 percent.

SFH averages the influence of interference. As a result all BSs experience more or less the same interference. Because the interference is distrib-

uted equally it is possible to reconstruct erroneous bits by channel coding. The
SFH algorithms used in the GSM are discussed in a subsequent section.

In this chapter, we first describe radio link measurements in GSM. In
section 6.3 we present the details of APC, DTX, and SFH. Section 6.4 is
devoted to future techniques (i.e., channel borrowing and smart antenna) that
may be used in GSM to reduce interference and improve system performance.

6.2 RADIO LINK MEASUREMENTS

In GSM the MS uses the BS identity code (BSIC) (see Figure 6.1) to distinguish between neighboring BSs.

The signal level values (RXLEV) and signal quality level values
(RXQUAL) used in GSM are listed in Tables 6.1 and 6.2, respectively. Intercell
handover from the serving cell to a neighbor cell occurs when RXLEV and/or
RXQUAL is low on the serving cell and better on the neighbor cell. Intracell
handover from one channel/time slot to another channel/time slot in the same
cell occurs when RXLEV is high but RXQUAL is low.

The MS monitors the signal strength of neighbor BSs and maintains a
list of the six strongest nonserving BSs. A new BS is selected from the list if

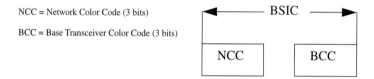

NCC = Network Color Code (3 bits)

BCC = Base Transceiver Color Code (3 bits)

Fig. 6.1 GSM BSIC

Table 6.1 Signal Level Values in GSM

RXLEV	dBm
0	< −110
1	−110 to −109
2	−109 to −108
3	−108 to −107
•	•
•	•
62	−49 to −48
63	> −48

Table 6.2 Quality Level Values in GSM

RXQUAL	Bit Error Rate (BER) (%)
0	< 0.2
1	0.2 to 0.4
2	0.4 to 0.8
3	0.8 to 1.6
4	1.6 to 3.2
5	3.2 to 6.4
6	6.4 to 12.8
7	> 12.8

☞ The path loss criterion for the serving BS is not met for 5 seconds.

☞ The signaling link with the serving BS fails.

☞ The serving BS becomes barred.

☞ Nonserving cell access signal is greater than that of the serving BS for 5 seconds, and by at least the CELL_RESELECT_HYSTERSIS value in dB

6.3 RADIO LINK FEATURES OF GSM

6.3.1 Dynamic Power Control

The GSM network is designed so that the MS is instructed to use only the minimum power level necessary to achieve effective communication with the BTS. GSM defines eight power classes for the BTS transmitter (see Table 6.3) to cover all five classes of MSs (0.8 W to 20 W).

The MS measures the receive power level of the serving BS, the quality of receive signal, the receive power level, and ID codes for up to six neighbor BSs. The BS measures the receive power level and signal quality of each MS, the distance to the MS, and the transmit power of the MS and BS.

Signal power level is determined by averaging the incoming signal level over a specified period of time. The receive power level in dBm is mapped to a value between 0 and 63. The value 0 corresponds to <-110 dBm (see Table 6.1). Quality level is determined by computing bit error rate (BER) (see Table 6.2). The BER is mapped to 8 levels where 0 is the best quality (BER $< 2 \times 10^{-3}$). Once the BS determines the minimum required transmit power, it sends this information to the MS.

In the BS, transmit power control may be employed, but it is optional. For the BTS, the power output is nominally controlled in 2-dB steps to provide

Table 6.3 GSM BS Power Classes

BS Power Class	Maximum RF Power (W)
I	320
II	160
III	80
IV	40
V	20
VI	10
VII	5
VIII	2.5

better cochannel interference performance. This allows a better QoS or a greater frequency reuse. The setting of the power step at 0 for the BTS corresponds to the relevant power class of the BTS transmitter.

Both MS and BTS power control is performed in 2-dB steps (see Figure 6.2), down from the level of the power class to a minimum of +13 dBm (refer to Table 6.4). The power output level of the MS is controlled in a monotonic sequence of 15 steps of 2 dB on the command through SACCH from the BTS. Power levels are sent to MSs via a 5-bit transmit power (TXPWR) field in the downlink SACCH message block. The MS sends confirmation to the BTS via MS_TXPWR_CONF field in the uplink SACCH message.

The use of minimum transmitting power to access the network helps to increase the battery life of the mobile set and reduce interference. By carefully controlling the ramp-up of the transmitter as well as the power level, the spectral interference with other GSM equipment can be minimized.

6.3.2 Discontinuous Transmission (DTX)

If we assume that the transmitter is active only 50 percent of the time, then the average interference can be reduced by 3 dB. A more practical method that is directly applicable to the capacity assessment procedure is to relate the DTX activity factor v of the transmitter to the occupancy; e.g., if v = 0.4, P_{occup} = 0.8, then the effective channel occupancy P_{active} = 0.4 x 0.8 = 0.32. P_{occup} is used to determine the capacity, whereas P_{active} is used to calculate the interference. The improvement in the overall S/I ratio of the system allows the use of a smaller reuse distance and, therefore, increases the spectrum efficiency.

Another advantage of DTX is the reduction in the power consumption of the mobile phone; this is especially important for the handheld unit. The drawback of DTX is the introduction of clipping to the speech due to imperfect

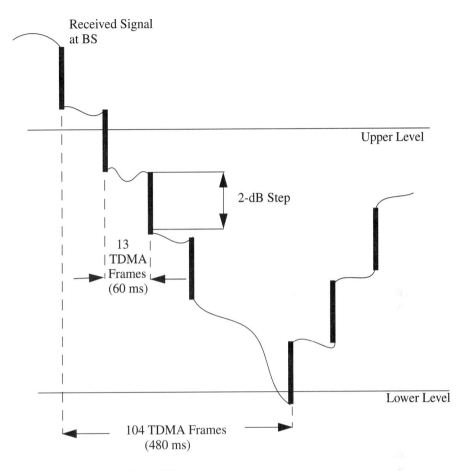

Fig. 6.2 Power Control in GSM

operation of the voice activity detector (VAD) which can sometimes fail to detect the exact times of the start and end of the speech bursts. Another disadvantage of DTX is the noise contrast between the active and silent periods. The use of DTX on the downlink is not possible. Therefore, it is potentially possible to have more capacity on the uplink with DTX than the downlink. However, it probably has little practical significance as the traffic in both links is usually equal for such applications. Consequently, DTX can be used to improve the quality of the uplink only without additional gains in capacity. The capacity advantage is achieved from decreasing reuse distance because DTX has a greater impact when the system is interference limited.

Discontinuous transmission is a GSM feature in which speech is transmitted only when there is speech available to transmit. This helps to reduce RF interference in MSs. A VAD is used to initiate the switching process. Com-

Table 6.4 GSM MS Power Classes and Power Control (dBm)

Power Control Level	Mobile Power Class				
	I (20 W)	II (8 W)	III (5 W)	IV (2 W)	V (0.8 W)
0	43	—	—	—	—
1	41	—	—	—	—
2	39	39	—	—	—
3	37	37	37	—	—
4	35	35	35	—	—
5	33	33	33	33	—
6	31	31	31	31	—
7	29	29	29	29	29
8	27	27	27	27	27
9	25	25	25	25	25
10	23	23	23	23	23
11	21	21	21	21	21
12	19	19	19	19	19
13	17	17	17	17	17
14	15	15	15	15	15
15	13	13	13	13	13

fort noise is introduced at the receiver to help to maintain a high level of intelligibility during the intervals when the speech is cut. Comfort noise is low-level background noise based on the statistics of the acoustic noise at the transmitter to assure the listener that the radio link is still active. The use of DTX saves power and helps to provide a longer battery life.

GSM transmission will cease four speech block periods (20 ms each) (see Figure 6.3) after speech activity has stopped. However, the mobile will periodically send a signal called a silence indicator (SID) every 480 ms to provide comfort noise level information to the BS so that the person on the far end can hear some low-level noise and not conclude that the link is down.

6.3.3 SFH

SFH is used in GSM to improve performance in the multipath fading environment and to reduce the required S/I ratio. The mobile radio channel is a frequency-selective fading channel. Fades occur when there is a loss in signal

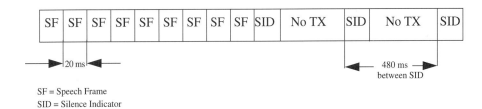

SF = Speech Frame
SID = Silence Indicator

Fig. 6.3 DTX in GSM

power due to variations in terrain such as valleys or hills or due to objects such as buildings or even large metal objects such as aircraft interfering with the signal path, causing the original signal to be attenuated or canceled out. When the mobile passes through areas of fade and poor reception, there is less chance of losing the radio link in these areas by invoking SFH. This is especially important while performing a handover to another cell.

GSM uses SFH to improve signal quality. In SFH, the hop rate is less than the message bit rate, whereas, in a fast hop system, FH occurs at a rate that is much higher than the message bit rate. Fast FH is generally used in the SS systems.

In GSM, the operating frequency is changed only with every TDMA frame. The hopping rate is 216.7 hops per second. This corresponds to 1/(frame duration) or $1/(4.1615 \times 10^{-3})$. A mobile transmits at one frequency during a time slot and hops to a different frequency before the next time slot. A frequency synthesizer is used to change the frequency and to settle on the new frequency within a fraction of one time slot (577 μs).

FH provides frequency diversity to overcome Rayleigh fading due to multipath propagation. Rayleigh fading may cause fades of 40 to 50 dB deep on the received signal, and the radio link may be lost. FH allows the maintenance of the radio link by shifting onto another frequency before the link is totally lost.

FH also provides interference diversity. At any time, the amount of interference on various channels in a given cell varies from channel to channel. A receiver on a channel with strong interference will not suffer excessive errors because the receiver will not carry successive bursts on the same high-interference channel.

FH reduces the S/I ratio required for good communications. For a non-hopping radio link, the minimum required S/I ratio is about 12 dB, whereas FH reduces the requirement to 9 dB. With the reduced S/I requirement, the system capacity will be improved.

Different hopping algorithms can be assigned to the MS with a given channel set. One of the algorithms is cyclic hopping in which FH is performed through the assigned frequency list, i.e., from first frequency, to second frequency, to third frequency, and so on until the frequency list is repeated. The other algorithm is the random hopping in which FH takes place in a random fashion through the frequency list. A BTS with m frequencies available can theoretically perform up to $m!$ different nonrepeating hopping sequences.

For a set of N_f frequencies, GSM allows 64 sequences to be built. The difference between the MS and BTS in frequency hopping mode is that in the MS only three out of the eight time slots are available to receive, transmit, and monitor, whereas the BTS uses all eight time slots, since it is capable of supporting eight MSs in one frame. The BTS must also be capable of receiving and transmitting in all eight time slots. The FCCH, SCH, and BCCH are not allowed to hop. The SCH is used for synchronization with the system as the MS initially seeks service or gets ready to move to another cell. The FCCH is used to allow the MS to accurately tune to a BS. The BCCH is used to provide general information on a per-BTS basis required by the MS for registration into the system.

Two different implementation schemes of SFH are used in BSs. These are RF hopping and baseband hopping. The RF hopping needs agile transceivers, as in the MS, except that two or three synthesizers are often required to allow one synthesizer to be tuned while the others are being used. The tuning time for each individual synthesizer is a minimum of one time slot. The main disadvantage of the RF hopping is that a hybrid combiner must be used since there needs to be non-frequency-selective signal combining. Also, a continuous transmission of the BCCH is required. RF hopping is suitable for BTS configurations with few (two to three) transceivers.

The baseband hopping (see Figure 6.4) is suitable when a large number of transceivers are used in one BTS. It uses a fixed-frequency transceiver and multiplexes a number of baseband processing systems to use the appropriate transmitter for the defined hopping sequence. Bseband hopping avoids the wideband combiner because the resulting frequency of each transceiver is fixed, so a selective combiner can be used. Baseband hopping requires one transceiver

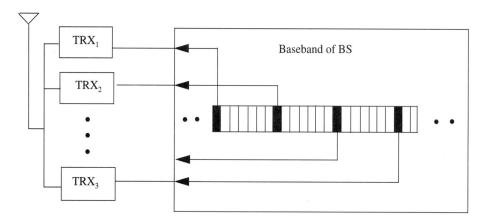

Fig. 6.4 Baseband Frequency Hopping Implementation

to be allocated for one frequency. This implementation is only cost effective in large systems that already have a number of transceivers at the BTS.

Hybrid hopping is a combination and compromise of the two implementation schemes. For the receive path, RF hopping is used because the need for wideband filters over the GSM frequency range does not present a problem for the selectivity of the BTS. Baseband hopping is used for the transmit path to reduce the output losses. The intermodulation requirements are met by providing extra selectivity by selection combiners.

GSM allows 64 x N_f different FH sequences that are generated for N_f different frequencies. Two parameters—*mobile allocation index offset* (MAIO) and *hopping sequence number* (HSN)—are used to describe them. The MAIO takes as many values as the number of frequencies in the set. Its value lies in between 0 to $N-1$ (represented by 6 bits). This determines the next frequency where the mobile will hop. The HSN is $0 \leq HSN \leq 63$. Cyclic hopping results when HSN = 0; otherwise the hopping is random.

Figure 6.5 shows the SFH on three different frequencies for GSM. The hopping sequences are C_0 to C_2, C_2 to C_1, and C_1 to C_0.

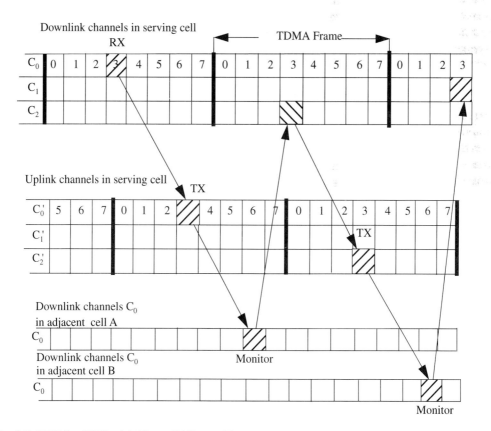

Fig. 6.5 SFH for GSM with Three Different Frequencies

The algorithm used in GSM (see Figure 6.6) requires input parameters MA (actual allocation of frequencies $1 \leq N_f \leq 64$ to mobile), MAIO, HSN, $FN(T_1)$, $FN(T_2)$, and $FN(T_3)$. The values of FNs are received over the SCH. If HSN = 0, the algorithm chooses a cyclic hopping path and the mobile allocation index (MAI) is calculated as

$$MAI = (FN + MAIO)modN \qquad (6.1)$$

If HSN is not equal to 0, M, M', and T' are calculated. If $M' \leq N$, the intermediate parameter S is set to M', otherwise $S = (M' + T')$ mod N. Finally, MAI is calculated (see Figure 6.6). The same hopping sequence is used in both uplink and downlink.

6.4 FUTURE TECHNIQUES TO REDUCE INTERFERENCE IN GSM

The demand for spectrum to serve mobile users has rapidly increased. Therefore techniques that enhance system capacity are of prime interest. There are two main approaches that can be used in GSM for that purpose:

☞ Channel borrowing or effective channel management
☞ Advanced antenna technology

In the following section, we present brief descriptions of these approaches. We first discuss these channel borrowing schemes:

☞ Dynamic Channel Allocation (DCA)
☞ Hybrid Channel Assignment (HCA)
☞ Channel Borrowing without Locking (CBWL)

We then present the advanced antenna technology (smart or intelligent) to reduce interference.

6.4.1 Channel Borrowing

In DCA [2], a central pool of all channels is used. A channel is borrowed from the pool by a BS for use on a call. When the call is completed the channel is returned to the pool. The basic DCA has a self-organizing channel assignment algorithm based on dynamic real-time measurements of interference levels. These measurements are usually performed at the MS in order to reduce the computational load and the complexity of the system. All BS radios have access to the whole channel set, even if each BS is equipped with a number of transceivers less than the number of channels.

In the call setup phase, the BS assignment is done on the strongest signal from neighboring BSs (preferred BS). The channel assignment is based on interference consideration. The interference level of the idle channels is mea-

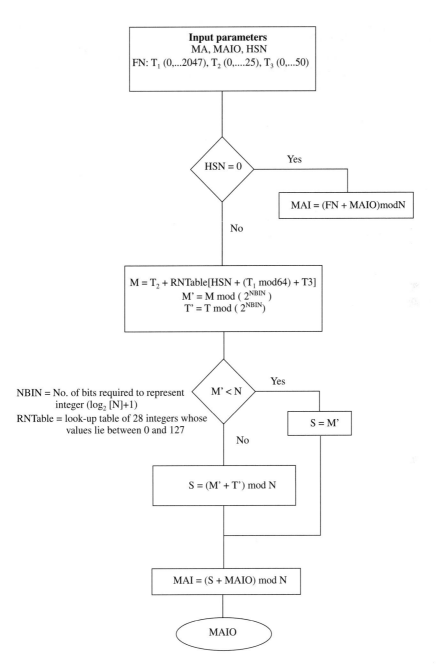

Fig. 6.6 Hopping Algorithm in GSM

sured and, by means of the signal level from the preferred BS, the resulting S/I ratio is estimated. If the S/I ratio exceeds the selected threshold value, the channel is considered a suitable channel. Different DCA algorithms differ in the selection of the preferred channel among the suitable channels. If the suitable channel set is found empty (no good-quality channels available at the preferred BS), the system looks for a new BS, received with a sufficient signal level, and searches for a good-quality channel. If no suitable channel is found, the call is blocked. During the call, the system monitors the quality of channel under use by measuring BER. If the quality threshold is exceeded, a handover request is initiated. A suitable channel is searched for among the channels of the current BS that show a better quality. This is a typical DCA operation, namely an intracell handover. If no channel is found, an intercell handover is used, searching again for a channel with suitable quality. If even this does not work, the call is forced to terminate.

Several variations of DCA have been proposed and some of them have been implemented. Simulation results indicate that Adaptive Channel Allocation (ACA), a variation of DCA, significantly increases the capacity of a TDMA system as compared to the traditional fixed channel assignment (FCA) (about 100 percent improvement). The ACA scheme with a subset of total available channels reserved for use of handovers shows significantly better results in terms of lost calls during handovers. However, this gives higher blocking at the initial access state.

In HCA [6], some channels are permanently assigned to each BS as in FCA, and others are kept in a central pool for borrowing as in DCA. *Channel locking* is used to prevent an increase in cochannel interference; that is, BSs within the required minimum channel reuse distance from a BS that borrows a channel cannot use the same channel.

Channel locking has some disadvantages. One is that the number of channels available for lending to a BS is limited, since the channel can be borrowed by a BS only when it is idle in all of the BSs within the required channel reuse distance of the borrowing BS. Another disadvantage is the difficulty in maintaining cochannel reuse distance at the minimum required value everywhere in the system. Because of this difficulty, DCA and HCA generally perform less satisfactorily than FCA under high loads [2,6]. Their other disadvantages relate to physical complexity. The transmitter of each BS must be able to transmit not only on the channel allocated permanently to that BS, but also on any of the channels that belong to the central pool. Also, to implement channel borrowing in a given BS, information must be known about channel usage at the BSs within the channel reuse distance of the given BS. This causes some complexity in the management of system resources. Reference [1] uses the generalized FCA and collision-type request channels to improve system performance.

The CBWL proposed in [5] has most of the advantages of other channel borrowing schemes and overcomes their disadvantages. In the CBWL, each BS is allocated channels as in FCA. If all channels of the BS are occupied and a new call arrives, channel borrowing is used. A channel can be borrowed only from an adjacent BS. The borrowed channel cannot be used by the original lending BS but can still be used in any nearby cochannel BSs. Thus, there is no channel locking. To prevent the increase of cochannel interference, borrowed channels are used with reduced transmitted power. Therefore, they can be accessed only in part of the borrowing cell. To determine whether a mobile is in the region that can be served by a borrowed channel, each BS transmits a signal with the same reduced power as that on a borrowed channel. The signal is called *borrowed channel sensing signal* (BCSS). If the BCSS is not above some suitable threshold at an MS, a borrowed channel cannot be used.

CBWL offers advantages in comparison with DCA and HCA. In CBWL, only a fraction of the total channels of the system need to be accessible at each BS. Without channel locking, channel reuse distance is always kept at a desired minimum. The CBWL exhibits better performance in light as well as in heavy traffic loads. In CBWL, channel borrowing at a BS does not require global information about channel usage in the system, so the control and management tasks are simplified. In CBWL, by appropriately organizing lending channel groups, it is possible to avoid using adjacent channels in the same BS, even with channel borrowing allowed. In comparison with generalized FCA, CBWL ensures good quality for borrowed and regular channels without increasing cochannel interference. In CBWL, a user can borrow channels from any of the adjacent BSs. There are six of these in the standard hexagonal layout geometry. The greater the number of channels that are potentially available to an arriving call, the more superior the performance.

The CBWL can be employed in existing cellular systems without additional infrastructure cost. Unlike cell splitting, CBWL does not require new BSs and additional antenna towers to increase system capacity. Simulations show that CBWL provides better channel utilization than a conventional cellular system with FCA, DCA, or HCA. In addition it can be beneficial in providing good performance in hard-to-reach (hot-spot) scenarios.

6.4.2 Smart Antenna [4]

Another way to reduce interference is to use a smart or intelligent antenna. A smart or intelligent antenna refers to a group of core RF technologies that control directional antenna arrays by means of sophisticated digital signal processing (DSP) algorithms. A smart antenna evaluates signal conditions continuously of each signal that is transmitted or received. The smart antenna then uses this information to determine how to manipulate the incoming signals to maximize performance. The smart antenna constructs a composite signal from multiple antenna feeds by optimizing signal characteristics. The optimization is accomplished by assigning specific weight to each of the incoming signals. A smart antenna functions autonomously and automatically and makes complex decisions in real time [3].

Smart antennas belong to two basic classes: switched beam and adaptive. A switched beam antenna combines signals according to a fixed number of beam patterns. One of the patterns will be considered a best fit for the signal on an individual channel at a given instance. The system logic may select another pattern as conditions change (i.e., the processor is said to switch between patterns as it tracks the signal). Pattern characteristics may be selected, but a beam may not be steered or swept on a continuous basis. For such a system to operate, signal processing occurs simultaneously over each of the hundreds of radio channels in a network. Such processing requires a powerful DSP engine, which must analyze the antenna signal across the entire frequency band occupied by the network, identify individual channels, and then apply appropriate processing.

The adaptive antenna essentially picks out the desired signal amid a field of interfering signals and thermal noise and self-regulates its performance to satisfy some preassigned criterion or criteria. Such criteria may include periodicity of switching and beam traversal. The adaptive array antenna effectively creates new spatial channels, forming areas of high signal gain while reducing the effect of other signals to a minimum. The adaptive array antenna increases the sensitivity in the direction of the desired signal, while reducing the sensitivity in certain angular directions corresponding to interfering signals. The adaptive antenna adjusts its directional beam pattern by using spatial filtering and internal feedback control, thereby maximizing S/I ratio. The adaptive array antenna consists of a linear or rectangular array of M homogeneous radiating elements. These elements are coupled together via some type of amplitude control and phase shifting mechanism to form a single output. The amplitude and phase control involve a set of complex weights.

The total array output in direction ψ_k is given as:

$$y_k(t) = \sum_{n=1}^{M} w_n \cdot e^{j(\omega t + \psi_{nk})} \tag{6.2}$$

where

w_n = complex weight applied to the output of the nth element and ω = frequency

With a suitable choice of weights, the array can be made to accept a desired signal from a direction ψ_k and nullify interference signals originated ψ_k for $k \neq i$. The weighting mechanism is optimized to steer the beam in a specific direction or directions. In reference [7] it has been shown that an M elements array has $M - 1$ degrees of freedom, which yields a maximum $M - 1$ independent pattern nulls. If the weights are controlled by a feedback loop to maximize the S/I ratio at the array output, the system behaves as an adaptive spatial filter.

The antenna elements can be arranged in various geometries, with uniform line and circular and planar arrays being very common. A circular array geometry provides complete coverage from a central BS as a beam can be steered through 360 degrees. The spacing between antenna elements is very critical in the design of antenna arrays.

Each MS is tracked in azimuth by a narrow beam for BS-to-MS transmission. The directive nature of the beams ensures that in a given system the mean interference power experienced by a single MS due to other active MSs is much less than the amount of interference that would be experienced if a conventional antenna, either omnidirectional or sectorized, were used. With adaptive antenna technology, many of the system parameters that are considered constraints in a conventional single antenna system are actually useful, flexible parameters that a system designer can manipulate to optimize system performance. This aspect of adaptive array antenna technology is quite attractive for the cellular/PCS system. Since TDMA-based cellular/PCS systems are designed to be interference limited, the use of adaptive antennas would be considerably beneficial in the following areas:

☞ **Coverage.** Adaptive beamforming can increase the cell coverage area substantially due to antenna gain and interference rejection. In a noise-limited environment, the cell coverage area is improved by a factor of $M^{1/\gamma}$, where M is the number of antenna elements in the array and γ is the propagation loss exponent. Generally, fewer cell sites are required to cover the given service area with the application of adaptive antennas at the BS.

☞ **Capacity.** Transmission bit rate is increased due to improvement in the S/I ratio at the output of the adaptive beamformer. In a noise-limited environment, the minimum rate improvement in S/I ratio that can be achieved is $10\log M$ dB. In addition, adaptive antenna technology provides the flexibility that allows a reuse factor of one; that is, a single frequency can be used in all cells. This is accomplished with centralized DCA. Also, the same channel, time slot, and frequency can be reused in the same sector.

☞ **Signal quality.** In a noise-limited environment, minimum receiver thresholds are reduced by $10\log M$ dB on average. In an interference-limited environment, an additional improvement in tolerable S/I ratio at a single element is achieved due to interference rejection afforded against directional interferers. The amount of improvement depends on the distribution of cochannel users in the neighboring cells.

☞ **Portable terminal transmit power.** If adaptive antennas were implemented in a system without changing other parameters such as cell size, the transmission power levels required for portable terminals would be reduced on average by at least $10\log M$ dB. The reduction in transmission power level, which results due to increase in antenna gain at the BS, improves the battery life. Another relevant consideration is increased fade margin for improved signal quality. For example, higher data rates lead to increased coverage area per cell, thus decreasing deployment costs.

E X A M P L E   6 – 1

Problem Statement

We consider a GSM system with the following data to show the advantage of adaptive array antennas.

- Coverage area: 60,000 mile^2
- One-way system bandwidth: 12.5 MHz
- Channel spacing: 200 kHz
- Frequency reuse factor: 4
- MS output power (W): 800 mW (29 dBm)
- BS antenna gain (G_{bs}): 20 dBi
- Receive cable /connector loss (L_c): 2 dB
- MS antenna gain (G_m): 0 dB
- Required S/I ratio: 12 dB
- Information rate: 271 kbps
- Receiver noise figure (F): 7 dB
- Propagation path-loss exponent γ: 4
- One-mile path-loss intercept (I_0): 80 dBm
- Lognormal fading margin (f_m): 10 dB

We first calculate the required minimum received power $p_{r_{min}}$:

$$p_{r_{min}} = p_{r_{threshold}} = kTB_wF + S/I = -174 + 7 + 10\log 200 \times 10^3 + 12 = -102 \text{ dBm}$$

where:
kT= –174 dBm/Hz
B= receiver noise bandwidth = 200 kHz
F= noise figure = 7 dB
S/I= required signal-to-interference ratio in the receiver bandwidth = 12 dB

Maximum allowable path loss:

$$p_{L_{max}} = W - p_{r_{min}} - L_c - f_m + G_{bs} + G_m =$$
$$29 - (-102) - 2 - 10 + 20 + 0 = 139 \text{ dB}$$

Cell radius in miles:

$$R \le [p_{L_{max}} - I_0]^{1/\gamma} = [139 - 80]^{1/4} = [59dBm]^{0.25} = 29.85 \text{ miles}$$

Number of cells required to cover the service area:

$$\frac{A_{cover}}{A_{cell}} = \frac{60000}{2.6 \times (29.85)^2} \approx 26$$

Next we calculate the improvement in S/I ratio from the adaptive antenna array (see Table 6.5).

We then calculate the maximum cell radius and number of cells to cover the service area (see Table 6.6). The results show that the number of cells required to cover the service area with an adaptive array antenna is significantly smaller than with the conventional one-element antenna.

Table 6.5 Improvement in S/I Ratio from Adaptive Array Antenna

No. of Antenna Elements	Improvement in S/I (dB)	Required S/I (dB)
2	$10 \log 2 = 3.0$	$12 - 3 = 9$
4	$10 \log 4 = 6.0$	$12 - 6 = 6$
6	$10 \log 6 = 7.78$	$12 - 7.78 = 4.22$

Table 6.6 Maximum Cell Radius and Number of Cells to Cover Service Area

No. of Antenna Elements	Maximum Allowable Path Loss (dB)	Cell Radius (miles)	No. of Cells Required to Cover Service Area
1	139	29.85	26
2	142	35.48	19
4	145	42.17	13
6	146.78	46.72	11

6.5 SUMMARY

In this chapter we discussed a number of mechanisms to reduce interference in the GSM system. We use adaptive power control in GSM to adjust the power of the radio transmitter between the BTS and MS. Both MS and BTS power control is performed in 2-dB steps down from the maximum power (for the power class) to a minimum of 13 dBm. The use of minimum transmitting power to access the network helps to increase battery life of the mobile phone and reduce interference.

DTX allows the transmitter to be powered on only when the MS user is actively talking. During the speech gaps nothing is transmitted. DTX helps to reduce RF interference and power consumption of the MS. For the interference-limited system, capacity improvement is achieved from decreasing the reuse factor.

SFH averages the influence of interference and reduces the S/I ratio required for good communication. With FH in GSM, about a 3-dB advantage can be achieved in S/I ratio.

Channel borrowing (or effective channel management) and advanced antenna technology are the two prime candidates for enhancing GSM system capacity.With the use of ACA it is possible to achieve almost a 100-percent gain in the system capacity, whereas with a two-element adaptive array antenna, a 3-dB advantage in S/I ratio is possible.

6.6 REFERENCES

1. Choudhury, G. L., and Rappaport, T. S., "Cellular Communication Schemes Using Generalized Fixed Channel Assignment and Collision Type Request Channels," *IEEE Transactions on Vehicular Technology* VT-31 (2), May 1982, pp. 53–65.

2. Cox, D. C., and Reudink, D. O., "Increasing Channel Occupancy in Large-Scale Mobile Radio Systems: Dynamic Channel Reassignment," *IEEE Transactions on Vehicular Technology* VT-22, November 1973, pp. 218–23.

3. Compton, R. T., *Adaptive Antenna—Concepts and Performance*, Prentice Hall, Upper Saddle River, NJ, 1988.

4. Garg, V. K., and Huntington, L., "Applications of Adaptive Array Antenna to a TDMA Cellular/PCS System," *IEEE Communications Magazine* 35 (10), October 1997, pp. 148–52.

5. Jiang, H., and Rappaport, T. S., "CBWL: A New Channel Assignment and Sharing Method for Cellular Communication Systems," *IEEE Transactions on Vehicular Technology*, 1993, pp. 189–93.

6. Kahwa, T. J., and Georganas, N. D., "A Hybrid Channel Assignment Scheme in Large-Scale, Cellular-Structured Mobile Communications Systems," *IEEE Trans., Communications*, COM-26, April 1978, pp. 431–38.

7. Litva, J., and Lo, T. K., *Digital Beam Forming in Wireless Communication*, Artech House, Norwood, MA, 1996.

GSM Logical Channels and Frame Structure

7.1 INTRODUCTION

In this chapter, we discuss the logical channels that are used in the GSM system. The logical channels carry user information and control signaling data. Different logical channels are used for different tasks. The information transmitted on a logical channel depends on a particular task. We also present the frame structure of GSM and provide details of the five different kinds of bursts that are used in GSM to carry user and control information. Mobility management and mobile identification procedures are also discussed.

7.2 GSM LOGICAL CHANNELS

A large amount of information is transmitted between the MS and the BS, particularly, user information (voice or data) and control signaling data. Depending on the type of information transmitted, different logical channels are used. These logical channels are mapped onto the physical channels (time slots). As an example, digital speech is carried by the logical channel called the traffic channel which during transmission can be allocated a certain physical channel. In the GSM system no RF carrier or time slot is assigned beforehand for the exclusive use of any particular task. Just about any time slot of any RF carrier can be used for a number of different tasks.

There are two basic logical channel types in the GSM: TCHs and CCHs. TCHs are used to carry either encoded speech or user data both in the uplink and downlink directions. TCHs support two information rates—full rate (TCH/F) and half rate (TCH/H). TCH/F carries user speech at 13 kbps (gross rate of 22.8 kbps) and data at 9.6 kbps, 4.8 kbps, and 2.4 kbps. The data rates are padded to achieve the gross rates to 12 kbps, 6 kbps, and 3.6 kbps, respectively.

A TCH/H carries user speech at the gross rate of 11.4 kbps. The user data rates for the TCH/H are 4.8 kbps and 2.4 kbps, respectively.

There are three types of control channels: BCCH, CCCH, and DCCH. The BCCHs are point-to-multipoint unidirectional channels. These channels are used for such functions as correcting mobile frequencies, frame synchronization, and CCH structure. These are downlink only channels. Other channels that belong to the BCCH group are the FCCH and SCH.

The FCCH is the downlink point-to-multipoint channel. It carries information for frequency correction of the MS. This channel is required for the correct operation of the radio subsystem and allows an MS to accurately tune to a BS. The FCCH sends all zeros in its burst to represent an unmodulated carrier.

The SCH carries information for frame synchronization of the MS and the identification of a BTS. The SCH has a 64-bit binary sequence that is previously known to the MS. The MS achieves the exact timing synchronization with respect to a GSM frame by correlating the bits with the internally stored 64 bits. The SCH carries BTS identification code (BSIC) and reduced TDMA frame number (RFN).

The CCCHs are point-to-multipoint bidirectional channels. They are primarily used to carry signaling information necessary for accessing management functions. These channels are used to establish connections between MSs and BSs before a DCCH is assigned to an MS. There are two downlink (BS to MS) and one uplink (MS to BS) CCCHs defined. The downlink channels are the PCH and AGCH. The PCH is used to page MSs. The AGCH is used to assign an MS to a specific DCCH. The RACH, which is an uplink channel, is used to request assignment of a DCCH.

The DCCHs are used for signaling and control after call establishment. There are two types of DCCHs—SDCCH and Associated Control Channel (ACCH). The SDCCH is a DCCH whose allocation is not linked to the allocation of a TCH. This is used before the MS is assigned a TCH. The SDCCH is used for authentication of the MS, for location updates, and for assignment to TCHs. The ACCHs are of two types—SACCH and FACCH. The SACCH is always associated with a TCH or SDCCH. It is used to carry general control information. The FACCH is similar to a blank-and-burst channel in that user information is precluded while data is being sent. FACCH is used to transmit handover orders.

Figure 7.1 shows all the logical channels that are used in the GSM. Each logical channel is allocated a half-rate convolutional code. The GSM system employs 8 channels (for full rate) or 16 channels (for half rate) in a TDMA frame. The channel coding is provided for the full-rate channel and based on handling 60 bits of user information at 5-ms intervals in accordance with the modified CCITT V.110 modem standard. Two hundred forty bits of user information are applied with 4 trailing bits to the punctured convolutional coder with length k = 5. The resulting 488 coded bits are reduced to 456 encoded information bits through puncturing (32 bits are not transmitted). They are divided into 114-bit information bursts that are applied in an interleaved manner to consecutive time slots. GSM control messages are 184 bits long; they are encoded using a shortened binary cyclic fire code, followed by a half-rate convolutional coder. This produces 184 message bits followed by 40 parity bits. Four trailing bits are added to yield a 228-bit information block. The block is applied to a half-rate, k = 5 convolutional coder. The resulting 456 encoded bits are interleaved in the same manner as the traffic channel speech data.

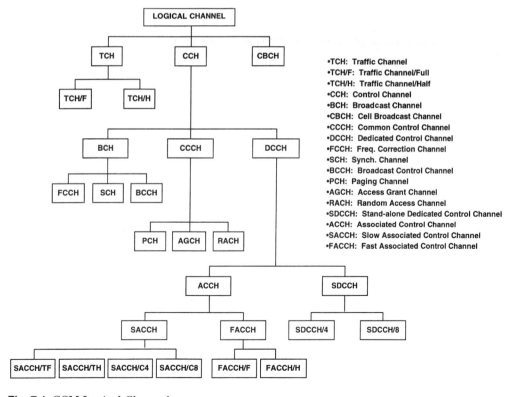

- TCH: Traffic Channel
- TCH/F: Traffic Channel/Full
- TCH/H: Traffic Channel/Half
- CCH: Control Channel
- BCH: Broadcast Channel
- CBCH: Cell Broadcast Channel
- CCCH: Common Control Channel
- DCCH: Dedicated Control Channel
- FCCH: Freq. Correction Channel
- SCH: Synch. Channel
- BCCH: Broadcast Control Channel
- PCH: Paging Channel
- AGCH: Access Grant Channel
- RACH: Random Access Channel
- SDCCH: Stand-alone Dedicated Control Channel
- ACCH: Associated Control Channel
- SACCH: Slow Associated Control Channel
- FACCH: Fast Associated Control Channel

Fig. 7.1 GSM Logical Channels

The 456 encoded bits in each 20-ms speech frame or control message are subdivided into eight 57-bit sub-blocks. The 8 sub-blocks that make up a single full-rate GSM frame are spread over 8 consecutive TCH time slots. If a burst is lost due to interference or fading, the channel coding ensures that enough bits will still be received correctly to allow the error correction to work.

7.3 ALLOWED LOGICAL CHANNEL COMBINATIONS

SDCCHs can share a physical channel with a BCH and a CCCH; in that case there can be 4 SDCCHs (referred to as SDCCH/4) mapped onto the same physical channel, or there can be 8 SDCCHs (referred to as SDCCH/8) that can share a physical channel with themselves.

An SACCH can be sent along with full- or half-rate TCHs. It is then referred to as SACCH/TF and SACCH/TH, respectively. It can also be sent along with SDCCH/4 and SDCCH/8. It is then referred to as SACCH/C4 and SACCH/C8, respectively.

FACCH is used to send preemptive signaling on full- or half-rate TCHs. It is then referred to as FACCH/F and FACCH/H, respectively.

The allowed logical channel combinations that can share the same physical channel in the GSM system are

☞ For traffic time slots

✗ TCH/F or FACCH/F and SACCH/TF traffic channel

✗ TCH/H(0) and TCH/H(1) or FACCH/H and SACCH/TH

✗ TCH/H(0) or FACCH/H(0) and SACCH/TH(0) + TCH/H(1) or FACCH (1) and SACCH/TH(1)

☞ For control time slots

✗ BCH, CCCH, SDCCH/4, and SACCH/4

✗ BCH and CCCH

✗ BCCH and CCCH

✗ SDCCH/8 and SACCH/C8

where:
BCH = FCCH and SCH and BCCH
CCCH = PCH or AGCH and RACH

7.3.1 TCH Multiframe for TCH/H

With one TCH/F, user information is sent in 24 out of 26 TDMA frames, with an SACCH frame and an idle frame occurring once every 26 TDMA frames (see Figure 7.2). With two TCH/H channels, user information requires only 12 out of 26 TDMA frames per TCH. In addition, there are two SACCH frames, one per user TCH, that require one TDMA frame out of every 26 frames as shown in Figure 7.3. Thus, two users can share the same physical channel by having their TCH frames and SACCH frames multiplexed onto the multiframe structure.

7.3.2 CCH Multiframe

The BCH and CCCH forward control channels are implemented only on certain Absolute Radio Frequency Channel Number (ARFCN) channels and are allocated time slots in a specific manner. BCH data is transmitted in time slot 0; the other time slots in a frame for that ARFCN are available for TCH data or DCCH data or are filled with dummy bursts. Furthermore, all time slots on all other ARFCNs within the cell are available for TCH or DCCH data.

The multiframe structure used for the first time slot of the radio channel (i.e., containing the downlink BCCH signal) is shown in Figure 7.4. In the downlink path, the FCCH is sent during the frequency correction frames (F),

0	1	2	3	4	5		12	13		23	24	25
T_0	T_1	T_2	T_3	T_4	T_5	— —	A	T_{13}	—	T_{23}	T_{24}	I

T: Frame for the ith TCH
A: SACCH
I: Idle Frame

Fig. 7.2 Speech Multiframe for TCH/F

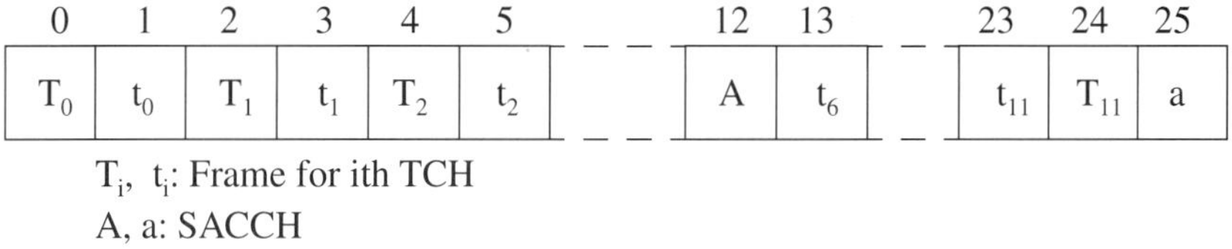

0	1	2	3	4	5		12	13		23	24	25
T_0	t_0	T_1	t_1	T_2	t_2	— —	A	t_6	—	t_{11}	T_{11}	a

T_i, t_i: Frame for ith TCH
A, a: SACCH

Fig. 7.3 Speech Multiframe for TCH/H

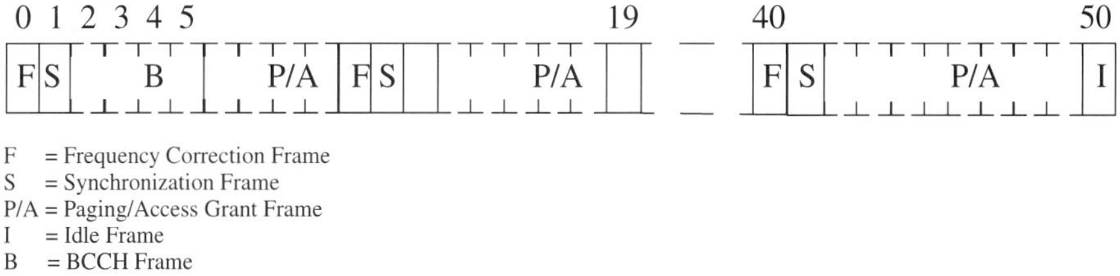

0 1 2 3 4 5   19   40   50

| F | S | B | P/A | F | S | P/A | | F | S | P/A | I |

F = Frequency Correction Frame
S = Synchronization Frame
P/A = Paging/Access Grant Frame
I = Idle Frame
B = BCCH Frame

Fig. 7.4 Downlink BCH + CCCH

the SCH during the synchronization frames (S), the BCCH during the BCCH frames (B), and PCH and AGCH of the CCCH during the PCH/AGCH frames (P/A). In the uplink path, all frames are used for the RACH (R) of the CCCH (see Figure 7.5). Idle frames are marked by I.

7.4 GSM FRAME STRUCTURE

The available forward (BS to MS, 935–960 MHz) and reverse (MS to BS, 890–915 MHz) bands are divided into 200-kHz channels referred to as ARFCNs. Each of the eight users for a full-rate channel utilizes the same ARFCN and occupies a unique time slot per frame. Each time slot consists of 156.25 bits, out of which 8.25 bits are used for guard time and 6 are the start and stop bits that are used to prevent overlap with adjacent time slots. Each time slot is 0.57692 ms. Figure 7.6 shows a normal burst (time slot). Only 148 bits are transmitted at a rate of 270.833 kbps. A single full-rate GSM frame contains 8 time slots with a time duration of 4.615 ms and 1250 bits (see Figure 7.6). The frame rate is 216.667 frames/s. The 13th and 26th frames are not used for traffic, but for control purposes.

7.5 GSM BURSTS

GSM uses five different types of the bursts: normal burst, synchronization burst, frequency correction burst, access burst, and dummy burst.

7.5.1 Normal Burst

The normal burst is used to carry information on TCH and control channels, except for RACH, SCH, and FCCH. This burst contains 156.25 bits. The encrypted bits are 57 bits of data or speech plus one bit "stealing Flag" to indicate whether the burst was stolen for FACCH signaling or not. The training sequence is a 26-bit pattern which is used by the equalizer to create a channel model. The tail bits always equal (0,0,0). They are used to provide start and stop bit patterns. The guard period is empty space and is used to prevent overlap between adjacent time slots during transmission (see Figure 7.7).

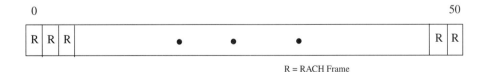

R = RACH Frame

Fig. 7.5 Uplink BCH + CCCH

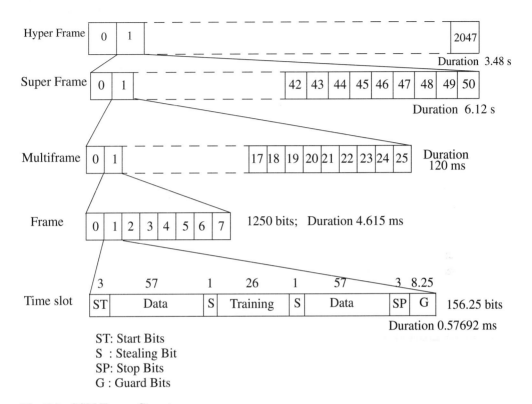

Fig. 7.6 GSM Frame Structure

3	57	1	26	1	57	3	8.25
TB	DATA	SF	T	SF	DATA	TB	G

TB: Tail Bits
SF: Stealing Bit
T: Training Bit
G: Guard Time

Fig. 7.7 Normal Burst

7.5.2 Synchronization Burst

The synchronization burst is used for time synchronization of the mobile. This burst contains a long synchronization sequence of 64 bits. The encrypted 78 bits are used to carry information of the TDMA frame number along with the BTS identification code (BSIC) as shown in Figure 7.8.

The TDMA frame is broadcast over an SCH to protect user information against eavesdropping. This is accomplished by ciphering the information

3	39	64	39	3	8.25
TB	Encrypted Bits	Extended Training Sequence	Encrypted Bits	TB	G

Fig. 7.8 Synchronization Burst

before transmitting it. The algorithm uses a TDMA frame number as an input parameter for calculating the ciphering key. By knowing the TDMA frame number, the MS will know what kind of logical channel is being transmitted on the CCH time slot 0. The BSIC is also used by the mobile to check the identity of the BTS during the measurements of signal strength (see Figure 7.9).

7.5.3 Frequency Correction Channel Burst

This burst is used for frequency synchronization of the MS. The fixed input bits are all zeros, causing the modulator to deliver an unmodulated carrier with an offset of 1625/24 kHz above the nominal frequency (see Figure 7.10).

7.5.4 Access Burst

The access burst is used for random access and has a longer guard period to protect for burst transmission from an MS that does not know the timing advance when it first accesses the system. This allows for a distance of 35 km from the BS to MS (see Figure 7.11).

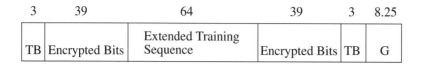

PLMN	BS	T_1 Superframe	T_2 Multiframe	T_3 Block Frame
color 3 bits	color 3 bits	index 11 bits	index 5 bits	index 3 bits

T_1 (11 bits) = FN div (26x51)   Range from 0 to 2047
T_2 (15 bits) = FN mod (26)   Range from 0 to 25
T_3 (13 bits) = [FN mod(51) - 1] div (10)   Range from 0 to 4

Fig. 7.9 Synchronization Channel Message Format

3	142	3	8.25
TB	Fixed All Zeros	TB	G

Fig. 7.10 Frequency Correction Burst

3	48	36	3	60
TB	Synchronization Sequence	Encrypted Message	TB	Additional Guard Time

Fig. 7.11 Access Burst

7.5.5 Dummy Burst

It carries no information and is sent from BTS on some occasions. The mixed bits are defined as modulating bit states (see Figure 7.12).

7.6 DATA ENCRYPTION IN GSM

Figure 7.13 shows the data encryption method used in GSM. Data is encrypted at the transmitter in blocks of 114 bits by taking 114-bit plain text data bursts and performing an "exclusive OR" logical function operation with a 114-bit cipher block. The decrypting function at the receiver is performed by taking the encrypted data block of 114 bits and performing the same "exclusive OR" operation using the same 114-bit cipher block that was used at the transmitter. The cipher block used at both ends of the transmission path for a given transmission direction is produced at the BS and MS by an encryption algorithm called A5. The A5 algorithm uses a 64-bit cipher key K_c, produced during the authentication process in call setup, and the 22-bit TDMA frame number (COUNT) which takes on (decimal) values from 0 through 4194304 and has a repetition time of about 5 hours which is close to the interval of the GSM hyperframe. The A5 algorithm produces two cipher blocks during each TDMA period, one for uplink path and the other for the downlink path.

7.7 MOBILITY MANAGEMENT

MSs periodically scan a list of PCHs and lock onto the channel with the strongest signal. In order to deliver an incoming call to an MS, a *page message* is broadcast on the PCH from the BS for which the MS is currently monitoring.

3	58	26	58	3	8.25
TB	Mixed Bits	Training Bits	Mixed Bits	TB	G

Fig. 7.12 Dummy Burst

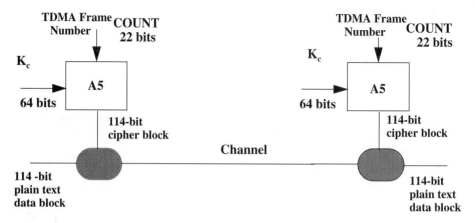

Fig. 7.13 Data Encryption in GSM

If the MS hears its identification code broadcast on the PCH, it responds with a *page response message.*

The difficulty in delivering a call to an MS is knowing which cell area should be paged and how many cell areas should be paged. One possibility would be to have all cells in GSM PLMN Area (GPA) or all cells in Europe page a mobile when delivering a call to a mobile subscriber. Involving too many cells in the paging process for call delivery can affect performance of the system.

Performance problems related to paging arise when too many cell areas are paged when attempting to deliver a call to an MS. As the page attempt rate increases to a given BS, eventually a resource becomes a bottleneck. Most likely the limiting resource would be the capacity of the PCH; however, it is also possible to encounter other bottlenecks such as BS real time. BS real time might become a bottleneck if a BS is unable to perform other call-handling functions because of the volume of pages that it is being required to broadcast. If the BS does not have adequate overload controls, it is possible that the number of successful pages can start to decrease as the page attempt rate increases beyond the saturation point. To keep the paging performance within a safe range, it is necessary to form clusters of cells and page only the cluster of cells for which the MS is known to be situated. These cell clusters are referred to as *location area* (LA).

The GPA is divided into LAs. Each LA is made up of one or more cell areas. An MS registers each time it enters a new LA. An MS is free to move around within a given LA without reregistering. When delivering calls to an MS, only one LA is paged. For example, if an MS is known to be within the ith LA, then only cells located in the ith LA are paged.

The LA identification includes Mobile Country Code (MCC)—3 digits; Mobile Network Code (MNC)—2 digits; and LA Code (LAC)—2 octets. The cell global identification consists of LAI plus Cell Identity (CI)—2 octets (refer to Figure 7.14).

An MSC coverage area contains one or more LAs. Each LA consists of one or more BSSs. Each BSS consists of one or more cell areas.

The *paging request message* may include more than one MS identification. The choice of the paging request message depends on the number of MSs to be paged and the types of identities that are used. A maximum of four mobiles can be paged in a paging request message when only TMSIs are used for identification. Upon receiving a paging request message, if the access to the network is allowed, the addressed MS initiates the assignment procedure by requesting a channel through RACH. Upon receiving the message, BSS sends an acknowledgment over the AGCH with the same random number as originally received and the channel number for SDCCH. The MS sends a Set Async Balanced Mode (SABM) frame containing the paging response (see Figure 7.15).

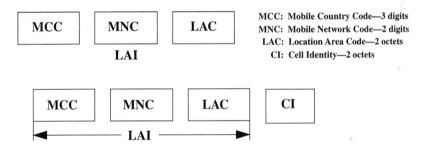

MCC: Mobile Country Code—3 digits
MNC: Mobile Network Code—2 digits
LAC: Location Area Code—2 octets
CI: Cell Identity—2 octets

Fig. 7.14 GSM Location Area and Cell Area Identification

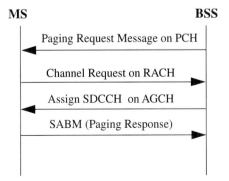

Fig. 7.15 Paging Procedure

7.7.1 Location Registration

GSM supports the following types of location registration:

☞ Geographic based

☞ Time based

☞ On/off based

In the geographic-based location registration, BSSs broadcast LAC information. The MS compares the new LAC with the last LAC. Registration takes place if LAC change is detected.

In the time-based location registration, the MS registers periodically. Updates depend on MS activity. Update intervals are broadcast by BSs. A minimum registration interval is 6 minutes (1 deci-hour) and a maximum registration interval is 25.5 hours (255 deci-intervals). A timer is reset when MS activity has taken place. The MS initiates location updating when time expires. The timer value of the MS is kept in memory when the MS is turned off.

In the on/off-based location registration, MS power-up is an *attach* operation and it causes a registration. MS power-down is a *detach* operation and it causes deregistration. Types of detach operations are

☞ Delete VLR entry

☞ Leave VLR entry, but flag MS as detached

Details about messages are given in chapter 9.

7.7.2 Mobile Identification

The mobile identification procedure is used to identify the MS when the VLR does not recognize the TMSI sent by the MS. This may result from the change in LAC. If MS identification is required, the VLR first sends a message (*MAP/ B Provide IMSI*) to the MSC (see Figure 7.16). On receiving this message, the MSC sends an *RIL3-MM Identity Request* message to the MS. The MS responds to this message by returning an *RIL3-MM Identity Response* message to the MSC. This message includes the MS's IMSI. On receiving this message, the MSC sends the *MAP/B IMSI Acknowledge* message to the VLR. If the VLR does not have the MS's IMSI, it requests the HLR for the user's profile which contains MS's IMSI. To do this, the VLR sends to the HLR a *MAP/D Update Location* message. The HLR responds with a *MAP/D Update Location Result* message, which is followed by a *MAP/D Insert Subscriber Data* message containing other MS data required by the VLR. The VLR acknowledges this message by the *MAP/D Insert Subscriber Data Result* message to the HLR.

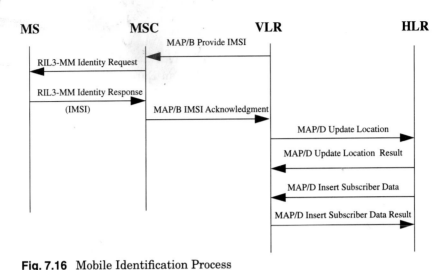

Fig. 7.16 Mobile Identification Process

7.8 SUMMARY

In this chapter, we presented the logical channel structure of the GSM system. The logical channels include traffic channels to carry user information and signaling channels to convey control data. Three types of logical channels are used in GSM. These are BCCH, CCCH, and DCCH. The BCCH is a point-to-multipoint unidirectional control channel from the BS to MSs. It is intended to broadcast a variety of information to MSs. The BCCH includes an FCCH and an SCH. A CCCH includes a PCH and an AGCH. Both of them are the downlink channels. The CCCH also includes the RACH which is an uplink channel. The DCCH is a point-to-point, unidirectional channel. Two types of DCCHs are used including SDCCH and ACCH. The SDCCH is employed for authentication and location updates of the MSs. The ACCH can be the FACCH or SACCH.

We also discussed five different types of bursts that are used in GSM including normal burst, synchronization burst, frequency correction burst, access burst, and dummy burst. We presented the structure of GSM frame, multiframe, superframe, and hyperframe. We also discussed the encryption method used in GSM.

We also provided details of the mobility management, paging, and mobile identification procedures. Necessary message flow diagrams were given.

7.9 REFERENCES

1. Mehrotra, A., *GSM System Engineering*, Artech House, Boston, 1997.
2. Haug, T., "Developing GSM Standards," Pan-European Digital Cellular Radio Conf., Nice, France, 1991.
3. Garg, V. K., and Wilkes, J. E., *Wireless and Personal Communications Systems*, Prentice Hall, Upper Saddle River, NJ, 1996.
4. Mouley, M., et al., "The GSM System for Mobile Communications," Mouly and Paulet, F-99120 Palaiseau, France, 1992.

CHAPTER 8

Speech Coding in GSM

8.1 INTRODUCTION

After the introduction of the ITU's 64-kbps PCM and 32-kbps ADPCM G.721 standards in 1986, the 13-kbps RPE codec [1-2] was selected for GSM. The VSELP codecs [3-4] operating at 8 and 6.7 kbps were chosen for the North American IS-54 and JDC wireless systems. More recently the 5.6-kbps half-rate GSM quadruple-mode VSELP speech codec was approved by ETSI [5]. In Japan the 3.45 kbps half-rate JDC speech codec using Pitch Synchronous Innovation (PSI) CELP was standardized [6]. In the low-delay category, a forward-adaptive 10-ms delay G.729 codec has recently been standardized for use in Europe and North America [7].

In this chapter, we first discuss speech coding methods and attributes of a speech codec. These are then followed by a brief discussion of the Linear-Prediction-based Analysis-by-Synthesis (LPAS) and a comparison of the ITU-T standards of the different codecs. We also concentrate on the GSM, full-rate and half-rate codecs.

8.2 SPEECH CODING METHODS

Speech coding is the process for reducing the bit rate of digital speech for transmission or storage, while maintaining a speech quality that is acceptable for the application.

Speech coding methods can be classified as:

☞ Waveform coding

☞ Source coding

☞ Hybrid coding

In Figure 8.1, the bit rate is plotted on a logarithmic axis vs. speech quality classes of poor to excellent, corresponding to the five-point mean-opinion score (MOS) scale values of 1 to 5, as defined by ITU. It may be noted that for low complexity and low delay, a bit rate of 32 to 64 kbps is required. This suggests the use of waveform codecs. However, for a low bit rate between 4 and 8 kbps, hybrid codecs should be used. These types of codecs tend to be complex with high delay.

8.3 SPEECH CODEC ATTRIBUTES

Speech quality as produced by a speech codec is a function of transmission bit rate, complexity, delay, and bandwidth. Therefore, when considering speech codecs it is essential to consider all these attributes. It should be noted that there is a strong interaction between all these attributes and that they can be traded off against each other. For example, low-bit-rate codecs tend to have more delay as compared to the higher-bit-rate codecs. They are generally more complex to implement and often have lower speech quality than the higher-bit-rate codecs.

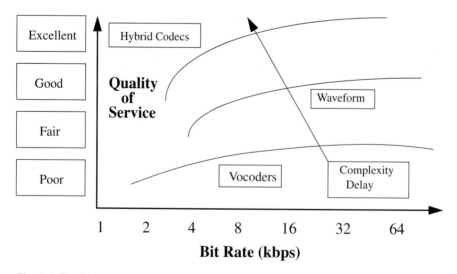

Fig. 8.1 Bit Rate vs. QoS

8.3.1 Transmission Bit Rate

Since the speech codec shares the communications channel with other data,
the peak bit rate should be as low as possible so as not to use a disproportion-
ate share of the channel. Codecs below 64 kbps were developed to increase the
capacity of equipment used for narrow bandwidth links. For the most part
codecs are fixed-bit-rate codec, meaning they operate at the same rate regardless of the input. In the variable-bit-rate codecs, network loading and voice
activity determine the instantaneous rate assigned to a particular voice channel. Any of the fixed-rate speech codec can be combined with a VAD and made
into a simple two-state variable-bit-rate system. The lower rate could be
either zero or some low rate needed to characterize slowly changing background noise characteristics. Either way, the bandwidth of the communications channels is used only for active speech.

8.3.2 Delay

The delay of a speech codec can have a great impact on its suitability for a particular application. For one-way delay of conversation greater than 300 ms,
the conversation becomes more like a half-duplex or push-to-talk experience
rather than an ordinary conversation. The components of total system delay
include frame size, look ahead, multiplexing delay, processing delay for computations, and transmission delay.

Most low-bit-rate speech codecs process one frame of speech data at a
time. The speech parameters are updated and transmitted for every frame. In
addition, to analyze the data properly it is sometimes necessary to analyze
data beyond the frame boundary. Hence, before the speech can be analyzed it
is necessary to buffer a frame of data. The resulting delay is referred to as
algorithmic delay. This delay component cannot be reduced by changing the
implementation. All other delay components can be reduced by changing
implementation. The second major contribution for delay comes from the time
taken by the encoder to analyze the speech and the decoder to reconstruct the
speech. This part of the delay is referred to as *processing delay*. It depends on
the speed of the hardware used to implement the codec. The sum of the algorithmic and processing delays is called the *one-way codec delay*. The third component of delay is due to transmission. It is the time taken for an entire frame
of data to be transmitted from the encoder to the decoder. The total of the
three delays is the *one-way system delay*. In addition, there is a frame interleaving delay which adds approximately an additional 20-ms delay to the
transmission delay. Frame interleaving is necessary to combat channel fading.

8.3.3 Complexity

Speech codecs are implemented in special-purpose hardware, such as digital
signal processors (DSPs). Their attributes are described by computing speed

in millions of instructions per second (MIPS), random-access memory (RAM) size, and read-only memory (ROM) size. For a speech codec, the system designer makes a choice about how much of these resources are to be allocated to the speech codec. Speech codecs with less than 15 MIPS are considered to be of low complexity; those requiring 30 MIPS or more are thought to be of high complexity. More complexity results in higher costs and greater power consumption; for portable applications, greater power consumption means reduced time between battery recharges or using larger batteries, which means more expense and weight.

8.3.4 Quality

Of all the attributes, quality has the most dimensions. Figure 8.1 is a typical plot that relates the performance of various coding schemes. In many applications there are large amounts of background noise (car noise, street noise, office noise, and so on). How well does the codec perform under these adverse conditions? What happens when there are channel errors during transmission? Are the errors detected or undetected? If undetected, the codec must perform even more robustly than when it is informed that the entire frames are in error. How well does the codec perform when speech is encoded and decoded twice? All these questions must be carefully evaluated during the testing phase of a speech codec. The speech quality is often based on the five-point MOS scale values of 1 to 5 as defined by ITU-T.

8.4 LPAS

LPAS methods [7, 8] provide efficient speech coding at rates between 4 and 16 kbps. In LPAS speech codecs, the speech is divided into frames of about 20 ms in length for which the coefficients of linear predictor (LP) are computed. The resulting LP filter predicts each sample from a set of previous samples.

   In analysis-by-synthesis speech codecs, the residual signal is quantized on a subframe-by-subframe basis (there are commonly 2 to 8 subframes per frame). The resulting quantized signal forms the excitation signal for the LP synthesis filter. For each subframe, a criterion is used to select the best excitation signal from a set of trial excitation signals.The criterion compares the original speech signal with trial reconstructed speech signals. Because of the synthesis implicit in the evaluation criterion, the method is called analysis-by-synthesis (ABS) coding. Various representations of excitation have been used [9]. For lower bit rates, the most efficient representation is achieved by using vector quantization. For each subframe, the excitation signal is selected from a multitude of vectors that are stored in a codebook. The index of the best-matching vector is transmitted. At the receiver this vector is retrieved from the same codebook. The resulting excitation signal is filtered through the LP synthesis filter to produce the reconstructed speech. Linear prediction ABS codecs using a codebook approach are commonly known as CELP codecs [10].

Parametric codecs are traditionally used at low bit rates. A proper understanding of the speech signal and its perception is essential to obtain good speech quality with a parametric codec. Parametric coding is used for those aspects of the speech signal that are well understood, while the waveform-matching procedure is employed for those aspects that are not well understood. Waveform-matching constraints are relaxed for those aspects that can be replaced by parametric models without degrading the quality of the reconstructed speech.

A parameter that is well understood in parametric coding is the pitch of the speech signal. Satisfactory pitch estimation procedures are available [11]. Piecewise linear interpolation of the pitch does not degrade speech quality. Pitch period is determined once every 20 ms and linearly interpolated between the updates. The challenge is to generalize the LPAS method such that its matching accuracy becomes independent of the synthetic pitch-period contour used. This is done by determining a time wrap of speech signal such that its pitch-period contour matches the synthetic pitch-period contour. The time wraps are determined by comparing a multitude of time-wrapped original signals with a synthesized signal. This coding scheme is called generalized ABS coding [12] and is referred to as Relaxed CELP (RCELP). The generalization relaxes the waveform-matching constraints without affecting speech quality.

8.5 ITU-T STANDARDS

Table 8.1 shows the different types of codecs that have been standardized by the ITU.

8.5.1 Bit Rate

For G.729 [13], the nominal bit rate of the codec is 8 kbps including the overhead for the codec-inherent error correction. The adaptive variable-bit-rate

Table 8.1 Comparison of ITU Speech Codecs

Codec	Bit Rate (kbps)	Frame Size (ms)	Subframe Size (ms)	Algorithmic Delay (ms)	MIPS (fixed-point DSP)	RAM (16-bit words)
G. 723.1	6.3	30	7.5	37.5	14.6	2.2 K
G. 723.1	5.3	30	7.5	37.5	16	2.2 K
G. 729	8	10	5	15	20	2.7 K
G. 729 A	8	10	5	15	10.5	2 K

capability allows the codec to operate in circuit multiplication equipment. Thus, during periods of congestion, operation continues at 6.4 kbps with a graceful degradation of speech quality. In contrast, when bandwidth is available, the bit rate increases to 9.6 kbps to improve the performance.

8.5.2 Delay

The requirement for the codec/decoder delay is important for two reasons. First, an excessive delay affects the quality of a conversation. Second, echo control devices are required if total one-way delay is above 25 ms. Algorithmic delay was set to be ≤ 16 ms as a requirement for G.729 and ≤ 5 ms as an objective. The corresponding limits for the total codec/decoder delay were set ≤ 32 ms and ≤ 10 ms, respectively. The algorithmic delay includes the frame size delay plus any other delay inherent in the algorithm such as a possible look-ahead time. Frame size was specified to be 10 ms. G.729 has a 5-ms look-ahead time. Assuming an 8-ms processing delay and an 8-ms transmission delay, the one-way system delay of G.729 is about 32 ms.

8.5.3 Complexity

The codec's complexity is measured in terms of MIPS and million operations per second (MOPS). From a service viewpoint, power consumption is used as the complexity metric to reflect what is important to the end user of a portable handset. G. 723.1 is of lower complexity as compared to G.729. ITU Study Group (SG) 14 requirements for G.723.1 codec are 10 MIPS, 2000 words of RAM, and 10,000 words of ROM.

8.6 WAVEFORM CODING

In general, waveform codecs are designed to be independent of signal. They map the input waveform of the encoder into a facsimile-like replica of it at the output of the decoder. Coding efficiency is quite modest. However, the coding efficiency can be improved by exploiting some statistical signal properties, if the codec parameters are optimized for the most likely categories of input signals, while still maintaining good quality for other types of signals as well. The waveform codecs can be further subdivided into time domain waveform codec and frequency domain waveform codec.

8.6.1 Time Domain Waveform Coding

The well-known representation of speech signal using time domain waveform coding is the A-law (in Europe) or μ-law (in North America) companded PCM at 64 kbps (Figure 8.2). Both use nonlinear companding characteristics to give a near-constant SNR over the total input dynamic range.

☞ A-Law

$$s(t) = \text{sgn}[i(t)]\frac{[1 + \ln(A|i(t)|)]}{1 + \ln A}, \ 1/A < |i(t)| < 1 \tag{8.1a}$$

or

$$s(t) = \text{sgn}[i(t)]\frac{[1 + A|i(t)|]}{1 + \ln A}, \ 0 < |i(t)| < 1/A \tag{8.1b}$$

where A = 87.6

☞ μ-Law

$$s(t) = \text{sgn}[i(t)]\frac{[1 + \mu|i(t)|]}{\ln(1 + \mu)}, \ -1 < i(t) < 1 \tag{8.2}$$

where μ = 255

The ITU G.721, 32-kbps ADPCM codec is an example of a time domain waveform codec. More flexible counterparts of the G.721 are the G.726 and G.727 codecs. The G.726 codec is a variable-rate arrangement for bit rates between 16 and 40 kbps. This may be advantageous in various networking applications to allow speech quality and bit rate to be adjusted on the basis of the instantaneous requirement. The G.727 codec uses core bits and enhancement bits in its bit stream to allow the network to drop the enhancement bits under restricted channel capacity conditions, while benefitting from them when the network is lightly loaded.

In differential codecs a linear combination of the last few samples is used to generate an estimate of the current one, which occurs in the adaptive predictor. The resultant difference signal (i.e., the prediction residual) is computed and encoded by the adaptive quantizing with a lower number of bits than the original signal, since it has a lower variance than the incoming signal. For a sampling rate of 8000 samples per second, an 8-bit PCM sample is represented by a 4-bit ADPCM sample to give a transmission rate of 32 kbps.

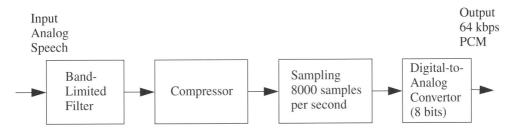

Fig. 8.2 PCM

Time domain waveform codecs encode the speech signal as a full-band signal and map it into as close a replica of the input as possible. The difference between various coding schemes is their way of using prediction to reduce the variance of the signal to be encoded to reduce the number of bits necessary to represent the encoded waveform.

8.6.2 Frequency Domain Waveform Coding

In frequency domain waveform codecs, the input signal undergoes a more or less accurate short-time spectral analysis. The signal is split into a number of frequency domain sub-bands. The individual sub-band signals are then encoded by using different numbers of bits to fulfill the quality requirements of that band based on its prominence. The various schemes differ in their accuracies of spectral analysis and in the bit allocation principle (fixed, adaptive, semiadaptive). Two well-known representatives of this class are sub-band coding (SBC) and adaptive transform coding (ATC).

8.6.3 Vocoders

Vocoders are parametric digitizing that use certain properties of the human speech production mechanism. Human speech is produced by emitting sound pressure waves that are radiated primarily from lips, although significant energy also emanates, in case of some sounds, from the nostrils, throat, and other areas. In human speech, the air compressed by the lungs excites the vocal cords in two typical modes. When generating voice sounds, the vocal cords vibrate and generate quasiperiodic voice sounds. In case of lower energy unvoiced sounds, the vocal cords do not participate in the voice production, and the source acts like a noise generator. The excitation signal is then filtered through the vocal apparatus, which behaves like a spectral shaping filter. This can be described adequately by an all-pole transfer function that is constituted by the spectral shaping action of gloat, vocal tract, lip radian characteristics, and so on.

In the case of vocoders, instead of producing a close replica of an input signal at the output, an appropriate set of source parameters is generated to characterize the input signal sufficiently close for a given period of time. The following steps are used in this process.

1. Speech signal is segmented in quasistationary segments of 5–20 ms.
2. Speech segments are subjected to spectral analysis to produce the coefficients of the all-zero analysis filter to minimize the prediction residual energy. This process is based on computing the speech autocorrelation coefficients and then using either the matrix inversion or iterative scheme.
3. The corresponding source parameters are specified. The excitation parameters as well as filter coefficients are quantized and transmitted to

the decoder to synthesize a replica of the original signal by exciting the all-pole synthesis filter.

The quality of this type of scheme is predetermined by the accuracy of the source model rather than the accuracy of the quantization of the parameters. The speech quality is limited by the fidelity of the source model used. The main advantage of vocoders is their low bit rate, with the penalty of relatively low synthetic speech quality. Vocoders can be classified into the frequency domain and time domain subclasses. However, frequency domain vocoders are generally more effective than the time domain vocoders.

8.6.4 Hybrid Coding

Hybrid coding is an attractive trade-off between waveform coding and vocoders, both in terms of speech quality and transmission bit rate, although generally at the price of higher complexity. It is also referred to as ABS codecs.

Most recent international and regional speech coding standards belong to a class of LPAS codecs. This class of codecs includes ITU G.723.1, G.728, and G.729, as well as all the current digital cellular standards, including

☞ Europe: GSM, full-rate, half-rate, and enhanced full-rate
☞ North America: Full-rate, half-rate, and enhanced full-rate for TDMA IS-136 and CDMA IS-95 systems
☞ Japan: PDC full-rate and half-rate

In an LPAS coder, the decoded speech is produced by filtering the signal produced by the excitation generator through both a long-term and a short-term predictor synthesis filter. The excitation signal is found by minimizing the mean-squared error over a block of samples. The error signal is the difference between the original and decoded signal. It is weighted by filtering it through a weighting filter. Both short-term and long-term predictors are adapted over time. Since the analysis procedure (encoder) includes the synthesis procedure (decoder), the description of the encoder defines the decoder. The short-term synthesis filter models the short-term correlations (spectral envelope) in the speech signal. This is an all-pole filter. The predictor coefficients are determined from the speech signal using LP techniques. The coefficients of short-term predictor are adapted in time, with rates varying from 30 to as high as 400 times per second.

The long-term predictor filter models the long-term correlations (fine spectral structure) in the speech signal. Its parameters are a delay and gain coefficient. For a periodic signal, delay corresponds to the pitch period (or possibly an integral number of pitch periods). The delay is random for nonperiodic signals. Typically, the long-term predictor coefficients are adapted at rates varying from 100 to 200 times per second.

An alternative structure for the pitch filter is the *adaptive codebook*. In this case, the long-term synthesis filter is replaced by a codebook that contains the previous excitation at different delays. The resulting vectors are searched, and the one that provides the best result is selected. In addition, an optimal scaling factor is determined for the selected vector. This representation simplifies the determination of the excitation for delays smaller than the length of excitation frames.

CELP coders use another approach to reduce the number of bits per sample. Both encoder and decoder store the same collection of codes (C) of possible length L in a codebook. The excitation for each frame is described completely by the index to an appropriate vector. This index is found by an exhaustive search over all possible codebook vectors, using the one that gives the smallest error between the original and decoded signals. To simplify the search it is common to use a gain-shape codebook in which the gain is searched and quantized separately. Further simplifications are obtained by populating the codebook vectors with a multipulse structure. By using only a few nonzero unit pulses in each codebook vector, efficient search procedures are derived. This partitioning of excitation space is referred to as an *algebraic codebook*. The excitation method is known as ACELP.

8.7 GSM VOCODERS

8.7.1 Full-Rate Vocoder

The GSM vocoder is referred to as a Linear Predictive Coding with Regular Pulse Excitation (LPC-RPE) [1]. Figure 8.3 shows the simplified block diagram. The encoder processes 20-ms blocks of speech. Each speech block contains 260 bits—the output rate is 13 kbps. The encoder has three major parts: linear prediction analysis, long-term prediction, and excitation analysis. The linear prediction analysis deals with an 8-tap filter characterized by 8 log-area ratios, each represented by 3, 4, 5, or 6 bits. In total, the 8 log-area ratios are represented by 36 bits. The long-term predictor estimates pitch and gain four times at 5-ms intervals. Each estimate provides a lag coefficient and gain coefficient of 7 bits and 2 bits, respectively. The remaining 188 bits are derived from the regular pulse excitation analysis as the long-term predictor operates over 5-ms intervals resulting in 47 bits per sub-block. The distribution of bits used in a GSM full-rate vocoder is given in Table 8.2.

8.7.2 Half-Rate Vocoder

The half-rate GSM vocoder with 5.6 kbps is a VSELP codec. This is a close relative of the CELP family. A slight difference is that VSELP uses more than one separate excitation codebook, which are separately scaled by their respective excitation gain factors. The half-rate GSM vocoder operates in one of four

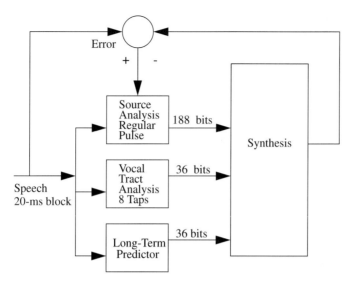

Fig. 8.3 GSM Full-Rate LPC-RPE Vocoders

Table 8.2 GSM Full-Rate Vocoder (13 kbps)

		Bits per 5 ms	Bits per 20 ms
LPC Filter	8 Parameters		36
LTP Filter	Delay Parameter	7	28
	Gain Parameter	2	8
Excitation Signal	Subsampling Phase	2	8
	Maximum Amplitude	6	24
	13 Samples	39	156
Total			260 or 13 kbps

different modes, depending on the grade of voice detected in the speech. Table 8.3 provides a summary for the synthesis modes 0–3. The speech spectral envelope is encoded by using 28 bits per 20-ms frame for the vector quantization of the LPC coefficients. The four different synthesis modes correspond to different excitation modes. Two bits per frame are assigned for excitation mode selection. The decision as to which excitation mode should be used is based on the Long-Term Predictor (LTP) gain. LTP gain is typically high for highly correlated voice segments and low for noise-like uncorrelated unvoiced segments.

In the unvoiced mode 0, speech is synthesized by superimposing the outputs of two 128-entry fixed codebooks to generate the excitation signal which

Table 8.3 GSM Half-Rate Vocoders

Parameter	Bits/Frame
LPC Coefficient	28
LPC Interpolation Flag	1
Excitation Mode Selection	2
Mode 0	
- Codebook 1 Index	$4 \times 7 = 28$
- Codebook 2 Index	$4 \times 7 = 28$
Mode 1, 2, 3	
- LTPD (subframe 1)	8
- $\Delta LTPD$ (subframe 2, 3, 4)	$3 \times 4 = 12$
- Codebook 3 index	$4 \times 9 = 36$
Frame Energy	5
Excitation Gain-Related quantity	$4 \times 5 = 20$
Total No. of Bits in each mode	112/20 ms

is the filtered through the synthesis filter and spectral postfilter. Both excitation codebooks 1 and 2 have a 7-bit address in each of the 5-ms subframes.

In modes 1–3 where the input speech exhibits different grades of voice, the excitation is generated by superimposing the 512-entry fixed trained codebook's output onto the adaptive codebook. The fixed codebook in mode 1–3 requires a 9-bit address, giving a total of 36 coding bits for the 20-ms frame. The adaptive codebook delay or long-term predictor delay (LTPD) is encoded using 8 bits to allow for 256 different LTP delay values. In order to minimize the variance of the LTP residual, the codec uses a range of integer and noninteger LTP delay positions. A further economy is achieved by encoding LTPD in the consecutive subframe differently than the previous subframe's delay. This is indicated by ΔLTPD in Table 8.3. Four encoding bits are used to allow for a maximum difference of positions with respect to the previous LTPD values. The energy of each subframe E_S is normalized with respect to the frame energy E_F which is vector quantized using 5 bits per 5-ms subframe to allow for 32 possible combinations. Thus, a total of 20 bits per frame are assigned for encoding the gain-related information.

8.8 SUMMARY

In this chapter, we presented the attributes necessary to select a speech codec, including transmission bit rate, delay, complexity of implementation, and speech quality. We discussed ITU standards for speech codecs and compared their performances. The chapter covered the details of the GSM full-rate (13-kbps) and the GSM half-rate (5.6-kbps) vocoders and gave the distribution of bits in these two vocoders.

8.9 REFERENCES

1. Kroon, P., and Deprettere, E. F., "Regular Pulse Excitation—A Novel Approach to Effective Multipulse Coding of Speech," *IEEE Trans. on Acoustics, Speech and Signal Processing,* 1986, pp. 1054–63.

2. Vary, P., Hellwig, K., Hofmann, R., Sluyter, R., Galland, C., and Rosso, M., "Speech Codec for the European Mobile Radio System," *Proc. ICASSP,* April 1988, pp. 227–30.

3. Gerson, I. A., and Jasiuk, M. A., "Vector Sum Excited Linear Prediction (VSELP) Speech Coding at 8 kb/s," *IEEE Journal on Selected Areas in Communications,* 1990, pp. 461–64.

4. Gerson, I. A., and Jasiuk, M. A., "Vector Sum Excited Linear Prediction (VSELP)," in *Advances in Speech Coding,* Atal, B. S., Cuperman, V., and Gersho, A., editors, Kluwer Academic Publishers, 1991, pp. 69–80.

5. Gerson, I. A., Jasiuk, M. A., Muller, J. M., Nowack, J. M., and Winter, E. H., "Speech and Channel Coding for the Half-Rate GSM Channel," *Proceedings ITG-Fachbericht 130, VDE-Verlag, Berlin,* November 1994, pp. 225–33.

6. Ohya, T., Suda, H., and Miki, T., "5.6 kb/s PSI-CELP of Half-Rate PDC Speech Coding Standard," *Proceedings of IEEE Conference on Vehicular Technology,* June 1994, pp. 1680–84.

7. Chen, J., Cox, R., Lin, Y., Jayant, N., and Melchner, M., "Coder for the CCITT 16 kb/s Speech Coder Standard," *IEEE Journal on Selected Areas in Communications,* June 1991, pp. 830–49.

8. Atal, B. S., "Predictive Coding of Speech Signals at Low Bit Rates," *IEEE Transaction, Comm.* 30 (4), 1982, pp. 600–14.

9. Kroon, P., and Deprettere, E. F., "A Class of Analysis-by-Synthesis Predic Coders for High Quality Speech Coding at Rates between 4.8 and 16 kbps," *IEEE Journal of Selected Areas of Communications* 6, 1988, pp. 353–63.

10. Atal, B. S., and Schroeder, M. R., "Stochastic Coding of Speech at Very Low Bit Rate," *Proc. International Conf. Comm.,* Amsterdam, 1984, pp. 1610–13.

11. Hess, W., "Pitch Determination of Speech Signals," Springer Verlag, Berlin, 1983.

12. Kleijn, W. B., Ramachandran, R. P., and Kroon, P., "Generalized Analysis-by-Synthesis Coding and Its Application to Pitch Prediction," *International Conf. Acoust. Speech Signal Processing,* San Francisco, 1992, pp. 1337–40.

Messages, Services, and Call Flows in GSM

9.1 INTRODUCTION

In preceding chapters we have discussed the cellular concepts and how they are applied to a GSM system. We have also discussed the architecture of GSM and the detailed formats of the various logical channels. In this chapter we discuss the various telephony services that can be offered on a GSM system and compare them with ISDN services. We then cover the end-to-end call flows for several of the services. GSM is offered throughout the world but must make minor adaptations for use in various countries. Two variations of SS7 are used in the world: CCITT SS7 and ANSI SS7. Much of the world is CCITT SS7 based, but North America and some other countries are ANSI SS7 based. We conclude the chapter by discussing various call flows for GSM.

9.2 GSM PLMN SERVICES

A telecommunication service supported by the GSM PLMN is defined as a group of communication capabilities that the service provider offers to the subscribers. The basic telecommunication services provided by the GSM PLMN are divided into bearer services, teleservices, and supplementary services.

9.2.1 Bearer Services

These services give the subscriber the capacity required to transmit appropriate signals between certain access points (i.e., user-network interfaces).

The capabilities of GSM bearer services include

☞ Rate adapted subrate information—circuit-switched asynchronous and synchronous duplex data, 300–9600 bps.

☞ Access to Packet Assembler/Disassembler (PAD) functions—PAD access for asynchronous data, 300–9600 bps.

☞ Access to X.25 public data networks—packet service for synchronous duplex data 2400–9600 bps.

☞ Speech and data swapping during a call—alternate speech/data and speech followed by data.

☞ Modem selection—selection of 3.1-kHz audio service when interworking with ISDN.

☞ Support of Automatic Request for retransmission (ARQ) technique for improved error rates—transparent mode (no ARQ) and nontransparent mode (with ARQ).

Table 9.1 provides a summary of these services and compares them with services available with ISDN.

9.2.2 Teleservices

These services provide the subscriber with necessary capabilities including terminal equipment functions to communicate with other subscribers.

Table 9.1 A Comparison of Bearer Services Supported by GSM and ISDN

Service	GSM	ISDN
Data Services	x	x
Alternate Speech/Data	x	x
Speech Followed by Data	x	x
Clear 3.1 kHz Audio	x	x
Unrestricted Digital Information (UDI)	x	x
PAD	x	
3.1 kHz External to PLMN	x	
Others		x

The GSM teleservices are

☞ Speech transmission—telephony, emergency call
☞ Short message services—mobile terminating point-to-point, mobile origi-
    nating point-to-point, cell broadcast.
☞ Message handling and storage services
☞ Videotex access
☞ Teletext transmission
☞ Facsimile transmission

A summary of the GSM teleservices is given in Table 9.2. The table compares GSM and ISDN teleservices.

9.2.3 Supplementary Services

These services modify or supplement basic telecommunications services.
These services are offered together or in association with basic telecommunications services.

The GSM supplementary services are

☞ Number identification services

Calling Number Identification Presentation (CNIP). When this
service is active, the ISDN number of the incoming call will be presented
on a display of the GSM telephone.

Table 9.2 A Comparison of Teleservices Supported by GSM and ISDN

Service	GSM	ISDN
Circuit Speech (Telephony)	x	x
Emergency Call	x	x
Short Message Point-to-Point	x	x
Short Message Cell Broadcast	x	x
Alternate Speech/Facsimile Group 3	x	x
Automatic Facsimile Group 3 Service	x	x
Voice-band Modem (3.1 kHz audio)	x	x
Messaging Teleservices	x	
Paging Teleservices	x	
Others		x

Calling Number Identification Restriction (CNIR). When the calling party activates this service, the calling party's number will not be presented on the display of the telephone.

Connected Number Identification Presentation (CNOP). When this service is active, the ISDN number of the telephone where the call is completed will be displayed on the GSM telephone.

Connected Number Identification Restriction (CNOR). If a called party has restricted the identification of their number, the GSM phone will not display the ISDN number of the telephone where the call is completed.

Malicious Call Identification (MCI). This service permits a user to identify an incoming call as a malicious call. The network will record details of the calling and called party for later use.

☞ Calling offering services

Call Forwarding Unconditional (CFU). When this service is active, all calls to the GSM phone will be routed to another number.

Call Forwarding Mobile Busy (CFB). When this service is active and the GSM phone is busy, calls will be routed to another number.

Call Forwarding No Reply (CFNRy). When this service is active and the GSM phone is not answered, calls will be routed to another number.

Call Forwarding Mobile Not Reachable (CFNRc). When this service is active and the GSM phone is not available (typically off or out of range), calls will be routed to another number.

Call Transfer (CT). The current call can be transferred to another telephone.

Mobile Access Hunting (MAH). This service allows multiple phones to be called (in sequence). The phones can be wireline or wireless.

☞ Call completion services

Call Waiting (CW). When this service is active, a second incoming call will generate an indication in the GSM phone.

Call Holding (HOLD). This service permits a GSM phone to temporarily set aside (put on hold) a call in order to place or receive a second call.

Completion of Call to Busy Subscriber (CCBS). When this service is active, a GSM subscriber can request that the network monitor the status of a busy phone and complete a call to it when the phone becomes idle.

☞ Multiparty services

Three-Party Service (3PTY). This service permits a GSM telephone to join two calls for simultaneous communications between three phones.

Conference Calling (CONF). This service permits a GSM telephone to have simultaneous communications with three to five phones.

☞ Community of interest services

Closed User Group (CUG). When this service is active, several GSM users can form groups to and from which access is restricted. A specific user may be a member of one or more closed user groups. Members of a specific closed user group can communicate among each other but not, in general, with users outside the group.

☞ Charging services

Advice of Charge (AoC). This service provides information to the GSM phone about the charging associated with a call that is established. The costs presented on the display might not match exactly the costs for the call.

Freephone Service (FPH). A GSM phone with a freephone number will pay for all incoming calls. This service is similar to 800 service in North America.

Reverse Charging (REVC). This service allows the GSM phone to pay for an incoming call on a selective basis.

☞ Additional information transfer service

User-to-User Signaling (UUS). This service permits a GSM phone to send user data to another GSM or ISDN phone.

☞ Call restrictions services

Barring All Originating Calls (BAOC). When this service is active, the GSM phone cannot place any calls but can receive calls.

Barring Outgoing International Calls (BOIC). When this service is active, the GSM phone cannot place any international calls, but can receive calls and can place domestic calls.

BOIC except Home Country (BOIC-exHC). When this service is active, and the GSM phone is roaming, the GSM phone cannot place any international calls except to the home country, but can receive calls and place domestic calls.

Barring All Incoming Calls (BAIC). When this service is active, the GSM phone cannot receive calls but can place calls.

Barring Incoming Calls when Roaming (BIC-Roam). When this service is active and the GSM phone is roaming, the GSM phone cannot receive calls but can place calls.

Table 9.3 summarizes the GSM supplementary services and compares them with the supplementary services available with ISDN.

Table 9.3 A Comparison of Supplementary Services Supported by GSM and ISDN

Service	GSM	ISDN
Call Number ID Presentation	x	x
Call Number ID Restriction	x	x
Connected Number ID Presentation	x	x
Connected Number ID Restriction	x	x
Malicious Call Identification	x	x
Call Forwarding Unrestricted	x	x
Call Forwarding Mobile Busy	x	x
Call Forwarding No Reply	x	x
Call Forwarding Mobile Not Reachable	x	x
Call Transfer	x	x
Call Waiting	x	x
Call Hold	x	x
Completion of Call to Busy Subscriber	x	x
3-Party Service	x	x
Conference Calling	x	x
Closed User Group	x	x
Multiparty	x	x
Advice of Charge	x	x
Reverse Charging	x	x
Flexible Alerting		x
Mobile Access Hunting	x	x
Freephone	x	x
Barring All Originating Calls	x	x
Barring Outgoing Calls	x	x

Table 9.3 A Comparison of Supplementary Services Supported by GSM and ISDN (Continued)

Service	GSM	ISDN
Barring Outgoing International Calls	x	x
Barring All International Calls	x	x
Barring Outgoing International Calls—Except Home	x	x
Barring Incoming Calls when Roaming	x	x
Do Not Disturb	x	x
Message Waiting Notification	x	x
Preferred Language Service		x
Remote Feature Control		x
Selective Call Acceptance	x	x
Voice Privacy	x	x
Priority Access and Channel Assignment		x
Password Call Acceptance		x
Subscriber PIN Intercept	x	x
Subscriber PIN Access		x
Voice Mail Retrieval		x
Others		x

The GSM system offers a subscriber an opportunity to move freely through countries where a GSM PLMN is operational. Agreements are required between the various service providers to guarantee access to service offered to subscribers.

9.2.4 GSM Service Quality Requirements

The GSM standards place a variety of requirements on the quality of service delivered to the user. Some of these service requirements are:

- ☞ The time from switching to service ready equal to 4 seconds in the home system and 10 seconds in the visiting system
- ☞ A connect time of 4 seconds to called network
- ☞ A release time of 2 seconds to called network
- ☞ The time to alert a mobile of an inbound call equal to 4 seconds in the first attempt and 15 seconds in the final attempt
- ☞ A maximum gap due to handover of 150 ms if intercell and 100 ms if intracell

☞ A maximum one-way speech delay of 90 ms

☞ An intelligibility of speech equal to 90 percent

9.2.5 MSC Performance

The MSC performance will meet GSM Recommendations 02.08 and 03.05.

The reliability objectives of the MSC as per GSM Recommendations 3.05 and 3.06 are

☞ Cutoff call or call release failure rate probability: $P \ll 0.0002$

☞ Incorrect charging, misrouting, no tone, or other failures: $P \ll 0.0001$

☞ Mean Accumulated Intrinsic Downtime (MAIDT) for one termination (i.e., MAIDT (1) $\ll 30$ minutes/year)

☞ Probability of losing HLR/VLR messages: $P \ll 0.0000001$

The service availability of an MSC is expressed in terms of the frequency or duration of loss of service. The loss of service to particular circuits, groups of circuits, subsystems, or the complete MSC is determined by the faults in the MSC.

The average cumulative duration of service denial due to faults affecting more than 50 percent of the circuits will not exceed three minutes during the first year of operation and two minutes during each subsequent year. On the average, a fault that causes more than 50 percent of the established calls to be disconnected prematurely will occur less than once a year.

9.3 GSM MESSAGES

In this section we discuss the messages that flow between various network elements in the GSM system. For a discussion of messages that flow between a GSM network and the wireline network, see chapter 17 (SS7).

9.3.1 MS-BS Interface

Messages between the BS and the MS are sent in a layered protocol stack (Figure 9.1). A low-level set of primitives are defined for transmission of messages (Figure 9.2), and layer 3 messages are transmitted within the primitives. Upper layers on one side of the interface communicate with their corresponding layer on the other side of the interface by sending messages to the next lower layer. This step is repeated until the messages reach the physical layer. The primitives are

☞ **Request.** This primitive is used when a higher layer requests a service from a lower layer.

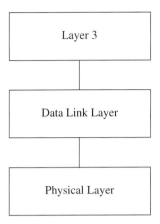

Fig. 9.1 Layering of MS-BS Interface

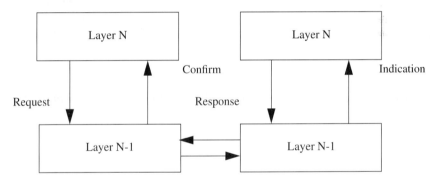

Fig. 9.2 Layer 2 Primitives

☞ **Indication.** A lower layer on the other side indicates the *request* to the next higher layer using the *indication primitive*.

☞ **Response.** The higher layer on the other side acknowledges the *indication* using the *response* primitive.

☞ **Confirm.** The lower layer on the requesting side confirms the request by sending a *confirm* primitive to the next higher layer.

Some Layer 3 messages can be transmitted only from BS to MS. Others can be transmitted only from MS to BS. A few messages can be transmitted in either direction. Some messages (call control and mobility management) flow directly from the MS to the MSC via the BS. These messages are described in Tables 9.5 and 9.6 since they do not originate or terminate in the BS (see chapter 5 and Figure 5.6 for more details). The layer 3 messages are segmented into three categories.

☞ RR messages are used to control the physical layer of the system. They are used for channel assignment, handover, ciphering, system information, paging, and MS information. The messages are summarized in Table 9.4.

Table 9.4 RR Messages

Message	BS->MS	MS->BS	Key Parameters	Description
Additional Assignment	x		Channel Description, Mobile Allocation	Assigns additional dedicated channels
Assignment Command	x		Channel Description, Mobile Allocation	Permits the BS to change the MS channel (frequency) to another channel in the same cell
Assignment Complete		x	Reason Code	MS has established main signaling link successfully
Assignment Failure		x	Reason Code	Indicates that the MS did not seize the new channel
Channel Mode Modify	x		Channel Mode (Voice or Data)	Changes the mode of the channel
Channel Mode Modify Acknowledge		x	Channel Mode (Voice or Data)	Acknowledges mode
Channel Release	x		Reason Code	Releases the indicated channel
Channel Request		x	Establishment Code (e.g., page, origination)	MS makes a request for a channel
Ciphering Mode Command	x		Start Ciphering	Indicates to MS to start ciphering
Ciphering Mode Complete		x	—	Indicates that the MS has started ciphering
Class Mark Change		x	MS Class Mark	Indicates that the class mark of the MS has changed
Class Mark Enquiry	x		—	BS requests class mark information from the MS (GSM Phase 2 only)
Frequency Redefinition	x		Channel Description, Mobile Allocation	Indicates that the frequencies and hopping sequence have changed

Table 9.4 RR Messages (Continued)

Message	BS->MS	MS->BS	Key Parameters	Description
Handover Access		x	—	Initial message sent on the new channel to inform new BS of the presence of the MS on that channel
Handover Command	x		Channel Description, Mobile Allocation, Frequency Channel Sequence	Changes the channel allocation of the MS
Handover Complete		x	Reason Code	Indicates that the MS has established the signaling link on the new channel
Handover Failure		x	Reason Code	Sent on the old channel to indicate that the MS has failed to seize the new channel
Immediate Assignment	x		Channel Description	Moves an idle MS to a dedicated channel
Immediate Assignment Extended	x		Channel Description (1), Channel Description(2)	Moves two idle MSs each to a dedicated channel
Immediate Assignment Rejected	x		Wait Indication	Sent to up to 5 MSs to indicate that no channel is available
Measurement Report		x	Measurement Results	Reports on signal strength measurements made on serving and neighboring cells
Paging Request: Type 1–3	x		MS Identity (1–4)	Pages up to 4 MSs, by their TMSI or IMSI Type 1: 1–2 MS Type 2: 2–3 MS Type 3: 1–4 MS
Paging Response		x	MS Identity, Classmark	Indicates that MS has responded to the page
Partial Release	x		Channel Description	Indicates that the MS should release some of the dedicated channels in use

Table 9.4 RR Messages (Continued)

Message	BS->MS	MS->BS	Key Parameters	Description
Partial Release Complete		x	—	Confirms that MS has deactivated some to the dedicated channels
Physical Information	x		—	Informs the MS to stop sending access bursts
RR Status	x	x	RR Cause	Sent by either side to indicate an error condition
Synchronization Channel Information	x		—	Sent by BS to synchronize the MS
System Information: Types 1–8	x		Call Channel Description, Call Selection Parameters, Neighbor Cell Description, RACH Parameters	Sent by BS to indicate the characteristics of the BS

☞ Mobility management messages are those messages that are unique to a phone that moves from one area to another. These messages are associated with determining the IMSI and assigning a TMSI, locating the MS, and authenticating the MS. The messages are summarized in Table 9.5.

Table 9.5 Mobility Management Messages

Message	BS->MS	MS->BS	Key Parameters	Description
Abort	x		Reason	BS requests that all mobility management connections be aborted (GSM Phase 2 only)
Authentication Reject	x		—	Indicates that authentication has failed
Authentication Request	x		RAND*	Network requests that the MS send authentication
Authentication Response		x	SRES†	MS sends the response to the authentication request
Circuit Mode Reestablishment Request		x	MS Identity, MS Classmark	MS requests at reestablishment of a failed connection

Table 9.5 Mobility Management Messages (Continued)

Message	BS->MS	MS->BS	Key Parameters	Description
Circuit Mode Service Abort		x	Message Type	MS requests that first mobility management connection be aborted (GSM Phase 2 only)
Circuit Mode Service Accept	x		—	Informs the MS that its request for service has been accepted
Circuit Mode Service Reject	x		—	Informs the MS that its request for service has been rejected
Circuit Mode Service Request		x	CM Service Type, MS Classmark, MS Identity	MS requests a service (e.g., connection establishment, supplementary service, SMS transfer)
Identity Request	x		Identity Type	Network requests that MS send its identity
Identity Response		x	MS Identity	MS sends its identity
IMSI Detach Indication		x	MS Classmark, MS Identity	MS requests that its identity be removed from the MSC/VLR
Location Update Accept	x		Location Area Identification, MS Identity	Indicates successful completion of update request
Location Update Reject	x		Reject Cause	Indicates failure of update request
Location Update Request		x	Location Area Identification, MS Classmark, MS Identity	MS requests that its identity be updated in the MSC/VLR
Mobility Management Status	x	x	Reject Cause	Sent by either side to indicate an error condition
TMSI Reallocation Command	x		Mobile Identity	Assigns a new TMSI
TMSI Reallocation Complete		x	—	Indicates that MS has accepted new TMSI

*Random number.
†Signed result.

☞ The third class of messages are circuit-switched call control messages. These messages are similar to those found in wireline networks and include call setup, call disconnect, call status, and call progress messages. The messages are summarized in Table 9.6.

Table 9.6 Circuit-Switched Call Control Messages

Message	BS->MS	MS->BS	Key Parameters	Description
Alerting	x	x	Facility, Progress Indicator, User-User Data	Alerting indication is sent from one side to the other; can also be used for user-to-user signaling
Call Confirmed		x	Bearer Capabilities	Confirms incoming call request
Call Proceeding	x		Bearer Capabilities	Confirms call establishment (Setup message)
Congestion Control	x	x	Congestion Level	Used to flow control user-user information
Connect	x	x	User-User Data	Indicates that the user has answered the call
Connect Acknowledge	x	x	—	Acknowledges the connect message
Disconnect	x	x	Cause, User-User Data	Indicates that network or the MS has released the call
Emergency Setup		x	Bearer Capability	Sent by MS to indicate an emergency call setup
Hold		x	—	MS requests that the current call be put on hold
Hold Acknowledge	x		—	BS confirms hold request of MS
Hold Reject	x		Cause	BS rejects hold request of MS
Modify	x	x	Bearer Capability	Sending side requests a change in bearer capabilities
Modify Complete	x	x	Bearer Capability	Confirms modify request
Modify Reject	x	x	Bearer Capability, Cause	Indicates failure of modify request

Table 9.6 Circuit-Switched Call Control Messages (Continued)

Message	BS->MS	MS->BS	Key Parameters	Description
Notify	x	x	Notification Indicator	Indicates that the other side has put the call on hold
Progress	x		Progress Indicator, User-User Data	Indicates that call progress information is handled by tones or announcements instead of ISDN messages
Release	x	x	Cause, Facility, User-User Data	Sending side requests that the connection be released
Release Complete	x	x	Cause, Facility, User-User Data	Confirms release message
Retrieve		x	—	Retrieves a call on hold
Retrieve Acknowledge	x		—	Acknowledges a retrieval request
Retrieve Reject	x		—	Indicates that the retrieval request cannot be processed
Setup		x	Mobility Identity, Calling Party Number, Facility, User-User Data	Starts a call
Start DTMF		x	Keypad Facility (digit)	Indicates that the network should convert the message into the start of a DTMF tone
Start DTMF Acknowledge	x		Keypad Facility (digit)	Acknowledges Start DTMF message
Start DTMF Reject	x		Cause	Indicates that the network cannot accept and thus send the DTMF tone
Stop DTMF		x	—	Indicates that the network should stop sending of a DTMF tone
Stop DTMF Acknowledge	x		—	Acknowledges Stop DTMF message
Status	x	x	Cause, Call State	Reports an error condition to the other side

Table 9.6 Circuit-Switched Call Control Messages (Continued)

Message	BS->MS	MS->BS	Key Parameters	Description
Status Enquiry	x	x	—	Request a status report from the other side
User Information	x	x	User-User Data	Sends user-to-user information

9.3.2 BS to MSC Messages on the A Interface

The signaling between the BS and the MSC is done over one or more E1 lines at a data rate of 2048 kbps. The traffic channel bit rate is less than 16 kbps but is transmitted over one 64-kbps channel in the E1 line. The transcoding from the radio link voice signal to the PCM signal needed for the wireline network can occur at either the BSC or the MSC but is considered part of the BSS. The signaling between the BS and the MSC is layered but is not identical to the corresponding named layer in the OSI reference model (see Figure 9.3).

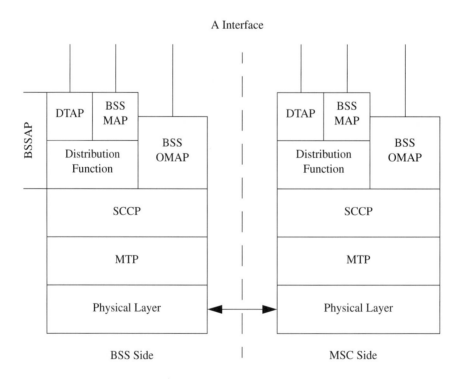

Fig. 9.3 Signaling Protocol Reference Model BS-MSC

The signaling protocol layers are

☞ Physical layer—one or more E1 (32 x 64 kbps = 2048 kbps) channels
☞ Message Transfer Part (MTP)
☞ Signaling Connection Control Part (SCCP)
☞ Base Station System Application Part (BSSAP)
☞ Base Station System Management Application Part (BSSMAP)
☞ Direct Transfer Application Part (DTAP)
☞ Base Station System Operations Maintenance Application Part (BSSO-MAP)

Table 9.7 summarizes the messages between the MSC and BSS.

Table 9.7 Layer 3 Messages between MSC and BSS

Message	BSS->MSC	MSC->BSS	Some Key Parameters	Description
Assignment Request		x	Radio Channel, Circuit Identity Code	Assigns radio resource at BSS
Assignment Complete	x		Radio Channel	Confirms radio resource assignment
Assignment Failure	x		Reason Code	Indicates that assignment failed
Block	x		Circuit Identity Code	Blocks use of a time slot
Blocking Acknowledge		x	Circuit Identity Code	Acknowledges the blocking of the time slot
Circuit Group Block	x		Circuit Identity Code, Circuit Identity Code List	Blocks use of several time slots (GSM Phase 2)
Circuit Group Blocking Acknowledge		x	Circuit Identity Code, Circuit Identity Code List	Acknowledges the blocking of several time slots (GSM Phase 2)
Circuit Group Unblock	x		Circuit Identity Code, Circuit Identity Code List	Unblocks several time slots and indicates that they may be returned to service (GSM Phase 2)

Table 9.7 Layer 3 Messages between MSC and BSS (Continued)

Message	BSS->MSC	MSC->BSS	Some Key Parameters	Description
Circuit Group Unblocking Acknowledge		x	Circuit Identity Code, Circuit Identity Code List	Acknowledges the unblocking of several time slots and returns them to service (GSM Phase 2)
Clear Command		x	Cause	Clears the indicated resource
Clear Complete	x		—	Indicates that the resource has been cleared
Clear Request	x		Cause	Requests that indicated resources be released
Unblock	x		Circuit Identity Code	Unblocks a time slot
Unblocking Ack		x	Circuit Identity Code	Acknowledges the unblocking of the time slot
Handover Candidate Enquire		x	Number of MSs	Request for number of MSs where handover is required
Handover Candidate Response	x		Number of MSs	Indicates the number of MSs where handover is required
Handover Request		x	Cell ID (serving), Cell ID (target)	Handover to this (new) BSS is requested
Handover Required	x		Cell Identifier List Preferred	The MS needs a handover to a new channel
Handover Required Reject		x	Cause	Indicates that a handover has not occurred
Handover Request Acknowledge	x		—	Indicates that the BSS can support the handover request
Handover Command		x	—	Indicates new channel that MS must retune
Handover Complete	x		—	Indicates successful handover to the new BS
Handover Succeeded		x	—	MSC sends indication to old BS indicating that the handover succeeded

Table 9.7 Layer 3 Messages between MSC and BSS (Continued)

Message	BSS->MSC	MSC->BSS	Some Key Parameters	Description
Handover Failure	x		Reason Code	Handover has been aborted due to resource allocation failure
Handover Performed	x		Cause Radio Channel	Indicates that the BSS performed a handover internal to the cell
Handover Detect	x		—	Indicates that the MS has successfully accessed the target cell
Resource Request		x	Cell Identifier	MSC requests the spare and optionally the total resources at a BS
Reset	x	x	Cause	Indicates that the sending side has lost all call state memory
Reset Acknowledge	x	x	—	Indicates that the sending side has cleared all calls and is ready to resume service
Resource indication	x		Cell Identifier, Resources	BS reports on the spare and optionally the total resources
Paging		x	IMSI, TMSI	Request that MS be paged
Overload	x	x	Cause Cell Identifier	Indicates that the processor is in overload
MSC Invoke trace		x	Trace Type, Mobile Identity	MSC invokes a trace record at the BSS (GSM Phase 2)
BSS Invoke trace	x		Trace Type	BSS invokes a trace record at the MSC (GSM Phase 2)
Trace Invocation	x	x	Trace Number	Invokes a trace record at the receiving entity (Phase 1)
Classmark Update	x		Class mark information	MS has changed classmark

Table 9.7 Layer 3 Messages between MSC and BSS (Continued)

Message	BSS->MSC	MSC->BSS	Some Key Parameters	Description
Classmark Request		x	—	MSC asks BS for the classmark (on the channel associated with the MS)
Cipher Mode Command		x	Encryption Information	Updates encryption information for the specified MS
Cipher Mode Complete	x		Algorithm	Indicates that successful cipher mode has been achieved on the air interface
Cipher Mode Reject	x		Cause	Cipher mode cannot be started
Complete layer 3 information	x		Layer 3 Information	BS sends layer 3 information when on receipt of initial layer 3 messages (GSM Phase 2)
Queuing Indication	x		—	Indicates a delay in assignment of the TCH (GSM Phase 2)
Service Access Point Identifier (SAPI) *n* Reject	x		Cause	Indicates that a message has been rejected
SAPI *n* Clear Command	x	x	—	Instructs receiving entity to clear resources
SAPI *n* Clear Complete	x	x	—	Indicates that all resources for SAPI *n* have been released
Reset Circuit	x	x	Circuit Identity	Indicates that the circuit state is unknown due to a failure
Reset Circuit Acknowledge	x	x	Circuit Identity	Indicates that the call using the circuit is cleared and the circuit is available for service
Confusion	x	x	Cause	Indicates that the message sent by the other side cannot be acted upon and no other failure message is available (GSM Phase 2)

Table 9.7 Layer 3 Messages between MSC and BSS (Continued)

Message	BSS->MSC	MSC->BSS	Some Key Parameters	Description
Unequipped Circuit	x	x	Circuit Identity Code, Circuit Identity Code List	Other side is attempting to use a circuit that the sending side does not recognize (GSM Phase 2)
Load Indication	x	x	Cell Identifier, Resource, Cause	Indicates that the resources are in overload or unavailable at the BSS (GSM Phase 2)
VGCS/VBS Setup		x	Group Call Reference	Sets up a voice call to a group of MSs (GSM Phase 2)
VGCS/VBS Setup Acknowledge	x		—	Confirms that the BSS will support the group call (GSM Phase 2)
VGCS/VBS Setup Refuse	x		Cause	BSS rejects the requested group call (GSM Phase 2)
VGCS/VBS Assignment Request		x	Group Call Reference, Circuit Identity Code, Encryption Information	Request to BSS to assign radio resources for the group call (GSM Phase 2)
VGCS/VBS Assignment Result	x		Channel Type, Cell Identifier, Chosen Channel	Indicates that radio resources for the group call have been assigned (GSM Phase 2)
VGCS/VBS Assignment Failure	x		Circuit Pool, Circuit Pool List, Cause	Indicates failure of assignment request (GSM Phase 2)
VGCS/VBS Queuing Indication	x		—	Indicates delay in assigning radio resources for group call (GSM Phase 2)
Uplink Request	x		—	Indicates that an MS has requested access to the uplink of a group call (GSM Phase 2)
Uplink Request Acknowledgment		x	—	Acknowledges uplink request (GSM Phase 2)

Table 9.7 Layer 3 Messages between MSC and BSS (Continued)

Message	BSS->MSC	MSC->BSS	Some Key Parameters	Description
Uplink Request Confirmation	x		Cell Identifier	Uplink has been successfully established (GSM Phase 2)
Uplink Release Indication	x		Cause	Uplink has been released (GSM Phase 2)
Uplink Reject Command		x	Cause	Request for Uplink has been rejected (GSM Phase 2)
Uplink Release Command		x	Cause	MSC has released uplink and it is available for allocation (GSM Phase 2)
Uplink Seized Command		x	Cause	Uplink is no longer available for allocation (GSM Phase 2)
Suspend	x		DLCI	BSS has detected an overload in the corresponding connection (GSM Phase 2)
Resume	x		DLCI	Overload situation no longer exists (GSM Phase 2)

9.3.3 MSC to VLR and HLR

The MSC must communicate with the HLR and the VLR to manage those functions that are unique to a mobile environment. We will discuss communication from the MSC to other switches in chapter 17 on SS7. In this section, we focus on the messages related to managing a mobile call.

The first set of messages are directly related to mobility management and are summarized in Table 9.8. The messages are transmitted by one side of the interface and confirmed by the other side using the *request*, *indication*, *response*, and *confirm* set of primitives as described in section 9.3.1 for the BS-MSC interface. The messages are segmented into several categories:

☞ **Location management services** update HLR and VLR entries so that the network can locate the MS.

☞ **Paging and search services** page the MS.

☞ **Access management services** support access to the network by the MS.

☞ **Handover services** support the movement (handover) of an MS from one cell to another cell.

☞ **Authentication management services** support the confirmation of the MS identity by using cryptographic techniques.

☞ **Security services** support privacy of the MS-BS link by using cryptographic techniques.

☞ **Subscriber management services** support HLR database updates when the subscriber changes service subscriptions (e.g., add/drop call waiting, call forwarding).

☞ **Identity management services** support the use of IMSI and TMSI by the MS.

☞ **Fault recovery services** support recovery from errors in MS network signaling.

The next set of services are for operations and maintenance (Table 9.9). For more details on Operation, Administration, and Maintenance (OA&M) refer to chapter 15.

Table 9.8 Mobility services

Service	Message	Key Parameters	Description
Location management services	UPDATE_LOCATION _AREA	IMSI, TMSI, LAI	MSC→VLR Used to update VLR entry when MS moves to new location area
	UPDATE_LOCATION	IMSI VLR Number	VLR→HLR Used to update HLR entry
	CANCEL_LOCATION	IMSI	HLR→VLR Used to cancel an entry in the VLR
	SEND _IDENTIFICATION	TMSI (New VLR) IMSI (Old VLR)	VLR→VLR Used by new VLR to obtain information from old VLR
	DETACH_IMSI	IMSI or TMSI	MSC→VLR Removes database entry in VLR
	PURGE_MS	IMSI, VLR Number	VLR→HLR Removes database entry in HLR
Paging and search	PAGE	IMSI, TMSI, LAI	MSC→VLR Used to initiate paging of an MS

Table 9.8 Mobility services (Continued)

Service	Message	Key Parameters	Description
	SEARCH_FOR_MS	IMSI	VLR→MSC Used for paging MS; used when VLR does not know which LAI the MS is in
Access management services	PROCESS_ACCESS _REQUEST	IMSI, IMEI, TMSI, LAI, MS-ISDN Number	MSC→VLR Used to initiate the processing of an MS access request
Handover services	PREPARE _HANDOVER	Target Cell ID, BSS-ADPU	MSC-A→MSC-B Used when inter-MSC handover is performed
	SEND_END_SIGNAL	BSS-APDU	MSC-A→MSC-B Used when inter-MSC handover is performed to release call records in MSC-B
	PROCESS_ACCESS _SIGNALING	BSS-APDU	MSC-B→MSC-A Used to pass A-interface information from MSC-B
	FORWARD_ACCESS _SIGNALING	BSS-APDU	MSC-A→MSC-B Used to pass A-interface information to MSC-B
	PREPARE _SUBSEQUENT _HANDOVER	BSS-APDU, Target MSC Number, Target Cell ID	MSC-A→MSC-B Used to indicate that a handover back to MSC-A or to a third MSC is needed
	ALLOCATE _HANDOVER _NUMBER	—	MSC→VLR Used to request a handover number from the VLR
	SEND_HANDOVER _REPORT	Handover Number	VLR→MSC-B Used to transfer handover number from MSC-B to MSC-A
Authentication management services	AUTHENTICATE	RAND (from VLR) SRES (from MS)	VLR→MSC Used to authenticate the MS

Table 9.8 Mobility services (Continued)

Service	Message	Key Parameters	Description
	SEND _AUTHENTICATION _INFO	IMSI (from VLR), Authentication Set (from HLR)	VLR→HLR VLR requests authentication information
Security management services	SET_CIPHERING _MODE	K_c	VLR→MSC Used to start encryption of the radio link
International mobile equipment identities management services	CHECK_IMEI	IMEI, Status	VLR→MSC & MSC→EIR Used to check the IMEI of an MS
	OBTAIN_IMEI	IMEI	VLR→MSC Used to obtain IMEI from the MSC
Subscriber management services	INSERT _SUBSCRIBER_DATA	IMSI, MS-ISDN Number, Services	HLR→VLR Used to update the VLR when the HLR data has changed because subscriber has changed service subscription (added services)
	DELETE _SUBSCRIBER_DATA	IMSI, MS-ISDN Number, Services	HLR→VLR Used to update the VLR when the HLR data has changed because subscriber has changed service subscription (removed services)
Identity management services	PROVIDE_IMSI	IMSI	VLR→MSC Used to obtain IMSI from an MS when the TMSI transmitted by the MS cannot be associated with a known system
	FORWARD_NEW _TMSI	IMSI, TMSI	VLR→MSC Used to allocate a new TMSI to an MS during an ongoing transaction (e.g., call set, request for supplementary service)

Table 9.8 Mobility services (Continued)

Service	Message	Key Parameters	Description
Fault recovery services	RESET	HLR Number, HLR ID List	HLR→VLR Indicates that the HLR has just been reset
	FORWARD_CHECK _SS_INDICATION	—	HLR→VLR Indicates to the MS that supplementary services may have been altered due to an HLR reset
	RESTORE_DATA	IMSI, Local MSI (LMSI), HLR Number	VLR→HLR Used when local roaming number for an MS maps to an unknown MS

Table 9.9 Operation and Maintenance Services

Service	Message	Key Parameters	Description
Subscriber Tracing Services	ACTIVATE_TRACE _MODE	IMSI, Trace Reference, Trace Type	HLR→VLR This message is used to activate a trace in the VLR on a questionable or nonfunctioning MS
	DEACTIVATE _TRACE_MODE	IMSI, Trace Reference, Trace Type	VLR→HLR This message is used to deactivate a trace on a questionable or nonfunctioning MS
	TRACE _SUBSCRIBER _ACTIVITY	IMSI, Trace Reference, Trace Type	VLR→MSC This message is used to activate a trace in the MSC on a questionable or nonfunctioning MS
Other Operation and Maintenance Services	SEND_IMSI	MS-ISDN Number, IMSI	VLR→HLR Used to obtain IMSI in visited system when only the MS-ISDN number is known

When MS makes an outgoing call or receives an incoming call, the MSC, VLR, and HLR must communicate with each other to obtain information about the MS. Some of the information needed includes the MS identity, MS services, MS temporary identities, and MS roaming number in the system. The messages sent between the MSC, VLR, and HLR are described in Table 9.10.

Table 9.10 Call-Handling Services

Message	Key Parameters	Description
SEND_INFO_FOR_INCOMING _CALL	Bearer Service, MS Roaming Number, MS-ISDN Number	MSC→VLR Used on incoming call to obtain information about an MS
SEND_INFO_FOR_OUTGOING _CALL	Bearer Service, Called Number	MSC→VLR Used on outgoing call to obtain information about an MS
SEND_ROUTING _INFORMATION	MS-ISDN Number (from MSC), IMSI (from HLR), MS Roaming Number (from HLR)	MSC→HLR Used to obtain information on an MS in order to route a call to the MS
PROVIDE_ROAMING _NUMBER	IMSI, MSC Number, MS-ISDN Number, Local MS Identity (LMSI) Roaming Number	HLR→VLR Used by HLR to obtain a roaming number from the VLR
COMPLETE_CALL	MS-ISDN Number, IMEI, Bearer Capability	VLR→MSC Used by VLR to request that an incoming or outgoing call be set up
PROCESS_CALL_WAITING	MS-ISDN Number, Bearer Capability	VLR→MSC Used by VLR to invoke call waiting at the MSC

Each MS can support a variety of services in addition to basic incoming and outgoing calls. These are called supplementary services and are described in section 9.2.3. The MSC, HLR, and VLR must communicate with each other to obtain information about the supplementary services that each MS is permitted to access. These messages are described in Table 9.11.

A GSM MS can function as a two-way pager. This service is an SMS and the messages to support this service are described in Table 9.12. For more details on SMS see chapter 10.

Table 9.11 Supplementary Services

Message	Key Parameters	Description
REGISTER_SS	Supplementary Service Code, Basic Service Group	MSC→VLR→HLR Used by MSC to register data at the HLR for supplementary services. The VLR relays the message.
ERASE_SS	Supplementary Service Code, Basic Service Group	MSC→VLR→HLR Used by MSC to remove data at the HLR for supplementary services. The VLR relays the message.
ACTIVATE_SS	Supplementary Service Code, Basic Service Group	MSC→VLR→HLR Used by MSC to activate a supplementary service. The VLR relays the message.
DEACTIVATE_SS	Supplementary Service Code, Basic Service Group	MSC→VLR→HLR Used by MSC to deactivate a supplementary service. The VLR relays the message.
INTERROGATE_SS	Supplementary Service Code, Basic Service Group	MSC→VLR→HLR Used by MSC to retrieve information about a supplementary service. The VLR relays the message.
INVOKE_SS	Supplementary Service Code, Basic Service Group	MSC→VLR→HLR Used by MSC to confirm information on a supplementary service during the invocation of the SS. The VLR relays the message.
REGISTER_PASSWORD	Supplementary Service Code, New Password	MSC→VLR→HLR Used by MSC to register a new password when the MS requests it. The VLR relays the message.
GET_PASSWORD	Password	HLR→VLR→MSC Used by HLR to obtain the password for a supplementary service. The VLR relays the message.

Table 9.11 Supplementary Services (Continued)

Message	Key Parameters	Description
PROCESS_UNSTRUCTURED _SS_REQUEST	Unstructured Supplementary Service Data Coding	MSC→VLR→HLR Used by MSC to relay information on a unstructured supplementary service. The VLR relays the message.
UNSTRUCTURED_SS _REQUEST	Unstructured supplementary service data coding	HLR→VLR→MSC Used by HLR to obtain information on an unstructured supplementary service. The VLR relays the message.
UNSTRUCTURED_SS_NOTIFY	Unstructured Supplementary Service Data Coding	HLR→VLR→MSC Used by HLR to notify an MS about an unstructured supplementary service. The VLR relays the message.

Table 9.12 SMS Management Services

Message	Key Parameters	Description
SEND_ROUTING_INFO _FOR_SM	MS-ISDN Number IMSI, LMSI	MSC→HLR Used to retrieve routing information needed for routing a short message
FORWARD_SHORT_MESSAGE	Originating address (IMSI, MS-ISDN Number, etc.), destination address user data	MSC→MSC Used to route short messages between MSCs
REPORT_SM_DELIVERY _STATUS	MS-ISDN Number	Gateway MSC→HLR Used to set the message waiting indicator in the HLR; also used to inform the HLR of successful short message transfer
READY_FOR_SM	IMSI, TMSI	MSC→VLR & VLR→HLR Used to indicate that the MS has available memory for the short message
ALERT_SERVICE_CENTER	Service Center Address	HLR→Interworking MSC Activated by HLR when message is waiting and MSC has available memory

Table 9.12 SMS Management Services (Continued)

Message	Key Parameters	Description
INFORM_SERVICE_CENTER	MS-ISDN Number	HLR→Gateway MSC Unconfirmed message used to inform service center of MS-ISDN number stored in the HLR database
SEND_INFO_FOR_MT_SMS	IMSI, or LMSI, MS-ISDN Number	MSC→VLR Invoked on mobile terminated short message to request subscriber information
SEND_INFO_FOR_MO_SMS	Service Center Address, MS-ISDN Number	MSC→VLR Invoked on mobile originated short message to request subscriber information

9.4 GSM CALL FLOW SCENARIOS

In this section, we discuss call flow scenarios used in GSM. In these call flow scenarios, we assume that the MS enters the new MSC area and requires a location update procedure involving registration, authentication, ciphering and equipment validation. The MS location registration update procedure is given in chapter 11. In this chapter we discuss the call flow scenarios involved with the call origination (wireless to wireline), call termination (wireline to wireless), call release, and handover (both inter- and intra-MSC). In these call flows, not every message on every link is shown. For details see GSM 09.02 [2].

9.4.1 Call Setup by an MS

The first call flow we will examine is the call setup by an MS (Figure 9.4).

1. The user dials a number and presses **SEND** or some other action to start a call. The MS sends a setup message to the MSC after it begins ciphering the radio channel. This message includes the dialed digits.

2. Upon receiving the setup message, the MSC requests the VLR to supply the subscriber parameters necessary for handling the call via the Send Info for Outgoing Call message. The message contains the called number and service indication.

3. The VLR checks for call barring conditions. If the VLR determines that the call cannot be processed, the VLR provides the reason to the MSC. In this case, we assume that the procedure is successful and the call can be

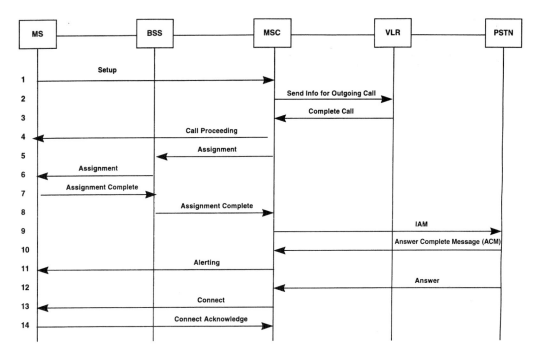

Fig. 9.4 Call Setup by a Mobile

processed. The VLR returns a Complete message to the MSC containing the service parameters for the subscriber.

4. The MSC sends a Call Proceeding message to the MS.

5. The MSC allocates an available trunk to the BSS currently serving the MS. The MSC sends an Assignment message to the BSS supplying it with the trunk number allocated and asks to assign a radio TCH for the MS.

6. The BSS allocates a radio channel and sends an Assignment message to the MS over SDCCH.

7. The MS tunes to the assigned radio channel and sends an Assignment Complete message to the BSS.

8. The BSS connects the radio TCH to the assigned trunk to the MSC and deallocates the SDCCH. The BSS informs the MSC with a (trunk and radio) Assignment Complete message. At this point a voice path is established between the MS and the MSC. The MS user hears silence since the complete voice path is not yet established.

9. The MSC sends the Initial Address Message (IAM) to the PSTN to request call setup. This message includes the digits dialed by the MS and details of the trunk that will be used for the call.

10. The PSTN sets up the call and notifies the MSC with an Address Complete message.

11. The MSC informs the MS that the destination number is being alerted. The MS hears the ringing tone from the destination local exchange through the established voice path. The MSC sends an Alerting message to the MS.

12. When the destination party goes off hook, the PSTN informs the MSC via the Answer message.

13. The MSC informs the MS that the connection has been established via the Connect message.

14. The MS sends a Connect Acknowledgment to the MSC, and the two parties can now talk.

9.4.2 Mobile-Terminated Call

In this scenario (Figure 9.5) it is assumed that the MS is already registered with the system and has been assigned a TMSI. It is also assumed that the MS is in its home system. A land subscriber dials the directory number of the mobile subscriber.

1. The PSTN routes the call to the MSC assigned this directory number. The directory number in the IAM message is the MS ISDN Number (MSISDN).

2. The MSC sends the Send Routing Information message to the HLR to provide the routing information for the MSISDN.

3. The HLR acknowledges the Send Routing Information message to the MSC. This message contains the MSRN. If the MS is roaming within the serving area of this MSC, the MSRN returned by the HLR will most likely be the same as MSISDN. In this scenario we assume that mobile is not roaming.

4. The MSC informs its VLR about the incoming call using a Send Info for Incoming Call message that includes MSRN.

5. The VLR responds to the MSC through a Page message that specifies the LAI and TMSI of the MS. If the MS is barred from receiving the calls, the VLR informs the MSC that a call cannot be directed to the MS. The MSC would connect the incoming call to an appropriate announcement (e.g., "The mobile phone that you have called is not permitted to receive calls").

6. The MSC uses the LAI provided by the VLR to determine which BSSs will page the MS. The MSC sends the Page message to each of the BSSs to perform the paging of the MS.

7. Each BSS broadcasts the TMSI of the MS in the Page Request message on the PCH.

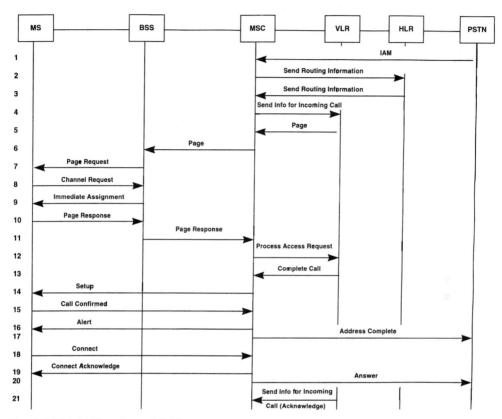

Fig. 9.5 Mobile-Terminated Call

8. When the MS hears its TMSI broadcast on the PCH, it responds to the BSS with a Channel Request message over the common access channel, RACH.

9. On receiving the Channel Request message from the MS, the BSS allocates an SDCCH and sends the Immediate Assignment message to the MS over the AGCH. It is over the SDCCH that the MS communicates with the BSS and MSC until a TCH is assigned.

10. The MS sends a Page Response message to the BSS over the SDCCH. The message contains the MS's TMSI and LAI.

11. The BSS forwards the Page Response message to the MSC.

12. The MSC sends a Process Access Request message to the VLR.

13. The VLR responds with a Complete Call message.

14. The MSC then sends a Setup message to the MS.

15. The MS responds with a Call Confirmed message.

16. The MSC then sends an Alert message to the MS.

17. The MSC sends an Address Complete message to the PSTN.

18. When the user answers, the MS sends a Connect message to the MSC.

19. The MSC sends a Connect Acknowledge to the MS.

20. The MSC sends an Answer message to the PSTN. The two parties can now talk.

21. The VLR closes the dialog with the MSC by sending a Send Info for Incoming Call (acknowledge) to the MSC.

9.4.3 Call Release

Under normal conditions, there are two basic ways a call is terminated: mobile initiated and network initiated. In this scenario (Figure 9.6), we assume that the mobile user initiates the release of the call.

1. At the end of the call the MS sends the Disconnect message to the MSC.

2. On receiving the Disconnect message, the MSC sends a Release message to the PSTN to release the call.

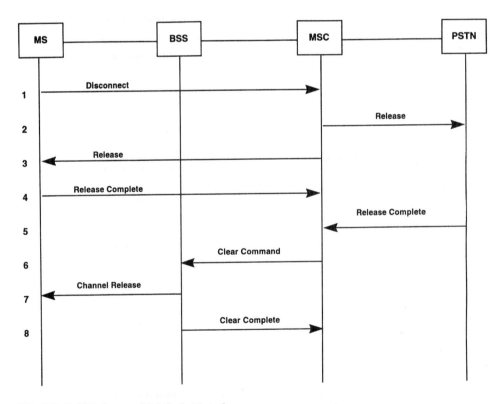

Fig. 9.6 Call Release—Mobile Initiated

3. The MSC asks the MS to begin its clearing procedure using the Release message.

4. After the MS has performed its clearing procedure, it informs the MSC through the Release Complete message.

5. The PSTN sends a Release Complete message to the MSC. The order of steps 4 and 5 may be reversed depending on timing in the network.

6. The MSC then sends the Clear Command message to the BSS to ask it to release all the allocated dedicated resources for a given SCCP connection.

7. The BSS sends the Channel Release message to the MS to release the TCH.

8. The BSS sends a Clear Complete message to the MSC informing it that all allocated dedicated resources have been released.

9.4.4 Handover

Basically there are two types of handover—internal and external. If the serving and target BTSs are located within the same BSS, the BSC for the BSS can perform handover without the involvement of the MSC. This type of handover is referred to intra-BSS handover. However, if the serving and target BTSs do not reside within the same BSS, an external handover is performed. In this type of handover the MSC coordinates the handover and performs the switching tasks between the serving and target BTSs. The external handovers can be classified as either within the same MSC (intra-MSC) or between different MSCs (inter-MSC). In the following call flow scenarios we focus only upon the external handovers.

9.4.4.1 Intra-MSC Handover (Figure 9.7) The MS constantly monitors the signal quality of the BSS-MS link. The BSS may also optionally forward its own measurements to the MS. When the link quality is poor, the MS will attempt to maintain the desired signal quality of the radio link by requesting a handover.

1. The MS determines that a handover is required; it sends the Measurement Report message to the serving BSS. This message contains the signal strength measurements.

2. The serving BSS sends a Handover Request message to the MSC. This message contains a rank-ordered list of the target BSSs that are qualified to receive the call.

3. The MSC reviews the global cell identity associated with the best candidate to determine if one of the BSSs that it controls is responsible for the cell area. In this scenario the MSC determines that the cell area is associated with the target BSS. To perform an intra-MSC handover, two resources are required: a trunk between the MSC and the target BSS

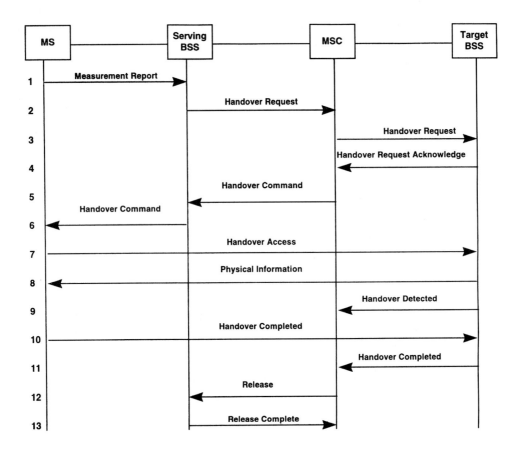

Fig. 9.7 Intra-MSC Handover

and a radio TCH in the new cell area. The MSC reserves a trunk and sends a Handover Request message to the target BSS. This message includes the desired cell area for handover, the identity of the MSC-BSS trunk that was reserved, and the encryption key (K_i).

4. The target BSS selects and reserves the appropriate resources to support the handover pending the connection execution. The target BSS sends a Handover Request Acknowledgment to the MSC. The message contains the new radio channel identification.

5. The MSC sends the Handover Command message to the serving BSS. In this message the new radio channel identification supplied by the target BSS is included.

6. The serving BSS forwards the Handover Command message to the MS.

7. The MS retunes to the new radio channel and sends the Handover Access message to the target BSS on the new radio channel.

8. The target BSS sends the Physical Information message to the MS.

9. The target BSS informs the MSC when it begins detecting the MS handing over with the Handover Detected message.

10. The target BSS and the MS exchange messages to synchronize/align the MS's transmission in the proper time slot. On completion, the MS sends the Handover Completed message to the target BSS.

11. At this point the MSC switches the voice path to the target BSS. Once the MS and target BSS synchronize their transmission and establish a new signaling connection, the target BSS sends the MSC the Handover Completed message to indicate that handover is successfully completed.

12. The MSC sends a Release message to the serving BSS to release the old radio TCH.

13. At this point the serving BSS releases all resources with the MS and sends the Release Complete message to the MSC.

It may be noted that GSM recommendations require that the open interval gap during a handover will not exceed 150 ms for 90-percent handovers. The open interval gap starts when the MS retunes to the new radio channel and ends after synchronization without any loss in voice/data transmission in the BSS or MSC.

9.4.4.2 Inter-MSC Handover (Figure 9.8) In this scenario we assume that a call has already been established. The serving BSS is connected to the serving MSC and the target BSS to the target MSC.

1. Same as step 1 in the intra-MSC handover.

2. Same as step 2 in the intra-MSC handover.

3. When a call is handed over from the serving MSC to the target MSC via PSTN, the serving MSC sets up an inter-MSC voice connection by placing a call to the directory number that belongs to the target MSC. When the serving MSC places this call, the PSTN is unaware that the call is a handover and follows the normal call routing procedures, delivering the call to the target MSC. The serving MSC sends a Prepare Handover message to the target MSC.

4. The target MSC sends an Allocate Handover message to its VLR to assign the TMSI.

5. The target VLR sends the TMSI in the Send Handover Report message.

6. Same as step 3 in the intra-MSC handover.

7. Same as step 4 in the intra-MSC handover.

8. The target MSC sends the Prepare Handover (acknowledge) message to the serving MSC indicating that it is ready for the handover.

9. The serving MSC sends the Send Handover (acknowledge) message to the target MSC.

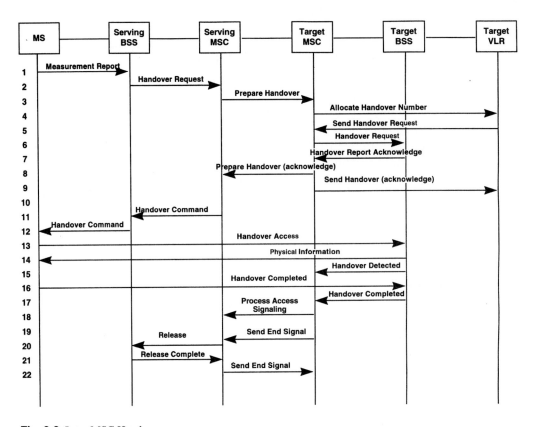

Fig. 9.8 Inter-MSC Handover

10. The target MSC sends a Send Handover Request (acknowledge) message to the target VLR.

11. Same as step 5 in the intra-MSC handover.

12. Same as step 6 in the intra-MSC handover.

13. Same as step 7 in the intra-MSC handover.

14. Same as step 8 in the intra-MSC handover.

15. Same as step 9 in the intra-MSC handover.

16. Same as step 10 in the intra-MSC handover.

17. Same as step 11 in the intra-MSC handover.

18. The target MSC sends a Process Access Signaling message to the serving MSC.

19. At this point the handover has been completed, and the target MSC sends a Send End Signal message to the serving MSC.

20. Same as step 12 in the intra-MSC handover.

21. Same as step 13 in the intra-MSC handover.
22. The serving MSC sends a Send End Signal (acknowledge) message to the target MSC.

9.5 SUMMARY

In this chapter we examined the basic and supplementary services that are provided to a GSM MS. GSM services are based on ISDN services, so we described each of the GSM services and compared them with the corresponding ISDN services. We briefly discussed the service quality and performance standards for GSM networks.

We described the messages that flow on the links between

☞ The MS and the BS
☞ The BS and the MSC
☞ The MSC and the VLR and HLR
☞ The HLR and the VLR

We then used the messages to show end-to-end call flows for mobile origination, termination, and release and for intra- and inter-MSC handovers.

9.6 REFERENCES

1. GSM Specification Series 01.02–1.06, "GSM Overview, Glossary, Abbreviations, Service Phases."
2. GSM Specification Series 02.01–2.88, "GSM Services and Features."
3. GSM Specification Series 03.01–3.88, "GSM PLMN Functions, Architecture, Numbering and Addressing Procedures."
4. GSM Specification Series 04.01–4.88, "MS-BSS Interface."
5. GSM Specification Series 05.01–5.10, "Radio Link."
6. GSM Specification Series 06.01–6.32, "Speech Processing."
7. GSM Specification Series 07.01–7.03, "Terminal Adaptation."
8. GSM Specification Series 08.01–8.60, "BSS-MSC Interface, BSC-BTS Interface."
9. GSM Specification Series 09.01–9.11, "Network Interworking, MAP."

Data Services in GSM

10.1 INTRODUCTION

In this chapter we examine data services in GSM. Analog cellular phones offered low-bit-rate data services as a never-planned-for option only after users starting using voice-band modems over the radio system. GSM data services were planned from the beginning, and offer many options similar to those offered over ISDN. In this chapter, we first examine the special needs for terminal equipment and MSs to support data. The standard 9-pin or 26-pin RS-232 connector is too large to include in a portable handset. Thus, other options are used in a digital system including GSM.

We then examine the standard data options supported in GSM: packet data, circuit data, and facsimile. Where appropriate we show protocol diagrams and discuss the special needs of sending data over a hostile radio link.

Since a GSM phone can function as a two-way pager, we discuss the special data service called Short Message Service (SMS). There are two SMS services that GSM supports. The first is a point-to-point service similar to standard paging with the addition of two-way capabilities. The second service is cell broadcast SMS which allows groups of MSs to be sent news, weather, sports, and stock market and other information. This is a one-way service.

Finally we examine a new service called GPRS. This service permits voice and circuit-switched data users to share facilities with packet-switched data users.

10.2 DATA INTERWORKING

Low-speed data is transmitted over the telephone network today using voice-band modems that range from 1200 baud to 56,000 baud. Some of the uses for this data include

☞ Accessing electronic mail
☞ Accessing remote computers
☞ File transfers
☞ Facsimile transmission
☞ Internet access
☞ Transaction services (e.g., credit card validation)

Transmission at higher rates is typically used for video conferencing, mainframe-to-mainframe communications, and other uses that would not initially be carried over a wireless mobile link.

Except for those PCS systems supporting PCM or ADPCM, the speech coding systems for transmission of voice have been designed for the voice transmission (using low bit rates) and have not been designed for the transmission of nonvoice signals such as voice-band modems. Therefore, both the MS and the PCS system must have capabilities to do protocol conversion to support the wide range of services currently supported using voice-band modems. The provision of the protocol conversion in the MS and the MSC is called interworking. Over the air interface, the data must be transmitted digitally since that is the only option available. If the air interface supports a data rate higher than the basic data rate of the voice-band modem, then interworking is possible. Interworking is not possible if the required data rate is higher than the needed data rate. For example, a raw data rate of 28.8 kbps is not possible over a 16-kbps data link. Those systems supporting PCM or ADPCM can interwork with voice-band modems without a data interworking platform. However, error performance may suffer because of the high error rate of the radio channel.

For the remainder of this section, we will focus on the digital GSM system and show the interworking functions needed to support data. The interworking function can support the data needs of the voice-band modem only when the air interface data rate is higher than the modem speed. Figure 10.1 shows one method for interworking by having a modem in the MS that communicates with a modem in the auxiliary device (fax machine, PC, credit card validation terminal, and so on) via an industry standard telephone jack (e.g., RJ-11). This method requires no modifications to the auxiliary device but does

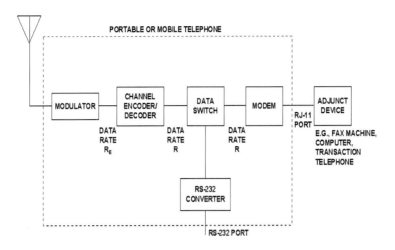

Fig. 10.1 Mobile Telephone Architecture for Data Interworking

require a different modem in the MS for each data service. Digital cellular phones are provided with a digital port on the phone (Figure 10.2). Since the RS-232 connector is large compared to most handheld phones, some special form of the connector is used on each phone. At this time, there does not appear to be a physical connector standard for the serial port on the phone. Thus each manufacturer uses its own type of connector. As the Infrared Data Association (IrDA) infrared serial port standard is used by more devices, that

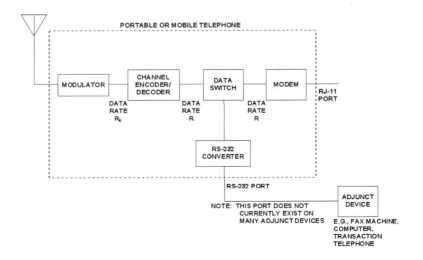

Fig. 10.2 Mobile Telephone Architecture for Data Interworking

may become the standard interface for phones as it has started to become in the PC industry. To support a direct serial port, the auxiliary device must have an option for bypass of the built-in modem and therefore a direct digital connection. While the RS-232 interface is standard on laptop computers, many devices (e.g., fax machines and transaction telephones) do not currently support a direct RS-232 connection in place of the internal modem.

Figure 10.3 shows the architecture of the PCS network side of the interworking function. In the network, a modem pool (Figure 10.4) must be installed for each service that needs interworking.

The GSM standards [11, 12, 13] define three types of MSs (Figure 10.5). Type 0 stations either do not support data or have all data functionality fully supported within the telephone (the Nokia 9000 combination GSM phone and PDA is an example). Type 1 MSs support the S interface with a separate ISDN terminal. Since most ISDN terminals do not include data terminal equipment, to support data applications, a data terminal must be connected to the ISDN terminal. In a type 2 MS, the R interface directly supports either an X series terminal (e.g., X.21, X.25) or a V series terminal (e.g., V.32). Since the terminal and the data path to the BTS may use different data rates, the MS must also perform a rate adaptation.

An example protocol stack (for X.25 service) is shown in Figure 10.6. The X.25 data is transmitted at layer 2 using Link Access Protocol-B (LAPB) and sent to the physical layer (layer 1) between the terminal and the MS. In the MS, LAPB is terminated and converted to the Layer 2 R(adio) Bit Oriented Protocol (L2RBOP) and to the RLP and sent to the MSC via the BSS. At the MSC, the RLP and L2RBOP are converted back to LAPB and the X.25 data is

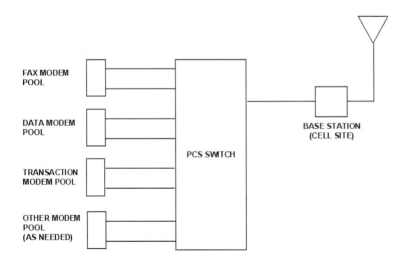

Fig. 10.3 PCS Network Architecture for Data Interworking

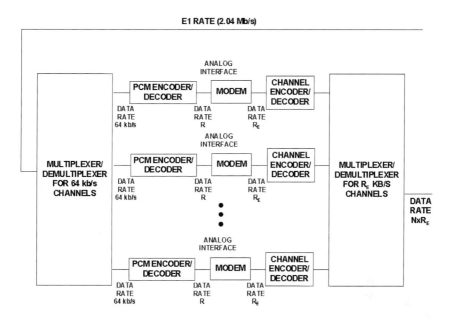

Fig. 10.4 Modem Pool for Data Interworking

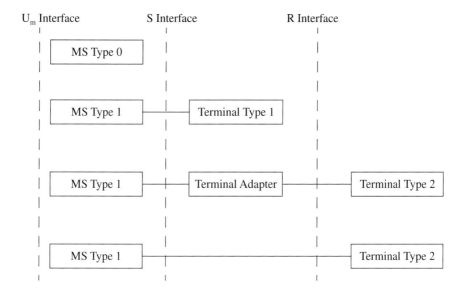

Fig. 10.5 Terminal Adaptation for GSM MSs

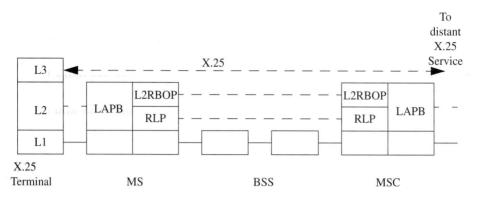

Fig. 10.6 Protocol Stack for X.25 Service

sent into the network. The L2RBOP protocol converts the LAPB data into fixed-length packets for transmission over the radio link. The purpose of the RLP is to combat the effects of bit errors on the radio link by number packets and retransmitting them when they are received in error by the other side. The physical layer also does error detection and correction but it is not sufficient to guarantee error-free performance for layer 2. RLP gives error-free performance for the layer 3 X.25 packets.

Control information from a terminal to a terminal adapter is transmitted using the AT command set [15]. The AT command set was first established for voice-band modems by the Hayes Microcomputer Company. It has since become an international standard for communicating with modems. It is used in GSM to communicate over the TE-TA interface (S interface). The S interface can be internal to the terminal or the MS. With the AT command set, the user (or application software) can control modem characteristics and dialing procedures.

10.3 GSM DATA SERVICES

GSM data services consist of circuit-switched and packet-switched data. Circuit-switched data can be to an analog modem, to an ISDN connection, or to a fax machine. Packet-switched data connects to a packet network. Current packet-switched data is via the signaling. In section 10.4, we will discuss a new GSM service of packet data on a TCH.

10.3.1 Interconnection for Switched Data

The physical layer of GSM offers a 13-kbps connection for the transmission of vocoded voice. When data is transmitted in place of voice, the maximum data rate that can be supported is, therefore, 13 kbps. To interconnect with a voice-band modem in the analog wireline network, there must be a modem in the MSC that functions at the correct rate (Figure 10.4). The GSM network supports circuit-switched data at rates from 300 to 9600 baud. At the MSC, the analog modem must support 300–9600 baud analog connections to modems in the wireline network. Currently, there are voice-band modems that will operate as high as 56 kbps. Support for data rates higher than 9600 bps requires that a GSM phone use multiple time slots and aggregate the slots to deliver the higher data rate. The use of multiple slots is equivalent to the use of multiple phones by the same user and therefore reduces the capacity of the cell.

Terminal adapters are needed to connect a terminal, laptop computer, and other items to the MS as described in section 10.1 and Figure 10.5.

On the radio link, a layer 2 functionality is added using L2R protocol and RLP. L2R provides the signaling functions needed by the modem (e.g., dialing, break) and communicates with RLP. The RLP provides the end-to-end error-free link over the radio path by segmenting the data into packets, numbering the packets, and error checking packets. This is the same as how RLP is used for other protocols.

Interconnection with ISDN is similar to that for an analog modem except that the data transmitted over the radio link must be rate adapted to 64 kbps and transmitted over a circuit-switched ISDN connection.

10.3.2 Group 3 Fax

Group 3 fax is a wireline-to-wireline service over analog voice-band or ISDN facilities. Since a GSM system does not transmit either analog voice or 64-kbps ISDN, an adapter function must be installed in the GSM network and at the GSM MS as described in section 10.2 (Figures 10.1–10.4). If a two-wire fax machine (analog type) is used, then a fax adapter (fax modem) must be installed between the MS and the fax machine. If a digital fax machine is used, then it can connect directly to the MS. See Figure 10.7 for examples of combinations of fax machines, fax adapters, and terminal adapters. It is possible, of course, to construct a completely integrated fax machine and MS. After the conversion from the fax machine protocol into the MS interface and protocol, the operation must be undone in the network. The network will have a modem pool to connect to the wireline network (Figure 10.4). If the connection is to another MS, then two modems will be necessary with a loop-around connection. The modem may be an analog fax modem or a digital ISDN modem depending on the other end of the connection (analog or ISDN).

The protocol stack for operation of the system is similar to the X.25 stack and is shown in Figure 10.8. The fax adapter protocol converts the T.30 data [18] into numbered frames for transmission over the radio link. It also communicates with the fax adapter on the other side for call control. The functions of L2RBOP and RLP are the same as for X.25 service.

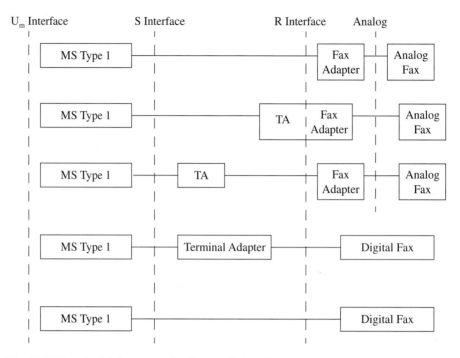

Fig. 10.7 Terminal Adaptation for Facsimile Machines

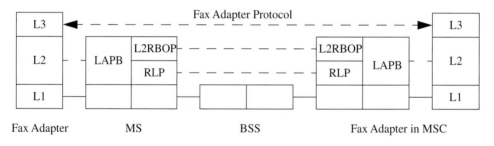

Fig. 10.8 Protocol Stack for Fax Service

10.3.3 Packet Data on the Signaling Channel

Packet Data on the Signaling (PDS) channel service [1,6] is a bearer service that enables data transfer in GSM networks of very small data packets on radio interface signaling channels for applications that use small amounts of data with a throughput in the range of 600–9200 bps and with a call duration of a few seconds. PDS may be applied to transfer data between a mobile user and the corresponding packet data network (e.g, X.25 or IP); it may also be used for data transmission between a mobile user and a host directly accessing an MSC or a PDSS2 support node.

The interworking between the GSM network and packet data network is not defined in the GSM specifications. There are two forms of PDS:

☞ **PDSS1.** In PDS Service 1 a terminal at the MS communicates with a packet network that is connected to the MSC. The service offers both mobile originating and mobile terminating connections as well as their release. The QoS offered and the data rates are comparable with the services offered by the circuit-switched air interface. In PDSS1, the service continues across handovers in the network and is offered to roaming MSs.

☞ **PDSS2.** In PDS Service 2 a terminal at the MS communicates with a node in the network called the PDSS2-Service Node (PDSS2-SN). The service offers the setup and release of mobile originating connections but not mobile-terminated service. A handover causes a termination of the PDSS2 connection, and the service can resume after a channel change or new channel establishment. The quality of service offered and the data rates are comparable with the services offered by the U_m interface. Roaming, handover, and security services are not offered in PDSS2 and must be done by the applications layer.

10.3.4 User-to-User Signaling

User-to-user signaling is an ISDN service that enables a user to send small amounts of data to another user in association with a call. The data is sent via signaling messages over the SS7 network. Many SS7 messages can have user-to-user data associated with them. The data can be sent during call setup or termination signaling, during alerting, and during the call via a user-to-user signaling message. The maximum length of the message associated with each signaling message is 128 octets.

In GSM the signaling messages between network elements can also have user-to-user information attached to them (see chapter 9). If the two parties that are in communication are both GSM users or if one is an ISDN user, then the user-to-user information is sent between the two phones. With a connection to an ISDN phone, the MSC must pass the user-to-user information via SS7 messages to and from the wireline ISDN network. For GSM to GSM user-to-user messages, the two MSCs will communicate directly or via the wireline network. The protocols, interpretation, and application layers of user-to-user signaling is left to the two ends. The GSM network and the ISDN network do not interpret the data but merely pass it from one phone to the other.

10.4 SMS

Many cellular phone users carry a pager in addition to a cellular phone. There are several reasons for the use of the pager. Older analog cellular phones do not operate for an entire day on a single battery. Therefore, the user must either leave the phone off for part of the day or must carry multiple batteries. With a pager, the user can apply power to the phone only when making calls; thus, the battery power is conserved. A second reason for using the pager is to receive calls. Cellular service charges for air time for both incoming and outgoing calls. A cellular user may not want to receive (and thus pay for) incoming calls from persons or organizations marketing products and services. If the phone is off, it cannot receive phone calls. Service providers either have moved to "calling-party-pays" numbers that permit the person calling the cellular phone to elect to pay for the air time charges through the use of a premium number service or do not charge the user for the first minute of incoming air time. A third reason for using the pager on analog cellular systems is to avoid fraud. If the phone is off except when making a call, the amount of time when the phone can be defrauded is minimized. Pagers can also be used to receive short messages at less cost than a phone call to a wireless phone.

When a GSM system supports paging by delivery of the page to a display on the phone, the service is called SMS. SMS allows the exchange of short alphanumeric messages between an MS and the GSM system and between the GSM system and an external device capable of transmitting and optionally receiving short messages. The external device may be a voice telephone, a data terminal, or a short message entry system. The SMS consists of message entry features, administration features, and message transmission capabilities. These features are distributed between a GSM system and the SMS Message Center (SMSMC) which together make up the SMS system. The SMSMC may be either separate from or physically integrated into the GSM system.

Short message entry features are provided through interfaces to the SMSMC and the MS. Senders use these interfaces to enter short messages, intended destination addresses, and various delivery options. MC interfaces may include features such as audio response prompts and DTMF reception for dial-in access from voice telephones, as well as appropriate menus and message entry protocols for dial-in or dedicated data terminal access. MS interfaces may include keyboard and display features to support message entry. Also, a GSM voice service subscriber can use normal voice or data features of the MS to call an SMS system to enter a message.

An SMS teleservice can provide the option of specifying priority level, future delivery time, message expiration interval, or one or more of a series of short, predefined messages. If supported by the teleservice, the sender can request acknowledgment that the message was received by the MS. An SMS recipient, after receiving a short message, can manually acknowledge the message. Optionally, the recipient can specify one of a number of predefined messages to be returned with acknowledgment to the sender.

SMS administration features include message storage, profile editing, verification of receipt, and status inquiry capabilities. The SMS transmission capabilities provide for the transmission of short messages to or from an intended MS and return of acknowledgments and error messages. These messages and acknowledgments are transmitted to or from the MS whether it is idle or engaged in a voice or data call. The GSM service provider may offer SMS transmission to its voice and data customers only or may provide an SMS-only service without additional voice or data transmission capability. All available MSs on a GSM PCH can receive a broadcast message. A broadcast message is not acknowledged by the MS. Broadcast messaging services may be made available to MSs on GSM PCHs as well as to MSs in a call on a GSM TCH.

A GSM phone can act like a two-way pager; thus, the user does not need a separate pager. The digital encryption on the radio link solves the fraud problem, and the use of premium numbers for calls to GSM phones solves the incoming call problems. The use of the sleep mode and paging slots by the BS extends battery life. The display on the phone can display messages in addition to names and numbers, and, with a proper keypad on the phone, the user can send messages also. Thus, a GSM phone user does not need a pager.

The GSM system supports two types of SMS. The first type is called *point-to-point service*. In this service (Figure 10.9), a user sending a short message must interact with a Short Message Entity (SME) to enter the message into the system. The user can interact with the SME via e-mail, by a phone call in an operator who takes the message and enters it into the system, or by a call to a voice response system that accepts DTMF tones. Once the message is in the system, the SME forwards the message to the MSC which then sends the message to the MS. If the mobile user is roaming, a gateway MSC may receive the message and forward it to the destination MSC.

A mobile user can send a short message to the SME by entering the message on the keypad of the phone and sending it into the system. The destination user must retrieve the message by calling into the SME. Alternately, the SME can forward the message to a destination e-mail address.

The SME is not defined in the GSM standards and is considered to be outside the GSM architecture.

Figure 10.10 shows the protocol stack for the SMS. The following layers are defined for SMS:

☞ **Application layer.** This layer depends on the specific service supported by the SME. The interpretation of the information transmitted depends

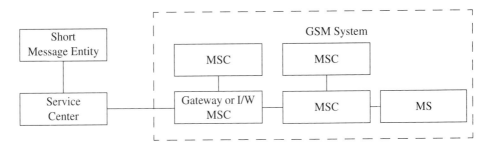

Fig. 10.9 Architecture of SMS

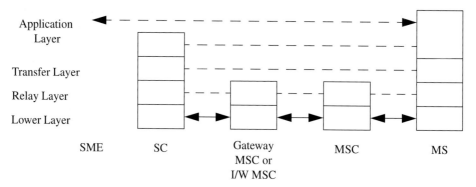

Fig. 10.10 Protocol Stack for SMS

on the software in the phone and on the user. For alphanumeric service, the display on the phone will present the letters and numbers transmitted, with the user determining the meaning of the message.

☞ **Transfer layer.** This layer transfers short messages from the Service Center (SC) to the MS. Protocol elements exist to transfer messages from the SC to the MS and obtain a delivery report; to transfer messages from the MS to the SC and obtain a delivery report; and to determine the status of messages sent in either direction.

☞ **Relay layer.** This layer is used to deliver services to the transfer layer. For details on the messages that are supported see chapter 9.

☞ **Lower layer.** This is the physical layer of GSM. See chapters 7 and 12 for the details.

A second type of service is the SMS cell broadcast (SMSCB) (see Figure 10.11). In this service, multiple MSs are sent the same message via a broadcast message. The purpose of this service is to provide the same information to many MSs in the system. The information could consist of news reports, sports

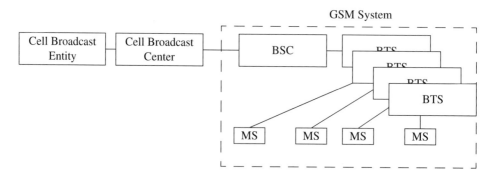

Fig. 10.11 Architecture of SMSCB

reports, stock data, weather alerts, and the like. The information service provider that originates the information is called the *cell broadcast entity*. The information is transferred to the *cell broadcast center* which relays the information to one or more BSCs. The BSCs in turn forward the message to the *base transceiver systems* for transmission over the air. The transmissions consist of the data and the identity of the mobiles that are to receive the data. The interface between the cell broadcast center and the BSC is not defined in the GSM standards but is left to a matter of agreement between the operators of each network element.

Messages to the MS are transmitted on the CBCH.

The MS and the SIM card have limited amounts of memory. Thus it is possible that messages sent directly to the MS or directly from the MS could be limited in number or length. Therefore, the MS may have a terminal connected to it that is capable of storing and generating the messages [14]. The interface uses mobile termination type 2 (MT2) as described previously in this chapter.

10.5 GSM GPRS

The GSM GPRS extends the packet capabilities of GSM to higher data rates and longer messages. The service supports sending point-to-point and point to multipoint messages. Two new nodes are added to the network to support GPRS (Figure 10.12). The serving GPRS support node communicates with MSs within its service area. The gateway GPRS support node communicates with packet networks that are external to the GSM network.

The protocol architecture of the GPRS system is shown in Figure 10.13. The application in the MS (or its data adjunct) communicates with the application in the distant packet terminal. The communication is through the higher layers and the network layers in the MS, where it is relayed through the BSS to the serving GPRS support node and on to the gateway GPRS node. From the gateway GPRS node, it is sent on to the packet switching network. As the data transverses the network, several protocols are used.

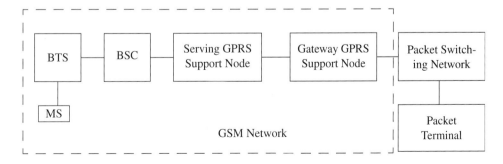

Fig. 10.12 GPRS Network Architecture

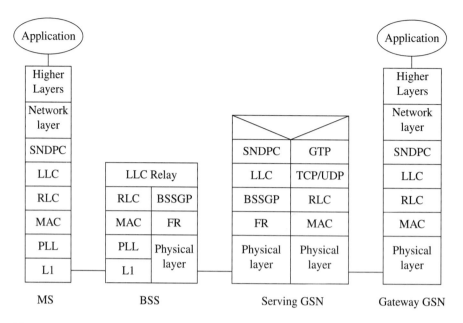

Fig. 10.13 Protocol Stack for GPRS

In the MS and the BSS:

☞ The **Sub-Network-Dependent Convergence Protocol (SNDCP)** performs header compression on the headers of the network layer.

☞ The **Logical Link Control (LLC)** provides the link layer control between the MS and GPRS serving node. It is based on LAPD.

☞ The **Radio Link Control (RLC)** transmits data blocks across the air interface, performs error detection, and performs error correction via an automatic repeat request process.

☞ The **Media Access Control (MAC)** operates similar to a slotted ALOHA channel.

☞ The **physical link layer** manages forward error correction, interleaving of frames, and radio channel congestion.

☞ The **radio frequency layer** manages the physical radio layer of the system, including frequency modulation.

Between the BSS and the serving GPRS node:

☞ **BSS GPRS Protocol (BSSGP).** This new protocol provides routing and QoS management.

☞ **Frame relay.** This standard wireline protocol supports packet communication between nodes.

☞ **Physical layer.** As needed between nodes (e.g., E1 link).

Between GPRS Nodes:

☞ **GPRS tunnel protocol.** This new protocol routes protocol data units through the network by adding packet routing information.

☞ **Transmission Control Protocol/User Datagram Protocol (TCP/ UDP) and Internet Protocol (IP).** These are the two standard protocols for the Internet.

☞ **Layer 2 (L2) and physical layer.** As needed between nodes. Some examples are ethernet, ISDN, and ATM.

On the radio channel, three new channels are added. These channels are similar to those used for circuit-switched connections.

☞ **Packet Broadcast CCH (PBCCH).** This channel transmits system information to all packet MSs in the area of a cell.

☞ **Packet Common CCH (PCCCH).** This channel has four subchannels. The Packet PCH (PPCH) is used to page MSs. The Packet Random Access Channel (PRACH) is used for MSs to access the network to initiate transmissions or respond to pages. The Packet Access Grant Channel (PAGCH) is used to send resource assignments to an MS. The Packet Notification Channel (PNCH) is used to send multicast information to MSs.

☞ **Packet TCH (PTCH).** This uplink and downlink channel is used to transmit data packets between the MS and the BS over the Packet Data TCH (PDTCH). The channel is also used to send control information to/ from MSs using the Packet Associated CCH (PACCH).

When each user has a steady flow of information to transmit (for example, a data file transfer or a fax transmission), fixed-assignment access methods are useful as they make an efficient use of communication resources. However, when the information to be transmitted is bursty in nature, fixed-assignment access methods result in wasting communication resources. Furthermore, in a cellular system where subscribers are charged based on a channel connection time, fixed-assignment access methods may be too expensive for transmitting short messages. Random-access protocols provide flexible and efficient methods for managing a channel access to transmit short messages. Random-access methods give freedom for each user to gain access to the network whenever the user has information to send. Because of this freedom, these schemes can result in contention among users accessing the network. Contention may cause collisions and may require retransmission of the information. The commonly used random-access protocols are pure ALOHA, slotted ALOHA, and CSMA/CD. GPRS is a packet radio system; therefore its characteristics are similar to ALOHA systems.

In the pure ALOHA scheme, each user transmits information whenever the user has information to send. A user sends information in packets. After sending a packet, the user waits a length of time equal to the round-trip delay for an acknowledgment (ACK) of the packet from the receiver. If no ACK is received, the packet is assumed to be lost in a collision and it is retransmitted with a randomly selected delay to avoid repeated collisions. The normalized throughput S (average packet arrival rate divided by the maximum throughput) of the pure ALOHA protocol is given as

$$S = Ge^{-2G} \qquad\qquad (10.1)$$

where G = normalized offered traffic load.

From Eq. (10.1) it should be noted that the maximum throughput occurs at traffic load G = 50 percent and is $S = 1/2e$. This is about 0.184. Thus, the best channel utilization with the pure ALOHA protocol is only 18.4 percent.

In the slotted ALOHA system, the transmission time is divided into time slots. Each time slot is made exactly equal to packet transmission time. Users are synchronized to the time slots so that, whenever a user has a packet to send, the packet is held and transmitted in the next time slot. With the synchronized time slots scheme, the interval of a possible collision for any packet is reduced to one packet time from two packet times, as in the pure ALOHA scheme. The normalized throughput S for the slotted ALOHA protocol is given as

$$S = Ge^{-G} \qquad\qquad (10.2)$$

where G = normalized offered traffic load.

The maximum throughput for the slotted ALOHA occurs at G = 1.0 (Eq. [10.2]), and it is equal to $1/e$ or about 0.368. This implies that, at the maximum throughput, 36.8 percent of the time slots carry the successfully transmitted packets, whereas the other 63.2 percent of the time slots remain empty.

GPRS has the characteristics of a slotted ALOHA system but uses queued requests and channel reservation techniques to grow the traffic to the maximum throughput and hold at the maximum.

Cai and Goodman simulated the performance of the GPRS system [19] and found these results:

☞ When a single time slot is reserved for both control and traffic, the maximum throughput is about 4 kbps per slot.

☞ When 8 time slots are used, the maximum throughput is about 5 kbps per slot or 40 kbps maximum.

☞ Since calls are gated into the system by queuing them, the blocking rate on a single user's packets is low (about 0.1 percent or less).

☞ When only one slot is available, blocking will increase to the order of 5 percent.

Cai and Goodman's work was on a draft version of the GPRS standard and therefor does not completely characterize the final GPRS system. It does however give an indication of its operation under load. We can see from the results that GPRS is a low-data-rate service.

10.6 SUMMARY

In this chapter we examined data transmission over a GSM network. GSM offers a rich set of data features that enable MSs and their adjuncts to send and receive data with a variety of wireless and wireline data networks. We examined circuit-switched and packet-switched data. Like the ISDN services that GSM emulates, a variety of channels can be used to send data. The user can send data over the CCH, over a TCH as a data call, or on either the CCH or the TCH using user-to-user information. A new service GPRS defines a packet channel for transmission of data.

Data services range from low-speed circuit-switched data to group 3 fax. For more details on data services, consult the references.

10.7 REFERENCES

1. GSM Specification Series 02.63, "Digital Cellular Telecommunications System (Phase 2+): Packet Data on Signalling Channels Service (PDS)—Stage 1."

2. GSM Specification Series 02.87, "Digital Cellular Telecommunications System (Phase 2+): User-to-User Signalling (UUS) Service Description—Stage 1."

3. GSM Specification Series 03.40, "Digital Cellular Telecommunications System (Phase 2+): Technical Realization of the Short Message Service (SMS) Point-to-Point (PP)."

4. GSM Specification Series 03.41, "Digital Cellular Telecommunications System (Phase 2+): Technical Realization of Short Message Service Cell Broadcast (SMSCB)."

5. GSM Specification Series 03.46, "Digital Cellular Telecommunications System: Technical Realization of Facsimile Group 3 Nontransparent."

6. GSM Specification Series 03.63, "Digital Cellular Telecommunications System (Phase 2+): Packet Data on Signalling Channels Service (PDS) Service Description—Stage 2."

7. GSM Specification Series 04.11, "Digital Cellular Telecommunications System (Phase 2+): Point-to-Point (PP) Short Message Service (SMS) Support on Mobile Radio Interface."

8. GSM Specification Series 04.12, "Digital Cellular Telecommunications System (Phase 2+): Short Message Service Cell Broadcast (SMSCB) Support on the Mobile Radio Interface."

9. GSM Specification Series 04.22, "Digital Cellular Telecommunications System (Phase 2+): Radio Link Protocol (RLP) for Data and Telematic Services on the

Mobile Station-Base Station System (MS-BSS) Interface and the Base Station System-Mobile-Services Switching Centre (BSS-MSC) Interface."

10. GSM Specification Series 04.63, "Digital Cellular Telecommunications System (Phase 2+): Packet Data on Signalling Channels Service (PDS) Service Description—Stage 3."

11. GSM Specification Series 07.01, "Digital Cellular Telecommunications System (Phase 2+): General on Terminal Adaptation Functions (TAF) for Mobile Stations (MS)."

12. GSM Specification Series 07.02, "Digital Cellular Telecommunications System (Phase 2+): Terminal Adaptation Functions (TAF) for Services Using Asynchronous Bearer Capabilities."

13. GSM Specification Series 07.03, "Digital Cellular Telecommunications System (Phase 2+): Terminal Adaptation Functions (TAF) for Services Using Synchronous Bearer Capabilities."

14. GSM Specification Series 07.05, "Digital Cellular Telecommunications System (Phase 2+): Use of Data Terminal Equipment-Data Circuit Terminating Equipment (DTE-DCE) Interface for Short Message Service (SMS) and Cell Broadcast Service (CBS)."

15. GSM Specification Series 07.07, "Digital Cellular Telecommunications System (Phase 2+): AT Command Set for GSM Mobile Equipment (ME)."

16. GSM Specification Series 09.06, "Digital Cellular Telecommunications System (Phase 2+): Interworking between a Public Land Mobile Network (PLMN) and a Packet Switched Public Data Network/Integrated Services Digital Network (PSPDN/ISDN) for the Support of Packet Switched Data Transmission Services."

17. GSM Specification Series 09.07, "Digital Cellular Telecommunications System (Phase 2+): General Requirements on Interworking between the Public Land Mobile Network (PLMN) and the Integrated Services Digital Network (ISDN) or Public Switched Telephone Network (PSTN)."

18. CCITT Recommendation T.30, "Procedures for Document Facsimile Transmission in the General Switched Telephone Network."

19. Cai, J., and Goodman, D. J., "General Packet Radio Service in GSM," *IEEE Communications Magazine* 35 (10), October 1997.

Privacy and Security in GSM

11.1 INTRODUCTION

The designers of the GSM system wanted to ensure that the system was relatively secure. Cellular fraud is extensive in analog cellular systems since the identity of the subscriber is sent to the network without encryption ("in the clear"). The GSM system, on the other hand, has security controls that virtually eliminate cloning fraud. The designers also wanted to ensure that the users' communications on the GSM system would be private, so the GSM system also has controls ensuring user privacy.

GSM system privacy and security is achieved using four primary mechanisms.

1. Each subscriber is identified using a cryptographic security mechanism. The algorithm is highly resistant to attacks by individuals attempting to make fraudulent phone calls.

2. The subscriber's security information is stored in a secure computing platform called a smart card or a SIM card.

3. The GSM operator maintains the secrecy of the cryptographic algorithms and the keys for authenticating the subscriber and providing voice privacy. The algorithms are stored in the SIM card and in the authentication center.

4. The cryptographic keys are not shared with other GSM administrations.

In this chapter we examine the privacy and security of GSM without revealing any details of the algorithms used by each administration. First, we review the necessary requirements for wireless security with the goal being security as effective as that of a wireline telephone. Next we study the smart card as implemented by ETSI for GSM (also called a SIM card). Stored in the smart card are various data parameters and security algorithms which we examine. The details of the algorithm are unique to each service provider and are a closely kept secret; therefore, we examine only the high-level capabilities of the algorithms. Finally, we study the call flows used for registration and unique challenge which use GSM security methods.

11.2 WIRELESS SECURITY REQUIREMENTS*

Security in a wireless system is of concern to the system operator and to the user. The system operator wants to ensure that the user requesting the service is a valid user whose actual identity is the same as the claimed identity—the operator must authenticate the user. The user wants to have access to services without compromising privacy. Users get disturbed when their locations, calling patterns, and details of their conversations are monitored. It is possible to authenticate users without providing privacy, and it is possible to provide privacy without authenticating users. However, most privacy mechanisms provide authentication as part of the process.

11.2.1 Privacy of Communications

Figure 11.1 is a high-level diagram of a PCS system showing areas where criminals or hackers can compromise the security of the system. At each interface compromise is possible, so designers must pay attention to each of these areas. A PCS personal terminal needs privacy in these areas:

☞ **Call Setup Information.** During call setup, the handset will communicate information such as calling number, calling card number, and type of service requested to the network. The system must send all this information in a secure way.

☞ **Speech.** The system must encrypt all spoken communications so that hackers cannot intercept the signals by listening on the airwaves.

☞ **Data.** The system must encrypt data communications so that hackers cannot intercept data by listening on the airwaves.

---

*The material in this section is based on [12] and is used with permission of the IEEE.

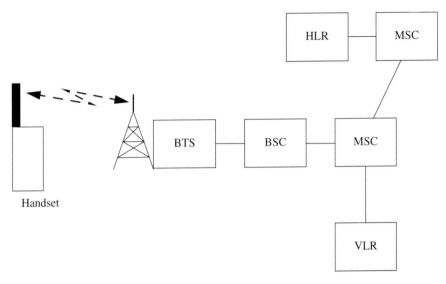

Fig. 11.1 Privacy Requirements

☞ **User location.** No information that a user might transmit should enable a listener to determine the user's location. The usual method to meet this need is to encrypt the user ID. Protection is needed against:

1. Radio link eavesdropping
2. Unauthorized access by outsiders (hackers) to the user location information stored in the network at the VLR and HLR
3. Unauthorized access by insiders to the user location information stored in the network. This is difficult to achieve, but not impossible.

☞ **User ID.** When a user interacts with the network, the user ID must be sent in a way that does not show the user ID. This prevents analysis of user calling patterns based on user ID.

☞ **Calling patterns.** No information must be sent from a handset that enables a listener of the radio interface to do traffic analysis on the PCS user. Typical traffic analysis information is

> Calling number
>
> Frequency of use of the handsets
>
> Caller ID
>
> Financial transactions

If the user transmits credit card information over any channel, the system must protect the data. For example, users may order items from mail order houses via a telephone that is wireless and may choose to speak their credit card numbers rather than dialing them via a keypad.

Or users may access bank voice response systems, where they send account data via tone signaling. Users may access calling card services of carriers and may speak or use tone signaling to send the card number.

All these communications need to be private. Since the user can send the information on any channel—voice, data, or control—the system must encrypt all channels.

11.2.2 Authentication Requirements

The system operator may or may not care if a call is placed from a stolen handset as long as the call is billed to the correct account. However, the owner of a handset *will* care. The network operator maintains a list of valid terminals in the EIR.

The terminal design should reduce theft of the handset by making reuse of a stolen handset difficult. Even if the handset is registered to a new legitimate account, the use of the stolen terminal should be stopped. The terminal design should also reduce theft of services by making reuse of a stolen handset's unique information difficult. To reduce theft, we need

☞ **Clone-resistant design.** In current wireless systems, cloning of handsets is a serious problem; methods must be put in place to reduce or eliminate fraud from cloning. Handset-unique information must not be compromised

> **Over the air.** Someone listening to the radio channel should not be able to determine information about the handset and then program it into a different handset.
>
> **From the network.** The databases in the network must be secure. No unauthorized people should be able to obtain information from those databases.
>
> **From network interconnect.** Systems will need to communicate with each other to verify the identity of roaming handsets. A fraudulent system operator could perpetrate fraud by using the security information about roaming handsets to clone handsets.
>
> **From fraudulent systems.** The communications scheme used between systems to validate roaming handsets should be designed so that theft of information by a fraudulent system does not compromise the security of the handset.
>
> **From security algorithms.** Any information passed between systems for the purpose of security checking of roaming handsets must have enough information to authenticate the roaming handset. It must also have insufficient information to clone the roaming handset.
>
> **From users cloning their own handsets.** Users themselves can perpetrate fraud on the system. Multiple users could use one account by cloning handsets.

☞ **A cryptographic system to reduce installation and repair fraud.**
Theft of service can occur at the time of installation of the service or
when a terminal is repaired. Multiple handsets can be programmed with
the same information (cloning).

☞ **Unique user ID.** More than one person may use a handset, so we must
uniquely identify the correct person for billing and other accounting
information.

☞ **Unique handset ID.** When all security information is contained in a
separate module (smart card), the identity of the user is separate from
the identity of the handset. Stolen handsets can then be valuable for
obtaining service without purchasing a new (full-price) handset. Therefore, the handset should have unique information contained within it
that reduces or eliminates the potential for stolen handsets to be re-reg-
istered with a new user.

11.2.3 System Lifetime Requirements

It has been estimated that computing power doubles every two years. An algorithm that is secure today may be breakable in 5–10 years. Since any system
being designed today must work for many years, it is reasonable to require
that the procedures last at least 20 years. Thus, the algorithm design must
consider the best available cracking algorithms available today and must have
provisions for being upgraded in the field.

11.2.4 Physical Requirements

Any cryptographic system used in a handset must work in the practical environment of a mass-produced consumer product.

11.3 SIM CARDS

The SIM card is a secure microprocessor-based environment implemented on a
credit-card-sized platform. Two types of SIM cards are used in GSM. An ID-1
card is the same size as a standard credit card (Figure 11.2). It looks like a
standard card and has embossing, picture, lettering, and a magnetic stripe
similar to a credit card, but it also includes a microprocessor. While some
larger GSM phones would use the ID-1 card, most GSM handsets use the plug-
in SIM card (Figure 11.3). An ID-1 card can be (permanently) converted to a
plug-in SIM card by removing the plug-in SIM from the plastic of the card.

The microprocessor platform is designed to be secure. Attempts to
reverse engineer the smart card will render the card inoperative and destroy
the data on the card. Certain data can be changed only by the manufacturer of
the card; other data can only be changed by entering a PIN. For more details
on the uses of smart cards and their detailed operation, see reference [13].

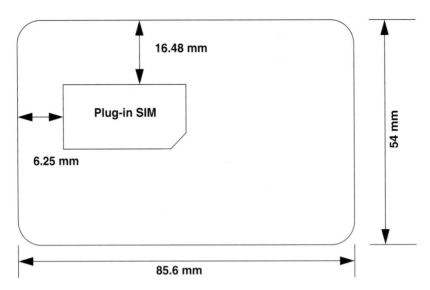

Fig. 11.2 ID-1 SIM Card

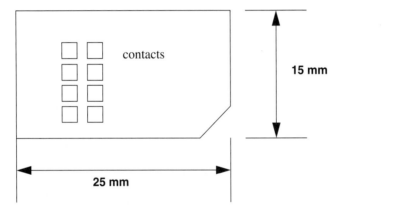

Fig. 11.3 Plug-in SIM Card

The smart card can contain both data and executable files and can support a wide range of access permissions. The file structure (Figure 11.4) is similar to the file structure used for other operating systems except that the naming convention is different. The root-level directory is called the Master File (MF). Under the MF are elementary files (EFs), which can contain data or executable files. The MF also contains directories called Dedicated Files (DFs) which can contain either additional DFs or EFs, up to the limit of the storage capacity of the SIM card.

If EF0 is present in a directory, it contains the PIN for the data and executable files in the directory. If EF0 is not present, the card uses the EF0 file

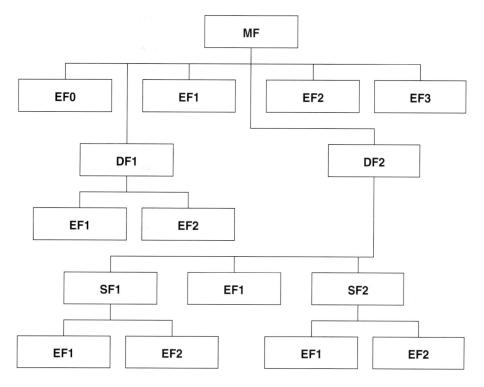

Fig. 11.4 File Structure of SIM Card

in the next highest directory. The EF0 file under the MF contains the PIN for all card functions except those directories (DFs) that have their own PIN. Each file/directory has access conditions (read and write). Several access conditions are defined (see Table 11.1) that enable executable and data files to be stored on the card. Since some users may not want to enter a PIN each time they access their phones, the smart card supports a second PIN (CHV2) that can be used to disable PIN checking. However, GSM administration may prevent the use of CHV2; thus each time the phone is powered up, the user must enter a PIN.

The SIM card provides storage capability for

☞ Administrative information—indicates mode of operation of the SIM (e.g., normal, type approval)

☞ ID card identification—a number uniquely identifying the SIM and the card issuer

☞ SIM service table—indicates which optional services are provided by the SIM

☞ IMSI

Table 11.1 File Access Conditions

Access Condition	Description
ALWays	The file can be accessed by any software resident on the card or off the card.
CHV1	The file can be accessed only after CHV1 has been correctly presented, or if CHV1 checking is disabled. The disabling of CHV1 checking may be inhibited by GSM administration.
CHV2	The file can be accessed only after CHV2 has been correctly presented. CHV2 is used to disable CHV1 checking and thus unlock the SIM card.
ADM	Reserved for use by GSM administration.
NEVer	Off-card software cannot access the data or program. On-card software may access the file.

☞ Location information—comprising TMSI, LAI, periodic location updating timer, and the location update status

☞ Encryption keys (K_c, K_i) and encryption key sequence number

☞ BCCH information—list of carrier frequencies to be used for cell selection

☞ Access control class(es)

☞ Forbidden PLMNs

☞ HPCMN search period—used to control the time interval between HPLMN searches

☞ Language preference—user-preferred language(s) of Man-Machine Interface (MMI)

☞ Phase identification

Location information, encryption key, and encryption key sequence number are updated on the SIM card after each call termination and when the handset is correctly deactivated in accordance with the manufacturer's instructions.

☞ The SIM card may also optionally provide storage capability for

☞ PLPN selector (for automatic PLPN selection)

☞ Cell broadcast message identifier selection

☞ Abbreviated dialing numbers/supplementary service control

☞ Fixed dialing numbers/supplementary service control

☞ MSISDN number(s) (for user number[s])

☞ Last number(s) dialed

☞ Capability configuration parameters (provide the parameters of required bearer capabilities associated with dialing numbers)

☞ Called party subaddress

☞ Short messages and associated parameters

☞ Accumulated call meter, accumulated call meter maximum value, price per unit, and currency table

GSM Phase 2+ defines a set of procedures to use to manage the interface between the SIM card and the handset [11]. The SIM toolkit identifies those procedures and enhances the capabilities of the handset/SIM combination. It is not required that a handset support the SIM toolkit.

11.4 SECURITY ALGORITHMS FOR GSM

GSM uses three security algorithms (see Figures 11.5 and 11.6):

1. **Authentication algorithm (A3).** Used by the handset to compute a signed response (SRES) to the random number (RAND) transmitted by the BS. SRES is transmitted to the BS during registration and other access messages (see section 11.5.1). The computation also uses a secret key (K_i) that is stored in the SIM card and is unique to each SIM card. The A3 algorithm is different for each GSM administration and is secret. Many GSM administrations use a common A3 that is available from the GSM Memorandum of Understanding (MoU) Security Group.

2. **Privacy key generation algorithm (A8).** Also uses RAND and K_i to generate a privacy key (K_c) that is used for voice and data privacy. A8 is also unique to each GSM administration. A common A8 is available from the GSM MoU.

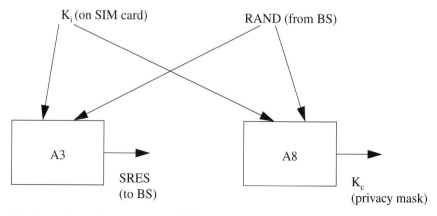

Fig. 11.5 Use of A3 and A8 in GSM Authentication

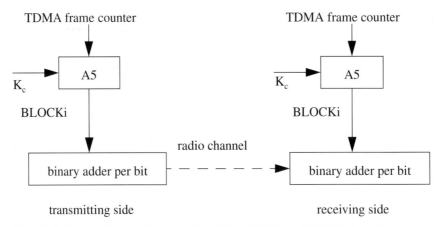

Fig. 11.6 Encryption and Decryption of the DCCH and TCH Using A5

3. **Encryption algorithm (A5).** Used to encrypt data transmitted on the DCCH and the TCH. The inputs to A5 are the privacy key (K_c) and the TDMA frame counter. The frame counter is 22 bits long and each frame is approximately 4.6 ms long. Therefore the encryption mask repeats approximately every 5 hours. For each frame, two outputs of A5— BLOCK1 and BLOCK2—are generated. BLOCK1 is used for encryption by the BS and BLOCK2 is used for encryption by the handset. The details of A5 are available from the GSM MoU. The secret key (K_i) is 128 bits long. The pseudorandom number transmitted by the BS (RAND) is also 128 bits long. The computed signed response (SRES) is 32 bits long.

11.5 TOKEN-BASED AUTHENTICATION

The designers of the GSM system wanted a security system that was under the control of the service provider and did not require the sharing of secret data between systems. They were also designing the system for Europe where each system is the size of a country. Thus, intersystem communications needs were minimal. Only when an MS, or handset, roamed into another country would an intersystem message be needed. The token-based system using security triplets meets this need.

The security key (K_i) and the details of the A3 and A8 algorithms are not shared between systems. Rather the GSM systems use a set of triplets consisting of a pseudorandom number (RAND); its corresponding response (SRES) generated by the authentication algorithm; and an encryption key (K_c) used for data, signaling, and voice privacy. The visited system requests the triplets from the home system. These triplets are computed and stored in the MS, in the home authentication center, and in the visited VLR. The comparison of the SRES values is done in the visited VLR.

All MSs are assigned an electronic serial number—the IMEI—when they are manufactured. At the time of service installation, the SIM card is assigned a 15-digit IMSI that is unique worldwide, K_i, and other data (see section 11.3). When the MS is turned on, it must register with the system. When it registers, it sends its TMSI and other data to the GSM network. The VLR in the visited system then queries the old VLR for the security data and location of the HLR and assigns a new TMSI to the MS. The MS uses the TMSI for all further access to that system. The TMSI provides anonymity of communications since only the MS and the network know the identity of the MS with a given TMSI. When an MS roams into a new system, the GSM system uses the TMSI to query the old VLR and then assigns a new TMSI; only when a communications failure occurs will the network request that the MS send its IMSI and then assign a new TMSI. The call flows for registration are described later on in this chapter.

The BS transmits a RAND on the DCCH that is received by the MS. When the MS accesses the system, it calculates SRES (described in section 11.4). It then transmits the desired message with its authentication to the network. The network does the same calculation and confirms the identity of the MS. All communications between the MS and the BS are encrypted to prevent a hacker from decoding the data and using the data to clone other phones.

Authentication is performed after the user identity (TMSI or IMSI) is known by the network and before the channel is encrypted.

Each system operator can choose its own authentication method. The MS and the HLR each support the same method and have common data. Each MS sends a registration request; then the network is sent a unique challenge. The MS calculates the response to its challenge and sends a message back to the network. The VLR contains a list of triplets—RAND (random number), SRES (signed response), K_c (privacy key); the network compares the triplet with the response it receives from the MS. If the response matches, the MS is registered with the network. The just-used triplet is discarded. After all triplets are used, the VLR must query the HLR for a new set. Each query typically results in three to five triplets.

Security of the MS is maintained by storing all data in the SIM which can be removed from the MS (see section 11.3).

11.5.1 Token-based Registration

The call flows for token-based registration are (see Figure 11.7):

1. The MS sends a registration message to the new system with the old TMSI and old LAI.
2. The new system queries the old VLR for data.
3. The old VLR returns security-related information (e.g., unused triplets and location of HLR).

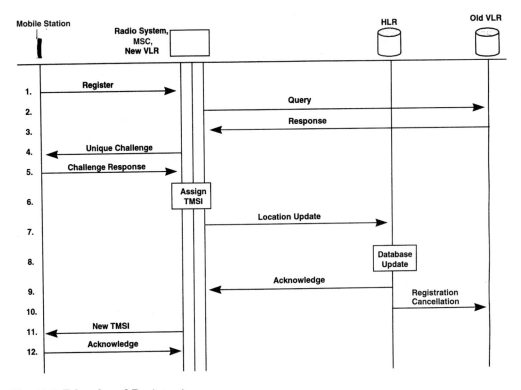

Fig. 11.7 Token-based Registration

4. The new system issues a challenge to the MS.

5. The MS responds to the challenge.

6. The new system assigns a new TMSI.

7. The new system sends a message to the HLR with MS location update information.

8. The HLR updates its location database with the new location of the MS.

9. The HLR acknowledges the message and may send additional security-related data (additional security triplets).

10. The HLR sends a registration cancellation message to the old VLR.

11. The new system sends an encrypted message to the MS with the new TMSI.

12. The MS acknowledges the message.

Steps 11 and 12 can occur any time after step 6 and are not time synchronized with steps 7–10.

If for any reason the old VLR is not reachable, the new visited network will request that the MS send its IMSI to the network; then the new network will establish communications with the HLR.

11.5.2 Token-based Challenge

The token-based challenge is integrated into the various call flows (e.g., registration, handover). We describe it separately here for clarity.

Since token-based systems must query the HLR for additional triplets when they are used, provisions are made to reuse the triplets. In those areas of the world where encryption of the radio link is not permitted or during times of network overload when encryption is disabled, the reuse of triplets can ultimately result in a security breach since it may be possible for other handsets to send a previously used challenge response pair and falsely gain access to the network. As token-based systems are more widely deployed, this type of fraud may be seen more and more.

The security-related information consisting of the RAND, SRES, and K_c triplet is stored in the VLR. When a VLR has used a token to authenticate an MS, it either deletes the token or marks it as used. When a VLR needs to use a token, it uses an unused set. If all sets are used, then the VLR may reuse a set that is marked used. The system operator defines how many times a token may be reused in the VLR. When a token is used the maximum number of times, it is deleted.

When a VLR successfully requests tokens from the HLR or an old VLR, it discards any tokens that are marked as used.

When an HLR receives a request for tokens, it sends any sets that are not marked as used. Those sets shall then be deleted or marked as used. Again, the system operator defines how many times a set may be reused before being discarded. When the HLR has no tokens, it will query the authentication center for additional tokens.

When a network must challenge an MS, the network will use one token from the current set of tokens and use the following call flow (see Figure 11.8):

1. The network transmits the (nonpredictable) RAND to the MS.
2. The MS computes the signature SRES of RAND using the encryption algorithm and the user authentication key (K_i).
3. The MS transmits the signature SRES to the network.
4. The MSC sends a message to the VLR requesting an authentication.
5. The VLR tests SRES for validity.
6. The VLR returns the status to the MSC.
7. The MSC sends a message to the MS with a success or failure indication.

11.6 SUMMARY

In this chapter we discussed how a GSM system maintains security and examined the high-level requirements for security in a wireless system. The network operator (GSM administration) is concerned with proper billing, and the user of the phone is concerned with privacy of communications. To avoid the

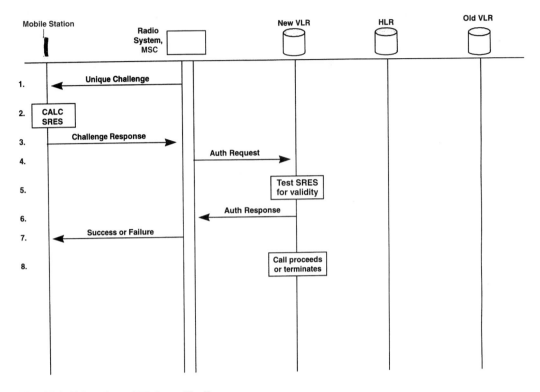

Fig. 11.8 Token-based Unique Challenge

high fraud rate of analog systems, the GSM designers have implemented the security data and algorithms into a secure computing platform—a smart card or SIM. We described the characteristics of the card and showed how the security of the data stored in the card is maintained.

We then examined the types of algorithms used by GSM to provide authentication, privacy keys, and privacy. Each of these algorithms is secret. A standard set is maintained by the GSM MoU and is available for use by operators, but the operator can implement any A3 or A8.

Finally we discussed call flows for registration and unique challenge. These two call flows provide the security mechanism of GSM calls.

11.7 PROBLEMS

1. A proposed wireless system stores security data in the handset rather than in a smart card. Describe how you would design a handset so that the security data stored there would be tamper resistant.

2. Computing power is estimated to double every two years or less. If a cryptographic algorithm takes 365 days to break using a readily available computer today (e.g., a desktop microcomputer), in how many years can the algorithm be broken in one hour of computing? In how many years can it be broken in one second and therefore be included in a digital scanner design?

3. Repeat the previous problem if a weaker algorithm is used that today can be broken in 30 days.

4. Describe the call flows that would be necessary for token-based registration when the old VLR cannot be queried for the TMSI and the network must request that the MS send its TMSI before messages can be exchanged with the HLR. If you have access to the GSM standards, compare your answer with the standards.

11.8 REFERENCES

1. ISO 7810, "Identification Cards—Physical Characteristics," 1985.

2. ISO 7811-1, "Identification Cards—Recording Technique, Part 1: Embossing," 1985.

3. ISO 7811-3, "Identification Cards—Recording Technique, Part 3: Location of Embossed Characters," 1985.

4. ISO 7816-1, "Identification Cards—Integrated Circuit(s) Cards with Contacts, Part 1: Physical Characteristics," 1987.

5. ISO 7816-2, "Identification Cards—Integrated Circuit(s) Cards with Contacts, Part 2: Dimensions and Locations of the Contacts," 1988.

6. ISO/IEC 7816-3, "Identification Cards—Integrated Circuit(s) Cards with Contacts, Part 3: Electronic Signals and Transmission Protocols," 1990.

7. prEN 726-3, "Terminal Equipment (TE); Requirements for IC Cards and Terminals for Telecommunication Use, Part 3: Application Independent Card Requirements."

8. prEN 726-4, "Terminal Equipment (TE); Requirements for IC Cards and Terminals for Telecommunication Use, Part 4: Application Independent Card Related Terminal Requirements."

9. ISO 7813, "Identification Cards—Financial Transaction Cards."

10. GSM 11-11, "Digital Cellular Telecommunications System (Phase 2+): Specification of the Subscriber Identity Module-Mobile Equipment (SIM-ME) Interface."

11. GSM 11-14, "Digital Cellular Telecommunications System (Phase 2+): Specification of the SIM Application Toolkit for the Subscriber Identity Module-Mobile Equipment (SIM-ME) Interface."

12. Wilkes, J. E., "Privacy and Authentication Needs of PCS," *IEEE Personal Communications,* August 1995.

13. Zoreda, J. L., and Oton, J. M., *Smart Cards*, Artech House, 1994.

Modulation and Demodulation

12.1 INTRODUCTION

The digital signals that are generated in the process of transmitting voice, data, and signaling information are generated at low data rates. These data rates, typically 1–50 kbps, are so low in frequency that their transmission directly from the transmitter to the receiver would require antennas that are thousands of meters long. Furthermore, the signals from one transmitter would interfere with the signals from other transmitters if they all used the same frequency band. Therefore, the baseband signals are modulated onto a radio frequency carrier for transmission from the transmitter to the receiver. As we will see in chapter 13, the radio environment at 800–2000 MHz is hostile. We must therefore choose modulation methods that are robust. In addition to the modulation methods we must also choose encoding algorithms that improve the performance of the system.

In this chapter we study three modulation methods, Minimum Shift Keying (MSK), Gaussian Minimum Shift Keying (GMSK), and π/4 Differential Quadrature Phase Shift Keying (π/4-DQPSK). GMSK is the modulation used by GSM, DCS-1800, PCS-1900, and DECT. MSK is introduced as a first step toward GMSK. PWT and PWT-E, the variations of DECT for the licensed and unlicensed 1900-MHz band in North America, use π/4-DQPSK. Since each of these methods is descended from phase shift keying (PSK), we will first study PSK and then show its relationship to the others.

We also examine the channel coding techniques that are used to combat the effects of fading over the radio channel.

12.2 INTRODUCTION TO MODULATION

The baseband data rates of a wireless transmitter are usually from a few kbps to as high as several hundred kbps. The wavelengths for those signals vary from a thousand meters to several hundred thousand meters. If we attempted to send these signals directly, the antennas would be very long and multiple transmitters would interfere with each other.

Obviously, when we want to send signals over any distance, baseband signaling is not sufficient. We must therefore modulate the signals onto an RF carrier. When we transmit the digital bit stream, we convert the bit stream into the analog signal, $a(t)\cos(\omega t + \theta)$. This signal has amplitude, $a(t)$, frequency, $\omega/2\pi$, and phase, θ. We can change any of these three characteristics to formulate the modulation method. The basic forms of the three modulation methods used for transmitting digital signals are

☞ Amplitude Shift Keying (ASK)
☞ Frequency Shift Keying (FSK)
☞ PSK

When ω and θ remain unchanged, we have ASK. When $A(t)$ and θ remain unchanged, we have binary (or M-ary) FSK. When $A(t)$ and ω remain unchanged, we have binary (or M-ary) PSK. Hybrid systems exist where two characteristics are changed with each new symbol transmitted. The most common method is to fix ω and change $A(t)$ and θ. This method is known as Quadrature Amplitude Modulation (QAM). Each of the modulation methods results in a different transmitter and receiver design, different occupied bandwidth, and different error rates. In the remainder of this chapter we will examine the methods used for GSM and DECT and calculate their error rates. Since all signals have a theoretical bandwidth that is infinite, all modulation methods must be band limited. Band limiting introduces detection errors, and the filter bandwidths must be chosen to optimize trade-offs between bandwidth and error rates.

The baseband outputs of the data transmitters are a series of binary data that cannot be sent directly over a radio link. The communications designer must choose radio signals that represent the binary data and permit the receiver to decode the data with minimal errors. For the simplest binary signaling system, we choose two signals denoted by $s_0(t)$ and $s_1(t)$ to represent the binary values of 0 and 1, respectively. Since no channel is perfect, the receiver will also have Additive White Gaussian Noise (AWGN), $n(t)$. The data receiver (Figure 12.1) will then process the signal and noise through a filter,

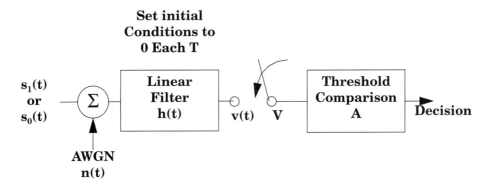

Fig. 12.1 Receiver Structure to Detect Binary Signals in AWGN

$h(t)$, and, at the end of the signaling interval, T, will make a determination of whether the transmitter sent 0 or 1.

The energies of $s_0(t)$ and $s_1(t)$ in a T interval are assumed to be finite and denoted as E_0 and E_1, respectively. For simplicity we assume that the noise has a probability density function of amplitude that is Gaussian and that the noise spectral density is flat with frequency (white noise) with the double-sided Power Spectral Density (PSD) of $N_0/2$.

When $s_0(t)$ is present at the filter as an input, its output at time $t = T$ is

$$V = S_0 + N, \text{ with } s_0(t) \text{ present} \qquad (12.1)$$

where:
S_0 = the output signal component at $t = T$ for the input $s_0(t)$, and
N = the output noise component.

Similarly, when $s_1(t)$ is present at the filter as an input, its output at $t = T$ is

$$V = S_1 + N, \text{ with } s_1(t) \text{ present} \qquad (12.2)$$

Since $n(t)$ is Gaussian with zero mean (implied by its constant power spectral density), N is also Gaussian. The variance of the noise, σ^2, can be determined as:

$$\sigma^2 = \int_{-\infty}^{\infty} |H(f)|^2 \frac{N_o}{2} \, df \qquad (12.3)$$

$$\sigma^2 = N_o \int_0^{\infty} |H(f)|^2 df = N_o B_N \qquad (12.4)$$

where:
$H(f)$ is the transfer function of the filter, and

$B_N = \int_0^{\infty} |H(f)|^2 df$ is the noise-equivalent bandwidth or simply the noise bandwidth of the receiver filter function $H(f)$.

Given that $s_0(t)$ is present at the receiver input, the Probability Density Function (PDF) of V is

$$p(v|s_0) = \frac{1}{\sqrt{2\pi}\sigma}e^{-\left(\frac{(v-S_0)^2}{2\sigma^2}\right)} \qquad (12.5)$$

Similarly, when $s_1(t)$ is present at the receiver input, the conditional probability function of V is

$$p(v|s_1) = \frac{1}{\sqrt{2\pi}\sigma}e^{-\left(\frac{(v-S_1)^2}{2\sigma^2}\right)} \qquad (12.6)$$

From Figure 12.2, given that $s_0(t)$ is present, the probability of error is

$$P(e|s_0) = \int_A^\infty p(v|s_0)dv \qquad (12.7)$$

If $s_1(t)$ is present, it is

$$P(e|s_1) = \int_{-\infty}^A p(v|s_1)dv \qquad (12.8)$$

If the a priori probability that $s_0(t)$ was sent is p, and the a priori probability that $s_1(t)$ was sent is $q = 1-p$, then the average probability of error is

$$P_e = pP(e|s_0) + qP(e|s_1) \qquad (12.9)$$

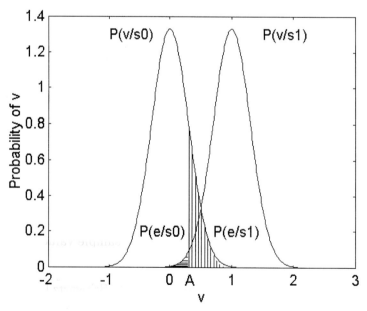

Fig. 12.2 Conditional Probability Density Function of the Filter Output at Time $t = T$

$$P_e = p\int_A^\infty \frac{1}{\sqrt{2\pi}\sigma}e^{-\left(\frac{(v-S_0)^2}{2\sigma^2}\right)}dv + q\int_{-\infty}^A \frac{1}{\sqrt{2\pi}\sigma}e^{-\left(\frac{(v-S_0)^2}{2\sigma^2}\right)}dv \qquad (12.10)$$

If we simplify Eq. (12.10), differentiate the result with respect to A, and then set the derivative equal to zero, we can determine the optimum choice for the threshold value, A, to minimize the probability of error, P_e.

$$A = A_{opt} = \frac{\sigma^2}{S_1 - S_0}\ln\frac{p}{q} + \frac{S_0 + S_1}{2} \qquad (12.11)$$

In most systems, the values of 0 and 1 are equally likely; if they are not, then the designer usually redesigns the encoding method to ensure that they are equally likely. Thus, $p = q$ and

$$A = A_{opt} = \frac{S_0 + S_1}{2} \qquad (12.12)$$

For the optimum value of A, with $p = q$, the probability of error is

$$P_e = \frac{1}{2}erfc\left[\frac{S_1 - S_0}{2\sqrt{2}\sigma}\right] = Q\left[\frac{S_1 - S_0}{2\sigma}\right] \qquad (12.13)$$

where $erfc(u)$ is the complementary error function $= 1 - erf(u) = 2Q(\sqrt{2}u)$. The error function, $erf(u)$ is defined as

$$erf(u) = \frac{2}{\sqrt{\pi}}\int_0^u e^{-t^2}dt \qquad (12.14)$$

and

$$Q(u) = \frac{1}{\sqrt{2\pi}}\int_u^\infty e^{-\frac{x^2}{2}}dx \qquad (12.15)$$

or

$$Q(u) \approx \frac{e^{-\frac{u^2}{2}}}{\sqrt{2\pi}u}, \; u \gg 1 \quad \text{Gaussian Integral} \qquad (12.16)$$

Next we want to find the filter that provides the minimum probability of error, as expressed by Eq. (12.13). At time t_0, the sample value consists of a signal-related component $g_0(t_0)$ and a noise component $n_0(t_0)$. This filter is known as a matched filter and has the transfer function $H_0(f)$ optimized to provide the maximum SNR at its output at time t_0. Schwartz [6] shows that this filter must be the conjugate match of the signal, $s(t)$. Since we have two signals, $s_0(t)$ and $s_1(t)$, we will need two filters in our receiver design.

If we transmit a signal, $s(t)$, then it has the Fourier transform, $S(\omega)$, which is a complex function. The optimum or matched filter must have a frequency response, $H(\omega)$, where

$$H(\omega) = S^*(\omega)e^{-j\omega t_0} \tag{12.17}$$

where:
  S^* is the complex conjugate of the Fourier transform of the signal.

In general this filter is not realizable, since analysis will show that it must have output before there is input, and we have not yet been able to design circuits that predict the future. However, we can design filters that approximate the ideal filter.

The SNR is defined as

$$\xi^2 = \frac{g_0^2(t_o)}{\sigma^2} \tag{12.18}$$

It can also be shown that the maximum value for SNR ξ is twice the energy of the input signal (E_g) divided by the single-sided input noise spectral density, regardless of the input signal shape.

$$\xi^2_{max} = \frac{2E_g}{N_o} \tag{12.19}$$

For a binary system, Eq. (12.19) becomes

$$\xi^2_{max} = \frac{1}{N_o} \int_0^T [s_1(t) - s_0(t)]^2 dt \tag{12.20}$$

Since the signals are zero outside the range $(0, T)$, the probability of error corresponding to the optimum receiver filter becomes

$$P_e = \frac{1}{2} erfc(\sqrt{z}) = Q(\sqrt{2z}) \tag{12.21}$$

where:

$$z = \frac{1}{4N_o} \int_0^T [s_1(t) - s_0(t)]^2 dt$$

If the transmitted pulses are allowed to take on any of "M" transmitted levels with equal probability, then the information rate per transmitted pulse is $\log_2 M$ bits. For a constant information rate, the bandwidth of the transmitted signal can be reduced by the same factor. With M-ary transmission, we will show that the error rates are higher, but if we have sufficient SNR then the higher error rates will not matter. Thus, we are using excess SNR to code the signal and reduce its bandwidth.

When we add additional levels to a baseband system, we are reducing the distance between detection levels in the receiver output. Thus, the error rate of a multilevel baseband system can be determined by calculating the appropriate reduction in the error distance. If the maximum amplitude is V, the error distance d_e between equally spaced levels at the detector is

$$d_e = \frac{V}{M-1} \qquad (12.22)$$

where:
M = number of levels.

Setting the error distance V of a binary system to that defined in Eq. (12.22) provides the error probability of the multilevel system.

$$P_e = \frac{1}{log_2 M}\left[\frac{M-1}{M}\right] erfc\left[\frac{V}{(M-1)\sqrt{2}\sigma}\right] \qquad (12.23)$$

where:

the factor $\left[\dfrac{M-1}{M}\right]$ reflects that the interior signal levels are vulnerable to both positive and negative noise,

the factor $\dfrac{1}{log_2 M}$ arises because the multilevel system is assumed to be coded, so symbol errors produce single bit errors ($log_2 M$ is the number of bits per symbol), and

the probability of multiple bit errors is assumed to be small and can be neglected.

Eq (12.23) relates error probability to the peak signal power V^2. To determine P_e with respect to average power, the average power of an M-level system is determined by averaging the power associated with the various pulse amplitude levels.

$$[V^2]avg = \frac{2}{M}\left[\left(\frac{V}{M-1}\right)^2 + \left(\frac{3V}{M-1}\right)^2 + ... + V^2\right] \qquad (12.24)$$

$$[V^2]avg = \frac{2V^2}{M(M-1)^2}\sum_{j=1}^{M/2}(2j-1)^2 \qquad (12.25)$$

where:
the levels $\dfrac{V}{M-1}[\pm 1, \pm 3, \pm 5, ..., \pm(M-1)]$ are assumed to be equally likely.

If T is the signaling interval for a two-level system, the signaling interval T_M for an M-level system providing the same data rate is determined as

$$T_M = T \, log_2 M \qquad (12.26)$$

For a raised cosine filter, the noise bandwidth is $B_N = \dfrac{1}{2T_M}$.

From Eq. (12.4), we get

$$\sigma^2 = \frac{N_o}{2T_M} \tag{12.27}$$

$$\sigma = \frac{1}{\sqrt{2}}\left[\frac{N_o}{T_M}\right]^{1/2} \tag{12.28}$$

Substituting Eq. (12.28) into Eq. (12.23), we get

$$P_e = \left[\frac{1}{log_2 M}\right]\left[\frac{M-1}{M}\right]erfc\left[\frac{V}{(M-1)\left(\frac{N_o}{T_M}\right)^{1/2}}\right] \tag{12.29}$$

The energy per symbol
$$E_s = E_b\,log_2 M = V^2\,T_M,$$

where:
E_b is the energy per bit

$$\therefore V^2 = \frac{E_b log_2 M}{T_M} \tag{12.30}$$

Substituting for V from Eq. (12.30) into Eq. (12.29), we get

$$P_e = \left[\frac{1}{log_2 M}\right]\left[\frac{M-1}{M}\right]erfc\left[\left(\frac{E_b}{N_o}\right)^{1/2}\frac{(log_2 M)^{1/2}}{M-1}\right] \tag{12.31}$$

$$SNR = \frac{\text{Signal Power}}{\text{Noise Power}} = \frac{E_b(log_2 M)(1/T_M)}{N_o\left(\frac{1}{2}T_M\right)} \tag{12.32}$$

$$\therefore SNR = 2log_2 M\left(\frac{E_b}{N_o}\right) \tag{12.33}$$

Another variation of baseband signaling is Antipodal Baseband Signaling (APBBS) where two signals of opposite polarities are sent. If $s_0(t) = -V$ and $s_1(t) = V$ for $0 \le t \le T$, then $s_1(t) - s_0(t) = 2\,V$.
We then calculate the value of z for Eq. (12.21) as

$$z = \frac{1}{4N_o}\int_0^T (2V)^2 dt = \frac{V^2 T}{N_o} = \frac{E_b}{N_o} \tag{12.34}$$

where E_b is the energy in either $s_0(t)$ or $s_1(t)$, or the bit energy.

$$P_e = \frac{1}{2}erfc\left[\sqrt{\frac{E_b}{N_o}}\right] = Q\left[\sqrt{\frac{2E_b}{N_o}}\right] \tag{12.35}$$

APBBS is used to modulate some signals; we will compare its SNR with other modulation methods.

12.3 PSK

For binary PSK (BPSK), we transmit two different signals. If the baseband signal is a binary 0, we transmit:

$$A \cos (\omega t + \pi) = -A \cos (\omega t) \tag{12.36}$$

and for binary 1, we transmit:

$$A \cos (\omega t) \tag{12.37}$$

BPSK can be considered a form of ASK where each nonreturn to zero (NRZ) data bit of value 0 is mapped into a −1 and each NRZ 1 is mapped into a +1. The resulting signal is passed through a filter to limit its bandwidth and then multiplied by the carrier signal cos ωt (see Figure 12.3).

We can also define PSK where there are M phases rather than two phases. In M-ary PSK, every n (where $M = 2^n$) bits of the binary bit stream are coded as a signal that is transmitted as $A \sin (\omega t + \theta_j), j = 1, M$.

The error distance of a PSK system with M phases is $V \sin (\pi/M)$ where V is the signal amplitude at the detector. A detection error occurs if noise of the proper polarity is present at the output of either of the two phase detectors. The probability of error is [2]

$$P_e = \left(\frac{1}{log_2 M}\right) erfc\left[\sin\left(\frac{\pi}{M}\right)(log_2 M)^{1/2}\left(\frac{E_b}{N_o}\right)^{1/2}\right] \tag{12.38}$$

The SNR is given as:

$$SNR = log_2 M\left(\frac{E_b}{N_o}\right)(\text{For } M > 2) \tag{12.39}$$

Each symbol has a length of T. Therefore

$$E_b = V^2 T/ \log_2 M \tag{12.40}$$

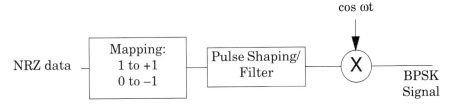

Fig. 12.3 BPSK Modulator

and the rms noise σ is:

$$N_o = \sigma^2\, 2T \tag{12.41}$$

for noise in a Nyquist bandwidth.

Also, as shown in reference [9], the bandwidth efficiency of the M-ary PSK is given as $\dfrac{R}{B_w} = \dfrac{log_2 M}{2}$ where R is the data rate and B_w is the bandwidth.

We will now examine several variations of PSK.

12.4 QUADRATURE PHASE SHIFT KEYING (QPSK)

If we define four signals, each with a phase shift differing by 90°, then we have QPSK. We have previously calculated the error rates for a general PSK signal with M signal points. For QPSK, $M = 4$, so substituting $M = 4$ in Eq. (12.38) we get

$$P_e = \left(\frac{1}{log_2 4}\right) erfc\left[\sin\left(\frac{\pi}{4}\right)(log_2 4)^{1/2}\left(\frac{E_b}{N_o}\right)^{1/2}\right]$$

$$P_e = \frac{1}{2} erfc\sqrt{\frac{E_b}{N_o}} = Q\left[\sqrt{\frac{2E_b}{N_o}}\right] \tag{12.42}$$

See Figure 12.4 for a plot of Eq. (12.42). The plot was generated by Matlab and the program can be found on the disk supplied with this book.

The input binary bit stream $\{b_k\}$, $b_k = \pm 1$; $k = 0, 1, 2, \ldots$, arrives at the modulator input at a rate $1/T$ bps and is separated into two data streams, $a_I(t)$ and $a_Q(t)$, containing even and odd bits, respectively (Figure 12.5). The modulated QPSK signal $s(t)$ is given as:

$$s(t) = \frac{1}{\sqrt{2}}a_I(t)\cos\left(2\pi ft + \frac{\pi}{4}\right) + \frac{1}{\sqrt{2}}a_Q(t)\sin\left(2\pi ft + \frac{\pi}{4}\right) \tag{12.43}$$

$$s(t) = A\cos\left[2\pi ft + \frac{\pi}{4} + \theta(t)\right] \tag{12.44}$$

where:

$$A = \sqrt{(1/2)(a_I^2 + a_Q^2)} = 1, \text{ and}$$

$$\theta(t) = -\text{atan}\frac{a_Q(t)}{a_I(t)}.$$

In Figure 12.6, we show an NRZ data stream of 01110001. We then show the in-phase (I) (0100) and quadrature (Q) (1101) signals that are generated from the NRZ data stream. Notice that the I and Q signals have bit lengths that are twice as long as the NRZ data bits. In the figure, there is no delay

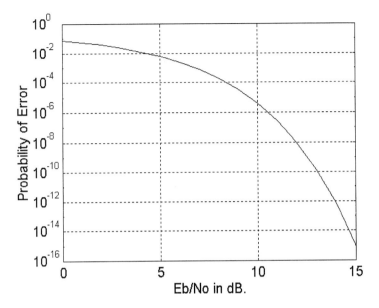

Fig. 12.4 Error Performance of QPSK

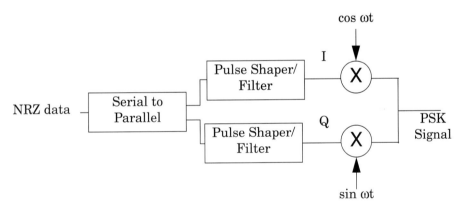

Fig. 12.5 QPSK Modulator

between the NRZ data and the I and Q data. In a real implementation there would be a 1- to 2-bit delay before the I and Q signals were generated. This delay accounts for the time for 2 bits to be received and decoded into the I and Q signals. Finally we show the generated QPSK signal. To make the figure clearer, we chose a carrier frequency that is four times higher than the data rate. In real systems, the carrier frequency would be many times that of the data rate. The plot was generated by Matlab and the program can be found on the disk supplied with this book.

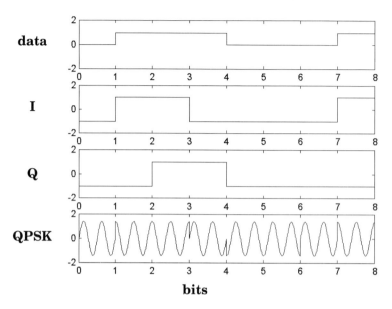

Fig. 12.6 QPSK Signals

The values of $\theta(t) = 0, -\pi/2, \pi/2, \pi$ represent the four values of $a_I(t)$ and $a_Q(t)$. On the I/Q plane, QPSK represents four equally spaced points separated by $\pi/2$ (Figure 12.7). Each of the four possible phases of carriers represents 2 bits of data. Thus, there are 2 bits per symbol. Since the symbol rate for QPSK is half of the bit rate, twice the information can be carried in the same amount of channel bandwidth as can be carried using BPSK. This is possible because the two signals I and Q are orthogonal to each other and can be transmitted without interfering with each other.

In QPSK, the carrier phase can change only once every $2T$ seconds. If from one $2T$ interval to the next one, neither bit stream changes sign, the carrier phase remains the same. If one component $a_I(t)$ or $a_Q(t)$ changes sign, a phase shift of $\pi/2$ occurs. However if both components, I and Q, change sign, then a phase shift of π or 180° occurs. When this 180° shift is filtered by the transmitter and receiver filters, it generates a change in amplitude of the detected signal and causes additional errors. Notice the 180° shift at the end of bit interval 4 in Figure 12.6.

If the two bit streams, I and Q, are offset by a half-bit interval, then the amplitude fluctuations are minimized since the phase never changes by 180° (Figures 12.8 and 12.9). This modulation scheme, called Offset Quadrature Phase Shift Keying (OQPSK), is obtained from conventional QPSK by delaying the odd bit stream by a half-bit interval with respect to the even bit stream. Thus, the range of phase transition is 0° and 90° and it occurs twice as often, but with half the intensity of the QPSK system. Comparing Figure 12.6

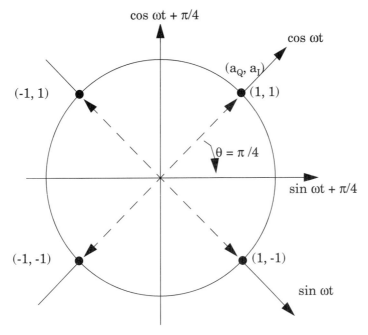

Fig. 12.7 Signal Constellation for QPSK

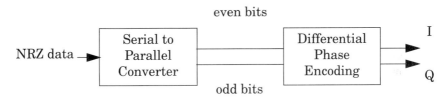

Fig. 12.8 Differential Encoding of π/4-DQPSK

with Figure 12.9, notice that the I signal is the same for both QPSK and OQPSK, but the Q signal is delayed by half a bit. Thus, the 180° phase change at the end of bit interval 4 of the QPSK signal is replaced by a 90° phase change at the end of bit interval 4 of the OQPSK signal. Also notice that phase changes occur more frequently with OQPSK system. While the phase changes will still cause amplitude fluctuations in the transmitter and receiver, the fluctuations will have smaller magnitude. The bit error rate and bandwidth efficiency of QPSK and OQPSK are the same as for BPSK.

In theory, QPSK or OQPSK systems can improve the spectral efficiency of mobile communication. They do, however, require a coherent detector. In a multipath fading environment, the use of coherent detection is difficult and often results in poor performance over noncoherently based systems. The

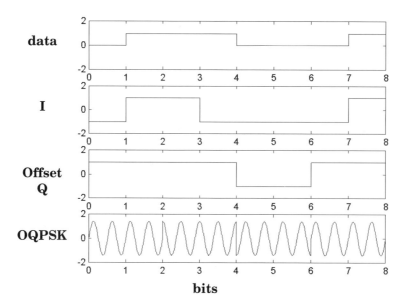

Fig. 12.9 OQPSK Signals

coherent detection problem can be overcome by using a differential detector, but then OQPSK is subject to intersymbol interference (ISI) which results in poor system performance. The spectrum of offset QPSK (OQPSK) is [5]

$$P_{QPSK}(f) = T\left[\frac{\sin \pi f T}{\pi f T}\right]^2 \qquad (12.45)$$

Eq. 12.45 is plotted in Figure 12.13.

12.5 π/4-DQPSK

We can design a PSK system to be inherently differential and thus solve detection problems. π/4-DQPSK is a compromise modulation method because the phase is restricted to fluctuate between $\pm\pi/4$ and $\pm 3\pi/4$ rather than the $\pm\pi/2$ phase changes for OQPSK. It has a spectral efficiency about 20 percent higher than the GMSK modulation used for DECT and GSM (see the next section).

π/4-DQPSK is essentially a π/4-QPSK with differential encoding of symbol phases. The differential encoding mitigates loss of data due to phase slips. However, differential encoding results in the loss of a pair of symbols when channel errors occur. This can be translated to an approximate 3-dB loss in E_b/N_o relative to coherent π/4-QPSK.

A π/4-QPSK signal constellation (Figure 12.10) consists of symbols corresponding to eight phases. Consider that these eight phase points are formed by superimposing two QPSK signal constellations, offset by 45° relative to each other. During each symbol period, a phase angle from only one of the two QPSK constellations is transmitted. The two constellations are used alternately to transmit every pair of bits (di-bits). Thus, successive symbols have a relative phase difference that is one of the four phases shown in Table 12.1.

Figure 12.10 shows the π/4-DQPSK signal constellation. When the phase angles of π/4-QPSK symbols are differentially encoded, the resulting modulation is π/4-DQPSK. This can be done either by differentially encoding the

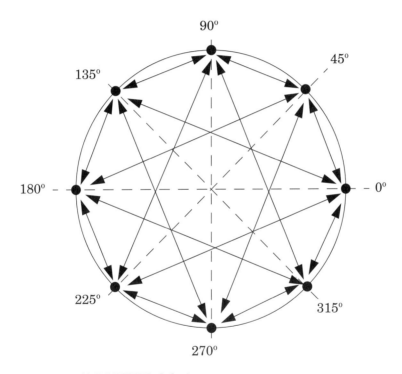

Fig. 12.10 π/4-DQPSK Modulation

Table 12.1 Phase Transitions of π/4-DQPSK

Symbol	π/4-DQPSK Phase Transition
00	45°
01	135°
10	−45°
11	−135°

source bits and mapping them onto absolute phase angles or, alternately, by directly mapping the pairs of input bits onto relative phase ($\pm\pi/4$, $\pm 3\pi/4$) as shown in Figure 12.10. The binary data stream entering the modulator $b_M(t)$ is converted by a serial to parallel converter into two binary streams $b_o(t)$ and $b_e(t)$ before the bits are differentially encoded (Figure 12.11).

$$I_k = I_{k-1}\cos\Delta\phi_k - Q_{k-1}\sin\Delta\phi_k \tag{12.46}$$

$$Q_k = I_{k-1}\sin\Delta\phi_k + Q_{k-1}\cos\Delta\phi_k \tag{12.47}$$

where:
I_k and Q_k are the in-phase and quadrature components of the $\pi/4$-DQPSK signal corresponding to the k-th symbol, and
the amplitudes of I_k and Q_k are ± 1, 0, $\pm\dfrac{1}{\sqrt{2}}$, respectively.

Since the absolute phase of $(k-1)$th symbol is ϕ_{k-1}, the in-phase and quadrature components can be expressed as:

$$I_k = \cos\phi_{k-1}\cos\Delta\phi_k - \sin\phi_{k-1}\sin\Delta\phi_k = \cos(\phi_{k-1}+\Delta\phi_k) \tag{12.48}$$

$$Q_k = \cos\phi_{k-1}\sin\Delta\phi_k + \sin\phi_{k-1}\cos\Delta\phi_k = \sin(\phi_{k-1}+\Delta\phi_k) \tag{12.49}$$

These component signals (I_k, Q_k) are then passed through baseband filters having a raised cosine frequency response as:

$$|H(f)| = \begin{cases} 1 & 0 \le |f| \le \dfrac{1-\alpha}{2T_s} \\[3mm] \sqrt{\dfrac{1}{2}\left\{1 - \sin\left[\dfrac{\pi T_s}{\alpha}\left(|f| - \dfrac{1}{2T_s}\right)\right]\right\}} & \dfrac{1-\alpha}{2T_s} \le |f| \le \dfrac{1+\alpha}{2T_s} \\[3mm] 0 & |f| \ge \dfrac{1+\alpha}{2T_s} \end{cases} \tag{12.50}$$

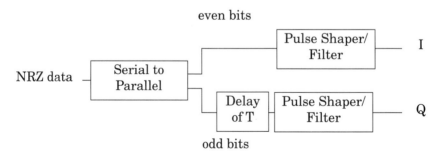

Fig. 12.11 OQPSK Encoding

where:
α is the roll-off factor, and
T_s is the symbol duration.

If $g(t)$ is the response to pulse I_k and Q_k at the filter input, then the resultant transmitted signal is given as:

$$s(t) = \sum_k g(t - kT_s)\cos\phi_k \cos\omega t - \sum_k g(t - kT_s)\sin\phi_k \sin\omega t \qquad (12.51)$$

$$s(t) = \sum_k g(t - kT_s)\cos(\omega t + \phi_k) \qquad (12.52)$$

where:
$2\pi\omega$ is the carrier frequency of the transmission.

The component ϕ_k results from differential encoding (i.e., $\phi_k = \phi_{k-1} + \Delta\phi_k$).

Depending on the detection method (coherent detection or differential detection), the error performance of π/4-DQPSK can either be the same or 3 dB worse than QPSK.

12.6 MSK

We previously showed that OQPSK is derived from QPSK by delaying the Q data stream by 1 bit or T seconds with respect to the corresponding I data stream. This delay has no effect on the error rate or bandwidth.

MSK is derived from OQPSK by replacing the rectangular pulse in amplitude with a half-cycle sinusoidal pulse. Figure 12.12 shows the data signal and the in-phase and quadrature pulses that are used to generate the MSK signal. Notice that the in-phase and quadrature signals are delayed by interval T from each other.

The MSK signal is defined as:

$$s(t) = a_I(t)\left|\cos\left(\frac{\pi(t - 2nT)}{2T}\right)\right|\cos 2\pi ft + a_Q(t)\left|\sin\left(\frac{\pi(t - 2nT)}{2T}\right)\right|\sin 2\pi ft \qquad (12.53)$$

$$s(t) = \cos\left[2\pi ft + b_k(t)\frac{\pi(t - 2nT)}{2T} + \phi_k\right] \qquad (12.54)$$

where:
$n = 0, 1, 2, 3, ...,$
$b_k = +1$ for $a_I \cdot a_Q = -1$,
$b_k = -1$ for $a_I \cdot a_Q = 1$,
$\phi_k = 0$ for $a_I = 1$, and
$\phi_k = \pi$ for $a_I = -1$.

Note that, since the I and Q signals are delayed by a 1-bit interval, the cosine and sine pulse shapes in Eq. (12.53) are actually both in the shape of a sine pulse.

MSK has the following properties:

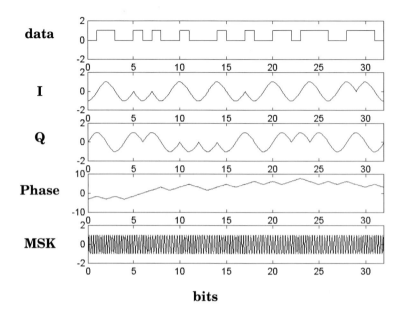

Fig. 12.12 Data, I, and Q Signals Used to Generate MSK Signal

1. For a modulation bit rate of R, the high frequency, $f_H = f + 0.25R$ when b_k =1, and the low frequency, $f_L = f - 0.25R$ when $b_k = -1$.
2. The difference between the high frequency and the low frequency is $\Delta f = f_H - f_L = 0.5R = 1/(2T)$, where T is the bit interval of the NRZ signal.
3. The signal has a constant envelope.

The error probability for an ideal MSK system is

$$P_e = \frac{1}{2} erfc \sqrt{\frac{E_b}{N_o}} = Q\left[\sqrt{\frac{2E_b}{N_o}}\right] \tag{12.55}$$

which is the same as for QPSK/OQPSK.

MSK modulation makes the phase change linear and limited to $\pm\pi/2$ over a bit interval T. This enables MSK to provide a significant improvement over QPSK. Because of the effect of the linear phase change, the PSD has low side lobes that help to control adjacent-channel interference. However, the main lobe becomes wider than QPSK (see Figure 12.13). Thus, it becomes difficult to satisfy the CCIR-recommended value of −60-dB side lobe power levels. The PSD for MSK is [5]

$$P_{MSK}(f) = \frac{16T}{\pi^2}\left[\frac{\cos 2\pi fT}{1 - 16f^2T^2}\right]^2 \tag{12.56}$$

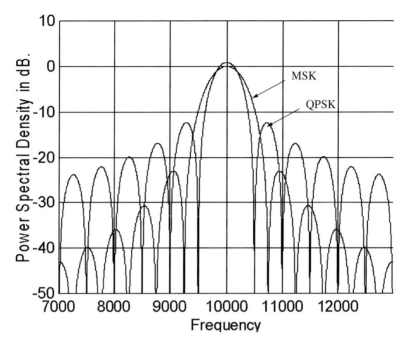

Fig. 12.13 Spectral Density of QPSK and MSK

Figure 12.13 shows the PSD for both MSK and QPSK. Notice that, while the first null in the side lobes occurs at a data rate of R for MSK and $R/2$ for QPSK, the overall side lobes are lower for MSK. Thus, MSK is more spectrally efficient than QPSK.

The detector for MSK is slightly different than for PSK (see Figure 12.14). We must generate the matched filter equivalent to the transmitted in-phase and quadrature signals. These two reference signals are

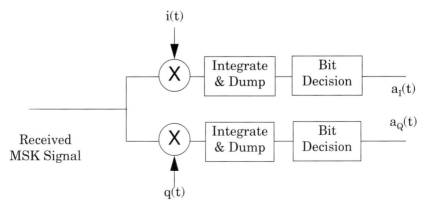

Fig. 12.14 Optimum MSK Detector

$$i(t) = \cos\left(\frac{\pi t}{2T}\right)\cos\omega t \qquad (12.57)$$

$$q(t) = \sin\left(\frac{\pi t}{2T}\right)\sin\omega t \qquad (12.58)$$

We multiply the received signal by $i(t)$ and $q(t)$ and perform an integration with detection at the end of the bit interval; we then dump the integrator output. This is the standard integrate-and-dump matched filter, with the reference signal $i(t)$ and $q(t)$ matched to the received waveform. At the end of the bit interval we make a decision on the state of the bit (+1 or –1) and output the decision as our detected bit. We do this for both the I channel and the Q channel with I and Q out of phase by T seconds to account for the differential nature of MSK.

12.7 GMSK

In minimal shift keying, we replace the rectangular data pulse with a sinusoidal pulse. Obviously other pulse shapes are possible. A Gaussian-shaped impulse response filter generates a signal with low side lobes and a narrower main lobe than the rectangular pulse. Since the filter theoretically has output before input, it can be approximated only by a delayed and shaped impulse response that has a Gaussian-like shape. This is the GMSK modulation.

The relationship between the premodulation filter bandwidth B and the bit period, T, defines the bandwidth of the system. If $B > 1/T$, then the waveform is essentially MSK. When $B < 1/T$, ISI occurs since each data pulse overlaps with the data pulse in the next position in the symbol time. However, ISI can be traded for bandwidth reduction if the system has sufficient SNR. GSM designers used a BT of 0.3 with a channel data rate of 270.8 kbps. DECT designers adopted a BT of 0.5 with a data rate of 1.152 Mbs. A choice of BT = 0.3 in GSM is a compromise between bit error rate and out-of-band interference since the narrow filter increases the ISI and reduces the signal power.

When GMSK was first proposed [3], the modulator was based on using FM (Figure 12.15). Since newer integrated circuits are available that enable easy construction of an I and Q modulator, there is a more modern method of generating the GSMK signal (shown in Figure 12.16).

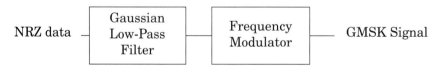

Fig. 12.15 GMSK Modulator Using Frequency Modulator

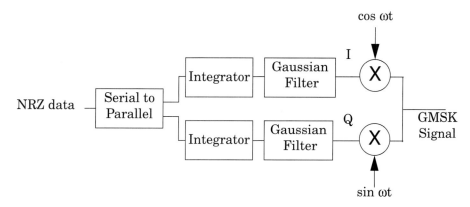

Fig. 12.16 GMSK Modulator Using Phase Modulator

Filtering of the NRZ data by a Gaussian low-pass filter generates a signal that is no longer constrained to a 1-bit interval. ISI is generated by the modulator.

The bit error rate performance of GMSK can be expressed [3] as:

$$P_e = \frac{1}{2} erfc \sqrt{\frac{\alpha E_b}{2N_o}} = Q\left[\sqrt{\frac{\alpha E_b}{N_o}}\right] \tag{12.59}$$

where α depends on the BT product. For GSM where BT = 0.3, α = 0.9; for DECT where BT = 0.5, α = 0.97; for BT = ∞, α = 2, which is the case for MSK (see Figure 12.17).

Demodulation of GMSK (Figure 12.18) requires multiplication by the in-phase and quadrature carrier signals followed by a low-pass filter with Gaussian shape. At the end of the bit interval we make a decision on the state of the bit (+1 or –1) and output the decision as our detected bit. As with MSK, we do this for both the I channel and the Q channel with I and Q out of phase by T seconds to account for the differential nature of GMSK. We ignore the ISI caused by the longer-than-one-bit-interval nature of the Gaussian-transmitted pulse.

12.8 SYNCHRONIZATION

The demodulation of a signal requires that the receiver be synchronized with the transmitted signal, as it is received at the input of the receiver. The synchronization must be for:

☞ **Carrier synchronization.** The receiver is on the same frequency as the transmitted signal, adjusted for the effects of doppler shifts.

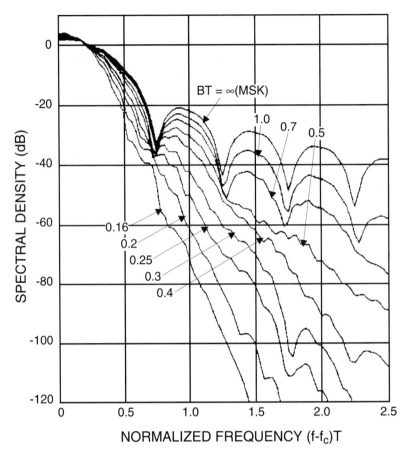

Fig. 12.17 Spectrum of GMSK
© *IEEE Transactions on Communications* 29 (7), July 1981. Used with permission.

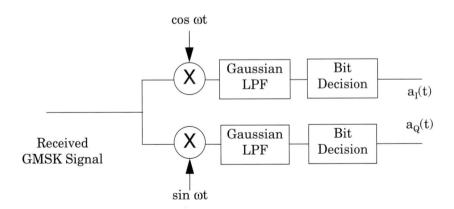

Fig. 12.18 GMSK Demodulator

☞ **Bit synchronization.** The receiver is aligned with the beginning and end of each bit interval.

☞ **Word synchronization (also known as frame synchronization).** The receiver is aligned with the beginning and end of each word in the transmitted signal.

If the synchronization in the receiver is not precise for any of these operations, then the bit error rate of the receiver will not be the same as described by the equations in the previous sections of this chapter. The design of a receiver is an area that standards have traditionally not specified. It is an art that enables one company to offer better performance in its equipment compared to a competitor. The methods of achieving synchronization discussed in this section are the traditional methods. A particular receiver may or may not use any of these methods, and proprietary methods are often used by many companies.

For PSK, the carrier signal changes phase every bit interval (Figure 12.19). If we multiply the received signal by an integer N, we can convert all of the phase changes in the multiplied signal to multiples of 360°. The new signal then has no phase changes and we can recover it using a narrowband phase-locked loop (PLL). After the PLL recovers the multiplied carrier signal, it is divided by N to recover the carrier at the proper frequency. By the suitable choice of digital dividing circuits, it is possible to get a precise 90° difference in the output of two dividers and thus generate both the cos ωt and the sin ωt signals needed by the receiver. There are also some down-converter integrated circuits that have a precise phase shift network contained within them. The carrier recovery is typically performed at some lower intermediate frequency rather than directly at the received frequency. For BPSK we would need an N of 2, but an N of 4 would be used to enable the sine and cosine term to be generated. For QPSK and its derivatives, an N of 4 is necessary; for π/4-DQPSK an N of 8 would be needed.

After we recover the carrier, we must reestablish the carrier phase to determine the values of the received bits. Somewhere in the transmitted signal must be a known bit pattern that we can use to determine the carrier phase. The bit pattern can be alternating 0s and 1s that we use to determine bit timing or it could be some other known pattern.

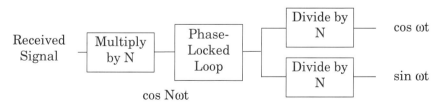

Fig. 12.19 Carrier Recovery for PSK

The advantage of differential keying (e.g., π/4-DQPSK) is that it is not important to know the absolute carrier phase. Only the change in carrier phase from one symbol to the next is important.

MSK is a form of FM; therefore, a different method of carrier recovery is needed. In Figure 12.20, the MSK signal has frequency f and deviation Δf = 1/2T. We first multiply the signal by 2, thus doubling the deviation and generating strong frequency components [4] at $2f + 2\Delta f$ and $2f - 2\Delta f$. We use two PLLs to recover these two signals.

$$s_1(t) = \cos(2\pi ft + 2\pi\Delta ft) \tag{12.60}$$

$$s_2(t) = \cos(2\pi ft - 2\pi\Delta ft) \tag{12.61}$$

We then take the sum and difference of $s_1(t)$ and $s_2(t)$ to generate the desired $i(t)$ and $q(t)$ signals.

$$i(t) = s_1(t) + s_2(t) = \cos(2\pi ft + \pi\Delta ft) + \cos(2\pi ft - \pi\Delta ft) = 2\cos 2\pi ft \cos \pi\Delta ft \tag{12.62}$$

$$q(t) = s_1(t) - s_2(t) = \cos(2\pi ft + \pi\Delta ft) - \cos(2\pi ft - \pi\Delta ft) = 2\sin 2\pi ft \sin \pi\Delta ft \tag{12.63}$$

The identical circuit can also be used for carrier recovery for a GMSK system [1, 3, 4].

The next step is to recover data timing or bit synchronization (Figure 12.21). Most communication systems transmit a sequence of 1s and 0s in an alternating pattern to enable the receiver to maintain bit synchronization. A PLL operating at the bit timing is used to maintain timing. Once the PLL is

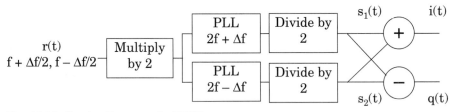

Fig. 12.20 Carrier Recovery for MSK

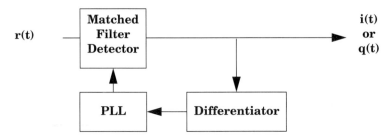

Fig. 12.21 Generalized Data Timing Recovery Circuit

synchronized on the received 101010... pattern, it will remain synchronized on any other patterns except for long sequences of all 0s or all 1s.

MSK uses an additional circuit to achieve bit timing. The $s_1(t)$ and $s_2(t)$ signals are multiplied together and low-pass filtered (Figure 12.22).

$$s_1(t)s_2(t) \ = \ \cos(2\pi ft + \pi\Delta ft) \times \cos(2\pi ft - \pi\Delta ft) \ = \ 0.5\cos 4\pi ft + 0.5\cos 2\pi\Delta ft \quad \text{(12.64)}$$

$$lowpassfiltered[s_1(t)s_2(t)] \ = \ 0.5\cos 2\pi\Delta ft \ = \ 0.5\cos\frac{\pi t}{T} \quad \text{(12.65)}$$

The output of the low-pass filter is a clock signal at one-half of the transmitted bit rate. The one-half bit rate clock is the correct rate for demodulation of the signal since the I and Q signals are at one-half of the bit rate.

Frame synchronization is determined by the receiver correlating the received signal with a known bit pattern. The receiver performs an autocorrelation function to determine when the bit pattern is received and then outputs a framing pulse (Figure 12.23).

12.9 EQUALIZATION

We show in chapter 13 that the received signal in a mobile radio environment travels from the transmitter to the receiver over many paths. The signal fades in and out and undergoes distortion because of the multipath nature of the channel. For a transmitted signal $s(t) = a(t) \cos[\omega t + \theta(t)]$, we can represent the received signal, $r(t)$, as:

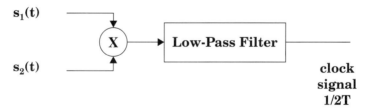

Fig. 12.22 MSK Data Timing Recovery Circuit

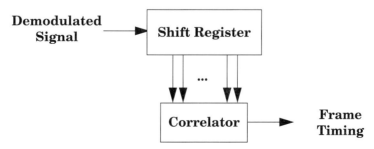

Fig. 12.23 Generalized Framing Recovery Circuit

$$r(t) = \sum_{i=0}^{n} x_i(t - \tau_i)a(t - \tau_i)\cos[\omega(t - \tau_i) + \theta(t - \tau_i)] \qquad (12.66)$$

$$+ y_i(t - \tau_i)a(t - \tau_i)\sin[\omega(t - \tau_i) + \theta(t - \tau_i)]$$

In chapter 13 we show that the received signal has Rayleigh fading statistics. But what are the characteristics of the x and y terms in Eq. (12.66)? If the transmitter signal is narrow enough compared to the fine multipath structure of the channel, then the individual fading components, $x_i(t)$ and $y_i(t)$ will also have Rayleigh statistics. If a particular path is dominated by a reflection off a mountain, hill, building, or similar structure, then the statistics of that path may be Rician rather than Raleigh. If the range of τ_i (delay spread) is small compared to the bit interval, then little distortion of the received signal occurs. If the range of τ_i is greater than a bit interval, then the transmissions from one bit will interfere with the transmissions of another bit, resulting in ISI.

SS systems transmit wide bandwidth signals and attempt to recover the signals in each of the paths and add them together in a diversity receiver. For our discussion here, however, we are transmitting narrowband signals, and the multipath signals are an interference to the desired signal. We need to construct a receiver that removes the effects of the multipath signal or cancels the undesired multipaths.

Another way to describe the multipath channel is to describe the channel as having an impulse response $h(t)$. The received signal is then written as:

$$r(t) = \int_{-\infty}^{\infty} s(t)h(t - \tau)d\tau \qquad (12.67)$$

We can then recover $s(t)$ if we can determine a transfer function $h^{-1}(t)$, the inverse of $h(t)$. One reason it is difficult to perform the inverse function is that it is time varying. The circuit that performs the inverse transfer function is called an equalizer (Figure 12.24).

Generally, we are interested in minimizing the ISI at the time when we do our detection (the sample time in a sample-and-hold circuit). Thus, we can model the equalizer as a series of equal time delays (rather than random as is the general case) with the shortest delay interval being a bit interval. We then construct a receiver that determines r_{eq}, the equalized signal:

$$r_{eq}(t) = \sum_{i=0}^{n} \alpha_i(t - \tau_n)r(t - \tau_n) \qquad (12.68)$$

We use our equalized signal r_{eq} as the input to our detector to determine the value of the i-th transmitted bit. We must adjust the values of α_i to achieve some measure of receiver performance. A typical measure is to minimize the mean square error between the value of the detected bit at the output of the

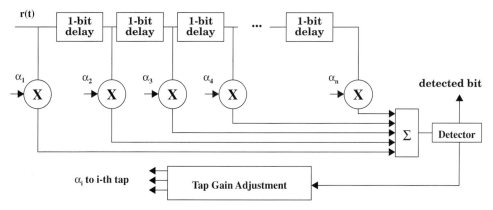

Fig. 12.24 Block Diagram of an Equalizer

summer and the output of the detector. Other measures for the equalizer are possible. For more details see reference [5].

12.10 CHANNEL CODING

Shannon showed that, by adding redundant bits to the source information, errors in a noisy channel can be reduced without sacrificing the information transmission rate, provided the information rate is less than the channel capacity. As discussed in chapter 8, the function of the speech codec is to reduce the average number of information bits per unit time. The transmitted information bits to the user over the channel must be the minimum number possible. The output of the speech codec is the input to the channel encoder. Channel coding is used to improve radio link performance in mobile communication by adding redundant bits to the source information. At the transmitter, a channel codec maps the digital information sequence generated by the speech codec or other data source into a coded version that can be decoded by the receiver with minimal errors.

There are two schemes used in error control—automatic repeat request (ARQ) and forward error correction (FEC). In ARQ schemes, the receiver detects whether the transmitted data block contains errors. If errors are detected, a retransmission request is sent to the sender. In ARQ schemes, error-free condition is achieved, but these schemes suffer with long transmission delays. Therefore, ARQ schemes are effective for data, facsimile, or still-image transmissions but not for voice transmissions. An example of an ARQ scheme is the radio link protocol used by the data services of GSM (see chapter 10).

When it is not possible for the transmitter to receive feedback from the receiver or when the delay is unacceptable, FEC schemes are used. Data is typically segmented into information blocks, with each information block containing unique added redundancy. The resultant output is a data stream of n-bit blocks where each block contains k information bits plus $(n - k)$ redundant bits. The addition of redundant bits is a trade-off between bit rate and the bit error rate.

For speech, the channel coding mechanism adds redundancy to the speech codec by adding extra code bits in a controlled manner so that the receiver can detect and correct errors caused by a noisy channel. The resulting channel codes are classified as **block codes** and **convolutional codes**. Block codes divide the sequence of source digits into sequential blocks of k digits. Each k-digit block is mapped into an n-digit block of output digits where $n > k$. The ratio of the total number of information bits to the total number of bits in the code word k/n is referred to as the **code rate** or **code efficiency.** The difference $(1 - k/n)$ is called **redundancy**. The encoder is said to produce an (n,k) code. Block codes are memoryless codes because each output code word depends on only one source k-bit block and not any preceding blocks of digits. The higher the rate, the more efficient the code is.

The minimum Hamming distance, d_{min} of a code is often used as a criterion for evaluating error detection or error correction capability. The Hamming distance, $d(u,v)$ between two code words u and v is defined as the number of different bits; the minimum Hamming distance is defined as the smallest value of $d(u,v)$. When the minimum Hamming distance is d_{min}, we can detect $(d_{min} - 1)$ bit errors, and we can correct Integer $[(d_{min} - 1) / 2]$ bit errors (where integer $[x]$ means the integer part of x). The additional coded bits lower the raw data transmission rate.

12.10.1 Linear Codes

A linear code is one in which the l-th component of $[V]$ can be expressed as a linear combination of the k information symbols. In a coding scheme where $m(x)$ or $[m]$ is the information code word block length, $g(x)$ or $[G]$ is the linear block code encoder, and $v(x)$ or $[V]$ is the linear block code block length, then we can write

$$v(x) = m(x)g(x) \tag{12.69}$$

$$\left[V \right] = \left[m \right]\left[G \right] \tag{12.70}$$

The generator matrix $[G]$ is given as

$$\left[G \right] = \left[r\ I \right] = \begin{bmatrix} r_1 & 1 & 0 & \ldots & \ldots & \ldots & 0 \\ r_2 & 0 & 1 & \ldots & \ldots & \ldots & 0 \\ \ldots & \ldots & \ldots & \ldots & \ldots & \ldots & \ldots \\ r_k & 0 & 0 & \ldots & \ldots & \ldots & 1 \end{bmatrix} \tag{12.71}$$

where:

$$r_j = \text{remainder of } \left[\frac{x^{n-k+j-1}}{g(x)} \right] \quad j = 1, \ldots\ldots k$$

12.10.2 Cyclic Codes

A cyclic code is one for which an end-around shift of a code word yields another code word. If $(y_1, y_2, y_3, ..., y_n)$ is a code word, it follows that, for a cyclic code, $(y_n, y_1, y_2, ..., y_{n-1})$ is also a code word. It should be noted that $(n - 1)$ code words are generated by cyclic shifts of a single code word. The cyclic code allows easy implementation of the encoder and decoder using shift registers. n-bit cyclic code words satisfy the following equation:

$$x^n + 1 = 0 \tag{12.72}$$

where calculations are based on modulo-2 operation.

A generator polynomial $G(x)$ for encoding is developed such that $G(x)$ is one of the factors of $x^n + 1$ and satisfies the following requirements:

☞ $G(x)$ is irreducible
☞ Root α^l of $G(x)$ should satisfy:

$$\alpha^l \neq 1 \text{ for k < n; and} \tag{12.73}$$

and

$$\alpha^l = 1 \text{ for k = n} \tag{12.74}$$

to guarantee that a k-bit-shifted code ($l < n$) becomes another code word. In general, the relationship between the order of generator polynomial p and the number of bits in a code word n and source bits in the code word k is given as:

$$n = 2^p - 1 \tag{12.75}$$

$$k = n - p \tag{12.76}$$

As an example, we consider $G(x) = x^5 + x^2 + 1$, a fifth-order generator polynomial, the number of bits in the code word $n = 2^5 - 1 = 31$, and the source bits in the code word $k = (31 - 5) = 26$.

12.10.2.1 BCH Code

BCH codes are the most popular codes in the class of cyclic codes. A BCH code is defined as (n, k, q) in which n = code-word length = $2^p - 1, p \geq 3$; k = information length; and q = number of errors that can be corrected, bounded by $q < (2^p - 1)/2$. The code rate $R = k/n$ varies over a wide range. The number of errors that can be corrected increases as the code rate decreases. The following constraints are imposed on a BCH code:

$$n - k \leq pq \tag{12.77}$$

$$d_{min} \geq pq + 1 \tag{12.78}$$

See Table 12.2 for some examples of BCH codes.

Table 12.2 Examples of BCH Codes

(n, k, q)	p	$g(x)$
(7,4,1)	3	$1 + x + x^3$
(15,11,1)	4	$1 + x + x^4$
(15,7,2)	4	$1 + x^4 + x^6 + x^7 + x^8$
(15,3,3)	4	$1 + x + x^2 + x^4 + x^5 + x^8 + x^{10}$

12.10.2.2 Hamming Codes   Hamming codes are linear cyclic codes. Hamming codes with $n = 2^p - 1$ and $k = n - p$ exist for any integer $p \geq 3$. The rate of this code is $R = k/n = (2^p - p - 1)/(2^p - 1)$, which approaches 1.0 as $p \to \infty$. Table 12.3 lists n, k, and R for the first eight Hamming codes.

Hamming codes are defined by a parity-check matrix containing $n - k = p$ rows and n columns. The columns consist of all possible nonzero p-component vectors.

$$\left[H\right] = \left[r^T \ I\right] \tag{12.79}$$

where:
I = unit matrix,
r_i = remainder of $\{x^{n-k+i-1}/[g(x)], i = 1, 2, ..., k\}$, and
$g(x)$ is the generator polynomial.

Table 12.4 gives Hamming codes generator polynomial up to block length $n = 2^{10} - 1$.

The Hamming weight of code word $[V]$ {of $v(x)$} is

Table 12.3 Examples of Hamming Codes

p	n	k	R
3	7	4	0.57
4	15	11	0.73
5	31	26	0.84
6	63	57	0.90
7	127	120	0.94
8	255	247	0.97
9	511	502	0.98
10	1023	1013	0.99

Table 12.4 Hamming Codes Generator Polynomial

p	$g(x)$
3	$1 + x + x^3$
4	$1 + x + x^4$
5	$1 + x^2 + x^5$
6	$1 + x + x^6$
7	$1 + x^3 + x^7$
8	$1 + x^2 + x^3 + x^4 + x^8$
9	$1 + x^4 + x^9$
10	$1 + x^3 + x^{10}$

$$W(V) = \sum_1^n 1_s^1 \text{ in V, and} \qquad (12.80)$$

$$d_{min} = min[W(v \subset V)] \qquad (12.81)$$

for any nonzero V.

If s denotes the number of errors detected and q is the number of errors corrected, then the detection capability of (n, k) codes will be $s = d_{min} - 1$, and error correction capability will be $q = (d_{min} - 1)/2$. Table 12.5 provides a summary of the first eight Hamming codes.

Table 12.5 Summary of Hamming Codes

	Hamming Single-Error Correction Codes	Hamming Double-Error Detection Codes
Code Length (n)	$n = 2^p - 1$	$n = 2^p - 1$
Message Length (k)	$k = 2^p - p - 1$	$k = 2^p - p - 2$
Parity-Check Length $(n - k)$	p	$p + 1$
Message	$m(x)$	$m(x)$
Generator Polynomial	$g(x)$	$h(x) = (1 + x)g(x)$
Code Word	$v(x) = m(x)g(x)$	$v(x) = m(x)h(x)$

12.10.2.3 Reed-Solomon Codes   Reed-Solomon codes are nonbinary BCH codes. They use input and output alphabets with 2^p symbols, $(0, 1, 2, ..., 2^p - 1)$. The block length of Reed-Solomon codes is $n = 2^p - 1$. The encoder for an (n, k, q) Reed-Solomon code produces one of n symbols from a Galois field (GF) (2^p) for each group of p successive uncoded data bits. For each block of k symbols, the Reed-Solomon code generates a block of n symbols. After the symbols are interleaved, each symbol is converted back to its binary equivalent for transmission. Reed-Solomon codes achieve the largest possible code minimum distance and error-correcting capability—q for any linear code with the same n and k values. Reed-Solomon codes are very effective in combating long strings of error (burst error) associated with channels that have memory.

Reed-Solomon codes are defined by generator polynomials, just as binary BCH codes are. For Reed-Solomon codes, the code generator polynomial has coefficients from the nonbinary alphabet $(0, 1, 2, ..., 2^p - 1)$ rather than binary alphabet $(0, 1)$. For Reed-Solomon codes, in performing polynomial multiplication and addition, modulo-2^p arithmetic is used. The properties of the Reed-Solomon codes are

☞ Block length: $n = 2^p - 1$
☞ Parity-check digits: $n - k = 2q$
☞ $d_{min} = 2q + 1$

12.10.2.4 Golay Code   The $(23, 12)$ Golay code has a minimum distance of 7 and corrects all patterns of three or fewer errors. The $(24, 12)$ Golay code has a minimum distance of 8 and, in addition to correcting all patterns of three errors, it also detects all patterns of four errors with a trivial reduction in code rate. The $(24, 12)$ Golay code is popular for many applications.

12.10.3 Convolutional Codes

The convolutional code is a powerful channel coding scheme for wireless mobile communications. In GSM, many types of logical channels are specified. The reason for supporting different dedicated channel types is that specific control channels are always available for signaling and link management information required to maintain the GSM radio link, even when the network is congested with speech. Each logical channel is allocated a convolutional code.

The convolutional encoder consists of shift registers and modulo-2 adders connected to some of the shift registers. The performance of convolutional code depends on two parameters—coding rate (r) and constraint length (K). The input sequence is fed to the K-stage shift registers, and output data is calculated using the contents of K-stage shift registers. The generator polynomials determine the encoding process.

GSM uses a 3-bit error redundancy code to assess the correctness of the bits that are more sensitive to errors in the speech frame. These bits are categorized as class Ia bits and include 50 bits. If one of these bits is in error, this may create a loud noise instead of a 20-ms speech slice. Detection of such errors allows the corrupted speech block to be replaced by something less disturbing (such as an extrapolation of the preceding block). The polynomial representing the detection code for category Ia bit is $g(x) = x^3 + x + 1$. At the receiving end, the same operation is performed and, if the remainder differs, an error is detected and the speech frame is eventually discarded.

The GSM convolutional code adds 4 bits (set of 0s) to the initial 185-bit sequence and then applies two different codes. The generator polynomials are $g_1(x) = 1 + x^3 + x^4$ and $g_2(x) = 1 + x + x^3 + x^4$. This results in 378 bits, or twice the 189-bit sequence (see Figure 12.25).

Convolutional decoding is performed using the Viterbi algorithm. The Viterbi decoder logically explores in parallel every possible user data sequence. It encodes and compares each one against the received sequence and selects the one that has the closest match—it is a maximum-likelihood decoder. To reduce the complexity (the number of possible data sequences doubles with each additional data bit), the decoder recognizes at each point that certain sequences cannot belong to the maximum-likelihood path and it discards them. The encoder memory is limited to K bits; in steady-state operation the Viterbi decoder retains only $2^K - 1$ paths. Its complexity increases exponentially with the constraint length K.

The GSM convolutional coding rate per data flow is 378 bits each 20 ms (i.e., 18.9 kbps). However, before modulating the signal, the 78 unprotected class II bits are added. Thus the GSM bit rate is 456 bits each 20 ms (i.e., 22.8 kbps).

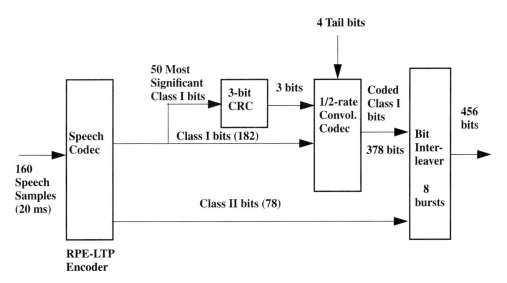

Fig. 12.25 Channel Coding in GSM (Full Rate)

12.11 INTERLEAVING AND DEINTERLEAVING

Because convolutional codes perform poorly on bursts of errors, interleaving is used in the transmitter to randomize the errors so that the convolutional codec can correct them.

The purpose of the interleaving algorithm is to avoid loss of the consecutive information bits. GSM blocks of full-rate speech are interleaved over 8 bursts; i.e., 456 bits of one block are divided into 8 subblocks, each containing 57 bits (see Figure 12.26). Each subblock is carried by a different burst and in a different TDMA frame. Thus a burst contains contributions from two successive speech blocks, i and i + 1. To avoid proximity relations between successive bits, the bits from block i use even positions in the burst and the bits from block i + 1 use odd positions. This ensures sufficient redundancy within the

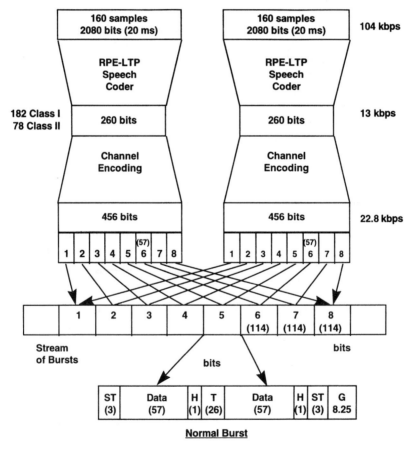

Fig. 12.26 Interleaving in GSM

interleaving process of the GSM signal structures to allow for one frame in five to be lost without a significant loss in voice quality.

Deinterleaving in the receiver is the reverse operation of interleaving. The major drawback of deinterleaving and interleaving is that they introduce delay. This delay amounts to the transmission time from the first burst to the last burst in a block and is equal to 8 TDMA frames (i.e., about 37 ms).

12.12 SUMMARY

In this chapter we studied the modulation and demodulation process applicable to GSM and DECT (and its North American variant PWT). Since baseband signals can be transmitted only over short distances with wires and require very long antennas to transmit them without wires, the baseband signals are modulated onto radio frequency carriers. First we studied ASK to determine its probability of error and methods for demodulation. From ASK we went on to phase modulation and the specific form (π/4-DQPSK) used for PWT. We showed that π/4-DQPSK has the same-shaped error rate curve as ASK but the SNR definition is different. We discussed MSK, a form of FM, as a first step toward GMSK which is used for DECT and GSM. Based on the literature, the error performance for GMSK also has the same-shaped curve as ASK with the proper definition of a correcting factor α. For GSM, where BT = 0.3, $\alpha \approx 0.9$; for DECT, where BT = 0.5, $\alpha \approx 0.97$.

We presented methods and block diagrams for recovery of the carrier frequency and phase, recovery of the bit timing, and recovery of the framing for various modulation methods. While our block diagrams were based on hardware approaches, they could just as easily be implemented in software in many cases. Many manufacturers have proprietary methods for achieving synchronization.

We looked at equalization techniques since multipath fading requires that an equalizer be used to lower the error rates on the channel.

We covered the characteristics of the linear and cyclic codes for error detection and error correction; BCH codes, Hamming codes, Reed-Solomon codes, and Golay codes; convolutional codes used in wireless communications and their applications in GSM channel coding for error detection and error correction. We concluded the chapter by presenting the concept of bit interleaving and deinterleaving to show how they have been used in the GSM to distribute the bit errors.

12.13 PROBLEMS

1. Determine the probability of error for an M-ary ASK system. Use as your signals $s_0(t) = 0$ and $s_n(t) = \dfrac{V_c}{(M-1)} \cos(2\pi f_c t)$.

2. If BPSK signaling is used through a channel that adds white noise with single-sided PSD, $N_0 = 10^{-8}$ W/Hz. Calculate the signal power required to give $P_e = 10^{-6}$ for a data rate of 1 Mbs.

3. Repeat problem 2 for the ASK and FSK modulation scheme.

4. Calculate the channel bandwidth required to transmit data at 1 Mbps for BPSK, 4-PSK, coherent BFSK, and noncoherent BFSK.

5. Find $\dfrac{E_b}{N_o}$ for M-ary PSK with $M = 8$ and $P_e = 10^{-6}$.

6. Design a M-ary PSK modulation scheme to provide $P_e = 10^{-6}$ with a data rate of 9.6 kbps and a channel bandwidth of 4.8 kHz. Assume $N_0 = 10^{-9}$ W/Hz.

7. Show that carrier recovery can be accomplished for $\pi/4$-DQPSK by multiplying the received signal by 8 and then dividing by 8.

12.14 REFERENCES

1. Buda, R. de, "Coherent Demodulation of Frequency Shift Keying with Low Deviation Ratio," *IEEE Transactions on Communications* 20, June 1972, pp. 466–70.

2. Garg, V. K., and Wilkes, J. E., *Wireless and Personal Communications Systems*, Prentice Hall, Englewood Cliffs, NJ, 1996.

3. Murota, K., and Hirade, K., "GMSK Modulation for Digital Mobile Radio Technology," *IEEE Transactions on Communications* 29 (7), July 1981.

4. Pasupathy, S., "Minimal Shift Keying: A Spectrally Efficient Modulation," *IEEE Communications Magazine*, July 1979.

5. Proakis, J. G., *Digital Communication*, McGraw Hill, New York, NY, 1989.

6. Schwartz, M., Bennett, W., and Stein, S., *Communications Systems and Techniques*, McGraw Hill, New York, NY, 1966.

7. Sklar, B., *Digital Communications: Fundamental and Applications*, Prentice Hall, Englewood Cliffs, NJ, 1988.

8. "Special Issue on Bandwidth and Power Efficient Coded Modulation," *IEEE Journal of Selected Areas in Communications* 7, August and December 1989 (2 parts).

9. "Special Issue on Bandwidth and Power Efficient Modulation," *IEEE Communications Magazine* 29, December 1991.

10. Stalling, W., *Data and Computer Communications*, 2d Ed., Macmillan Publishing Co., New York, NY, 1988.

11. Wozencraft, J. M., and Jacobs, I. M., *Principals of Communications Engineering*, John Wiley and Sons, New York, NY, 1965.

12. Ziemer, R. E., and Tranter, W. H., *Principle of Communications*, 3d Ed., Houghton Mifflin, Boston, MA, 1990.

13. Ziemer, R. E., and Peterson, R. L., *Introduction to Digital Communication*, Macmillan Publishing Co., New York, NY, 1992.

CHAPTER 13

Propagation Path Loss and Propagation Models

13.1 INTRODUCTION

In this chapter we discuss propagation and multipath characteristics of a radio wave. We also examine the concepts of delay spread, a cause of channel dispersion, and intersymbol interference. Since the analytical model of the propagation of radio waves in a real-world environment is complicated, we discuss empirical and semi-empirical models that have been developed by several authors for calculating path losses in urban, suburban, and rural environments and compare the results obtained with each model. The chapter also deals with Doppler spread and coherence bandwidth.

13.2 MULTIPATH CHARACTERISTICS OF A RADIO WAVE

Radio waves propagate through space as traveling electromagnetic (EM) waves. The energy of signals exists in the form of electrical (E) and magnetic (H) fields. Both electrical and magnetic fields vary sinusoidally with time. The two fields always exist together because a change in electrical field generates a magnetic field and a change in magnetic field develops an electrical field. Thus there is a continuous flow of energy from one field to the other.

Radio waves arrive at a mobile receiver from different directions with different time delays. They combine via vector addition at the receiver antenna to give a resultant signal with a large or small amplitude depending upon whether the incoming waves combine to strengthen each other or cancel each other. As a result, a receiver at one location may experience a signal strength several tens of dB different from a similar receiver located only a short distance away. As a mobile moves from one location to another, the phase relationship between the various incoming waves also changes. Thus, there are substantial amplitude and phase fluctuations, and the signal is subjected to fading. It should also be noted that, whenever relative motion exists, there is a **Doppler shift** in the received signal. In the mobile radio case, the fading and Doppler shift occur as a result of the motion of the receiver through a spatially varying field. Doppler shift also results from the motion of the scatterers of the radio waves (e.g., cars, trucks, vegetation). The effect of multipath propagation is to produce a received signal with an amplitude that varies quite substantially with location. At UHF and higher frequencies, the motion of the scatterers also causes fading to occur even if the mobile set or handset is not in motion.

The radio wave propagation laws are very similar to those known to us from our daily experience with light sources, whose beams propagate through direct line-of-sight (LOS) paths, reflection, diffraction, and so on. Of course there are differences due to their different wave lengths. The average received signal power at the MS diminishes when it moves away from the BS. This is known as **path loss**. For the sake of discussion, if we assume that the MS orbits around the omni-BS on a hypothetical orbit in the free space, the average received signal strength at the MS is constant, since no multipath propagation takes place due to lack of reflecting and scattering objects. In a real propagation scenario, when the BS transmits a sinusoidal radio frequency carrier and the MS travels in a tangential direction, the MS receives delayed, phase-shifted, and attenuated signals, which add vectorially at the MS's antenna. Signals sometimes add constructively and sometimes destructively. This results in rapid fluctuations in signal amplitude and phase. This is known as **fast fading** (or **Raleigh fading**) and often inflicts variations of up to 40 dB. The other characteristic of signal propagation is referred to as the **slow fading** (or **log-normal fading**). This obeys medium-term path loss variations due to terrain or obstruction due to large vehicles and the like, and it introduces shadowing effects.

Figure 13.1 illustrates the fading characteristics of a mobile radio signal. The rapid fluctuations caused by the local multipath are fast fading and the long-term variation in the mean level is slow fading. The slow fading is caused by movement over distances large enough to produce gross variations in the overall path between the BS and the MS. Fast fading is usually observed over distances of about half a wavelength. For VHF and UHF, a vehicle traveling at 30 miles per hour (mph) can pass through several fast fades in a second. Therefore, the mobile radio signal, as shown in Figure 13.1, consists of a short-term fast-fading signal superimposed on a local mean value (which

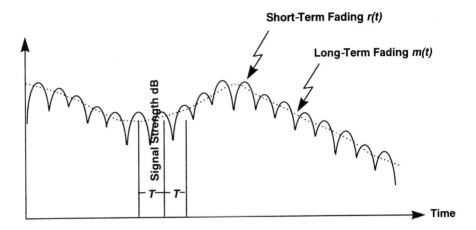

Fig. 13.1 Mobile Radio Signal Fading Representation

remains constant over a small area but varies slowly as the receiver moves). As noted above, even a stationary handset will see fading, though the signal fading rate is lower.

The received signal $s(t)$ can be expressed as the product of two parts: the signal subject to long-term fading $m(t)$ and the signal subject to short-term fading $r(t)$ as:

$$s(t) = m(t)r(t) \tag{13.1}$$

We refer to Figure 13.2 and assume that the z-axis is perpendicular to the surface of the earth and that the x-y plane lies on the surface of the earth. At every receiving point, we assume the existence of N plane waves of equal amplitude. The path-angle geometry for the i-th scattered plane wave is shown in Figure 13.2. If the transmitted signal is vertically polarized (i.e., the electrical field vector is aligned along the z-axis), the field components at the mobile are the electrical field E_z, the magnetic field component H_x, and the magnetic field component H_y. Using Clarke's model [1] these components at the receiving point, O, are expressed in the complex equivalent baseband form as:

$$E_z = E_0 \sum_{i=1}^{N} e^{j\phi_i} \tag{13.2}$$

$$H_x = -\frac{E_0}{\eta} \sum_{i=1}^{N} \sin\alpha_i e^{j\phi_i} \tag{13.3}$$

$$H_y = \frac{E_0}{\eta} \sum_{i=1}^{N} \cos\alpha_i e^{j\phi_i} \tag{13.4}$$

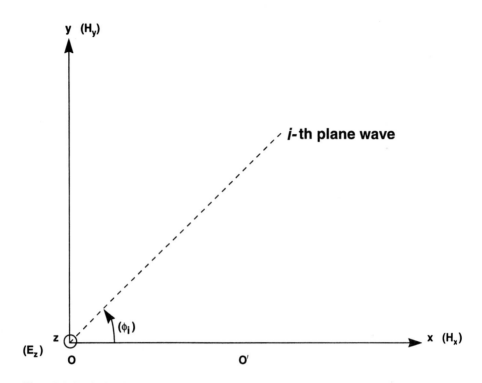

Fig. 13.2 Path Angle Geometry

where:
ϕ_i = phase angle relative to the carrier phase,
E_o = amplitude of the N plane wave, and
η = intrinsic wave impedance which is given as:

$$\eta = \sqrt{\frac{\mu_o}{\varepsilon_o}}$$
(13.5)

in which μ_o = free-space magnetic permeability($4\pi \times 10^{-7}$) H/m and
ε_o = free-space electric permittivity (8.854×10^{-12}) F/m.

The speed of light, c, is given as:

$$c = \frac{1}{\sqrt{\mu_o \varepsilon_o}} = 3 \times 10^8 \ \text{m/s}$$
(13.6)

For a vehicle moving in the x direction with a constant velocity, v, the received carrier is Doppler shifted by

$$f_d = f_m \cos\alpha = \frac{v \cos\alpha}{\lambda} = \frac{v_{eff}}{\lambda} = \frac{v_{eff} f_c}{c}$$
(13.7)

where:

$f_m = \dfrac{v}{\lambda}$ = maximum value of Doppler frequency, f_d at $\alpha = 0$,

v_{eff} is the effective velocity of the vehicle, and
f_c is the carrier frequency.

The Doppler frequency f_m is directly related to the phase change $\Delta\phi$ caused by the change in the path length. The Doppler shift is bounded to $\pm f_m$, which in general is much smaller than the carrier frequency f_c. The waves arriving from ahead of the vehicle experience a positive Doppler shift, whereas those coming from behind the vehicle have a negative Doppler shift. Thus, each component of the received signal is shifted by different values of Doppler frequency. For example, a vehicle traveling at 50 mph and receiving signals at a carrier frequency of 880 MHz will introduce a maximum Doppler

shift of $\dfrac{v}{\lambda} = \dfrac{50 \times 0.447 \times 888 \times 10^6}{3 \times 10^8} = 65.56$ Hz.

13.2.1 Short-Term Fading

By applying the central limit theorem and observing that α_i and ϕ_i are independent, it follows that $E_z, H_x,$ and H_y are complex Gaussian random variables for large N.

We consider the RF version of Eq. (13.2) for the field intensity E_z:

$$E_z = E_0 \sum_{i=1}^{N} e^{j(\omega_c t + \phi_i)} \tag{13.8}$$

The real part of E_z is given as:

$$Re[E_z] = E_0 \sum_{i=1}^{N} \cos\omega_c t \, \cos\phi_i - E_0 \sum_{i=1}^{N} \sin\omega_c t \, \sin\phi_i \tag{13.9}$$

Let $A_c = E_0 \sum_{i=1}^{N} \cos\phi_i$ and $A_s = E_0 \sum_{i=1}^{N} \sin\phi_i$, then Eq. (13.9) can be written as:

$$Re[E_z] = A_c\cos\omega_c t - A_s\sin\omega_c t \tag{13.10}$$

Since ϕ_i is uniformly distributed between 0 to 2π, the mean values of A_c and A_s are zero. The mean square values of A_c and A_s are $E(A_c^2) = E(A_s^2)$

$= \dfrac{E_0^2 N}{2} = P_0$, the mean received power at the MS.

Since A_c and A_s are uncorrelated and therefore independent, we can write $E[A_c A_s] = 0$. Thus, the density of A_c and A_s follows a normal distribution, and the envelope of A_c and A_s is given by:

$$r = (A_c^2 + A_s^2)^{1/2} \qquad (13.11)$$

and the phase, θ, is given as:

$$\theta = \operatorname{atan}\frac{A_s}{A_c} \qquad (13.12)$$

The square root of the sum of the square of two Gaussian functions is a Rayleigh distribution. Therefore, the probability density function for short-term or multipath fading is given by the Rayleigh distribution (refer to Figure 13.3):

$$p(r) = \frac{r}{P_0}e^{(-r^2)/(2P_o)} \qquad (13.13)$$

where:
$2P_0 = 2\sigma^2$ is the mean square power of the component subject to short-term fading, and r^2 is the instantaneous power.

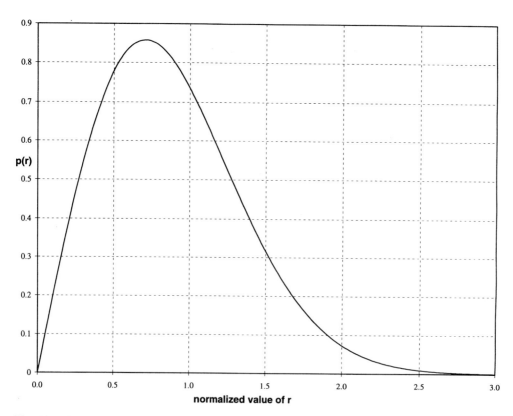

Fig. 13.3 Rayleigh Distribution—Short-Term Fading

The corresponding cumulative distribution function is

$$prob(r \le R) = P(R) = \int_o^R \frac{r}{P_0} e^{(-r^2)/(2P_0)} dr \qquad (13.14a)$$

$$P(R) = -e^{(-r^2)/(2P_0)} \Big|_0^R \qquad (13.14b)$$

$$P(R) = 1 - e^{-R^2/2P_0} \qquad (13.14c)$$

$$r_{mean} = E[r] = \int_0^\infty r p(r) dr = 1.2533\sqrt{P_0} = 1.2533\sigma \qquad (13.15)$$

$$\text{Mean Square: } = E[r^2] = \int_0^\infty r^2 p(r) dr = 2P_0 = 2\sigma^2 \qquad (13.16)$$

$$\text{Variance: } \sigma_r^2 = E[r^2] - (E[r])^2 = 0.4292 P_0 = 0.4292\sigma^2 \qquad (13.17)$$

The median value r_M, is defined as that for which $P(r_M) = 0.5$;

$$\therefore 1 - e^{(-r_M^2)/(2P_0)} = 0.5 \qquad (13.18a)$$

and

$$r_M = 1.1774\sqrt{P_0} = 1.1774\sigma \qquad (13.18b)$$

It is often convenient to write Eq. (13.13) and Eq. (13.14c) in terms of mean, mean-square value, or median rather than in terms of P_o.

Let $E[r] = \bar{r}$ and $E[r^2] = \overline{r^2}$.

In terms of the mean-square value

$$p(r) = \frac{2r}{\overline{r^2}} e^{\left[-\frac{r^2}{\overline{r^2}} \right]} \qquad (13.19a)$$

$$P(R) = 1 - e^{\left[-\frac{R^2}{\overline{r^2}} \right]} \qquad (13.19b)$$

In terms of the mean

$$p(r) = \frac{\pi r}{2\bar{r}^2} e^{\left[-\frac{\pi r^2}{(2\bar{r})^2} \right]} \qquad (13.20a)$$

$$P(R) = 1 - e^{\left[-\frac{\pi R^2}{(2\bar{r})^2} \right]} \qquad (13.20b)$$

In terms of the median

$$p(r) = \frac{2r\ln 2}{r_M^2} e^{\left(-\frac{r^2\ln 2}{2r_M^2}\right)} \tag{13.21a}$$

$$P(R) = 1 - 2^{-\left(\frac{R}{r_M}\right)^2} \tag{13.21b}$$

The Rayleigh probability density function describes the first-order statistics of the signal envelope over distances short enough for the mean level to be regarded as constant. First-order statistics are those for which distance is not a factor. The Rayleigh distribution gives information such as the overall percentage of locations (or time) for which the envelope lies below a specified value.

System engineers are interested in a quantitative description of the rate at which fades of any depth occur and the average duration of a fade below any given depth. This provides a valuable aid in selecting transmission bit rates, word lengths, and coding schemes in digital radio systems and allows an assessment of system performance. The required information is provided in terms of the number of fades per second and average fade duration below a specified level.

13.2.2 Level-Crossing Rate

The level-crossing rate, $N(R)$, at a specified signal level R is defined as the average number of times per second that the signal envelope crosses the level in a positive-going direction ($\dot{r} > 0$).

$$N(R) = \int_0^\infty \dot{r} p(R, \dot{r}) d\dot{r} \tag{13.22}$$

where:

$p(R, \dot{r})$ is the joint Probability Density Function (PDF) of R and $\dot{r}$, and a dot indicates the time derivative.

Using derivations given in references [6] and [9], the average level-crossing rate at a level R can be shown to be:

$$N(R) = \sqrt{\frac{\pi}{\sigma^2}} R f_m e^{\left(\frac{-R^2}{2\sigma^2}\right)} \tag{13.23}$$

Since $2\sigma^2$ = mean square value, therefore $\sqrt{2}\sigma$ is the root mean square (rms) value. The level-crossing rate for a vertical monopole antenna can then be given as:

$$N(R) = \sqrt{2\pi} f_m \rho e^{-\rho^2} = n_0 n_R \tag{13.24}$$

where:

$$\rho = \frac{R}{\sqrt{2}\sigma} = \frac{R}{R_{rms}} = \text{the ratio between the specified level and the rms amplitude of the fading envelope,}$$

$$f_m = \frac{v}{\lambda},$$

$$n_0 = \sqrt{2\pi}f_m,$$

$$n_R = \rho e^{-\rho^2},$$

n_R is the normalized level-crossing rate that is independent of wavelength and vehicle speed,

v = speed of vehicle, and
λ = carrier wavelength.

Figure 13.4 gives the n_R at a level R.

An approximate expression for $N(R)$ below a specified level, $\rho = \dfrac{R}{R_{rms}}$, can be given as:

$$N(R) \approx \sqrt{2\pi}\frac{v}{\lambda}\rho \qquad (13.25)$$

The above equations neglect the effect of the motion of the scatterers. When their motion is taken into account, the effect is to increase the fade rate.

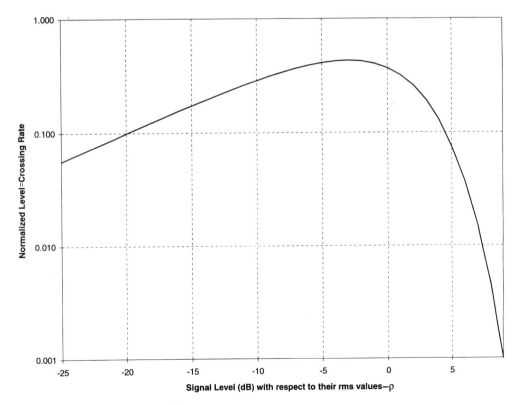

Fig. 13.4 n_R vs. Signal Level (ρ)

13.2.3 Average Fade Duration

The average fade duration is the average of τ_1, τ_2, ..., τ_n. The average duration of fades below the specified level R can be found from

$$E[\tau_R] = \tau(R) = \frac{prob[r \leq R]}{N(R)} \tag{13.26}$$

$$\tau(R) = \frac{e^{\rho^2} - 1}{\sqrt{2\pi}f_m\rho} = \frac{e^{\rho^2} - 1}{n_0\rho} \tag{13.27}$$

An approximate expression for $\tau(R)$ can be given as:

$$\tau(R) \approx \frac{\lambda}{v}\frac{\rho}{\sqrt{2\pi}} \tag{13.28}$$

E X A M P L E   1 3 – 1

Problem Statement

Calculate the level-crossing rate at a level of -10 dB and the average duration of fade for a cellular system at 900 MHz and a vehicle speed of 24 kilometers per hour (km/h). Assume the free-space speed of propagation for electromagnetic waves $= 3 \times 10^8$ m/s. Neglect the effects of the motion of the scatters. Compare the results obtained using the approximate expressions.

Solution

At 900 MHz, $\lambda = \dfrac{3 \times 10^8}{900 \times 10^6} = \dfrac{1}{3}$ m, $v = 6.67$ m/s, $f_m = \dfrac{6.67}{\frac{1}{3}} = 20$ Hz,

$$n_0 = \sqrt{2\pi}f_m = 50$$

From Figure 13.3, $n_R = 0.32$ at -10 dB.

$$N(R) = 0.32 \times 50 = 16.0 \text{ fades/s}$$

$$\rho e^{\rho^2} = n_R = 0.32$$

$$\rho = 0.316$$

$$\tau(R) = \frac{(1.105 - 1)}{50 \times 0.316} \approx 0.0066 \text{ s} = 6.6 \text{ ms}$$

Using the approximate expressions we get:

$$\text{Fading level} = \rho = -10 \text{ dB}$$

$$20 \log \rho = -10$$

$$\rho = 10^{-10/20} = 0.3162$$

$$N(R) \approx \sqrt{2\pi} \times \frac{6.67}{\frac{1}{3}} \times 0.3162 = 15.85 \text{ fades/s}$$

$$\tau(R) \approx \frac{1}{3 \times 6.67} \frac{0.3162}{\sqrt{2\pi}} = 0.0063 = 6.3 \text{ ms}$$

These results are quite close to those obtained using the exact expressions.

13.2.4 Rician Fading

When there is a dominant stationary (nonfading) signal component present, such as an LOS propagation path, the small-scale fading envelope distribution is Rician and is given as:

$$p(r) = \frac{r}{\sigma^2} \cdot e^{-\left[\frac{r^2 + A^2}{2\sigma^2}\right]} \cdot I_0[(Ar)/\sigma^2] \text{ for } A \geq 0, r \geq 0 \qquad (13.29)$$

where:
A = peak amplitude of the dominant signal and $I_0(...)$ = modified Bessel
Function of the first kind and zero order.

The Rician distribution is often described in terms of a parameter K, known as the Rician factor, and is expressed as:

$$K = 10\log\frac{A^2}{2\sigma^2} \text{ dB} \qquad (13.30)$$

As $A \rightarrow 0, K \rightarrow -\infty$ dB, and as the dominant path decreases in amplitude, the Rician distribution degenerates to a Rayleigh distribution.

13.2.5 Long-Term Fading

The probability density function for long-term fading is given by the lognormal distribution (refer to Figure 13.5):

$$p(m) = \frac{1}{m\sigma_m\sqrt{2\pi}}e^{[-(\log m - \bar{m})^2/(2\sigma_m^2)]}, \quad m > 0 \qquad (13.31)$$

where:
$\bar{m}$ is the mean of $\log m$, and
σ_m is the standard deviation.

Using $z = (\log m - \bar{m})/\sigma_m$, the cumulative distribution function is given as:

$$prob(z \leq Z) = P(z \leq Z) = \frac{1}{2} + \frac{1}{2}erf\left(\frac{Z}{\sqrt{2}}\right) \approx 1 - \frac{1}{\sqrt{2\pi}Z}e^{(-Z^2/2)} \qquad (13.32)$$

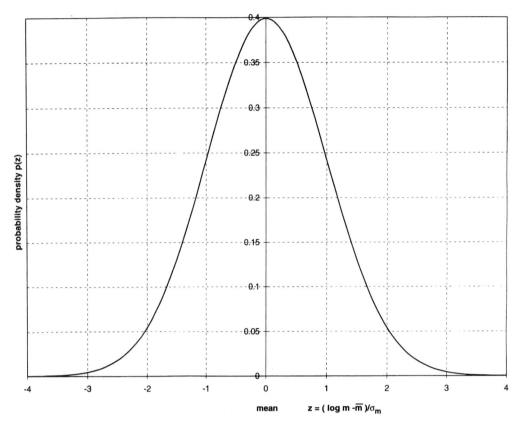

Fig. 13.5 Log-Normal Distribution–Long-Term Fading

13.3 FREE-SPACE PATH LOSS

We consider the system shown in Figure 13.6, where a cell-site transmitter is transmitting at an average power level of P_T. We want to find the received power level, P_R, at the receiving antenna (MS) located at a distance, d, from the transmitter.

For an isotropic antenna, in free space:

$$p_R = \frac{P_T}{4\pi d^2} \tag{13.33}$$

where:
P_T = average power level of transmitter,
d = distance between transmitter and receiver, and
p_R = power density at the receiver.

For an antenna radiating uniformly in all directions (spherical pattern), the power density, p_R, at the receiver is given by Eq. (13.33).

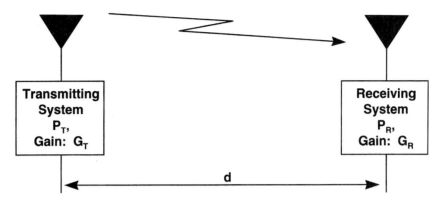

Fig. 13.6 A Simple Model for Path Loss in Free Space

When a directional transmitting antenna with a power gain factor, G_T, is used, the power density at the receiver site is G_T times Eq. (13.33).

The amount of power captured by the receiver is p_R times the aperture area, A_R, of the receiving antenna. The aperture area is related to the gain of the receiving antenna by (see references [9] and [10] for details):

$$G_R = \frac{4\pi A_R}{\lambda^2} \qquad (13.34)$$

where:

$\lambda = \dfrac{c}{f}$,

f = the transmission frequency in Hz,
$c = 3 \times 10^8$ m/s is the free-space speed of propagation for electromagnetic waves, and
A_R is the effective area, which is less than the physical area by an efficiency factor p_R.

Typical values for p_R range from 60 percent to 80 percent. The total received power, P_R, is:

$$P_R = A_R p_R \qquad (13.35)$$

Substituting the value of p_R and A_R from Eq. (13.33) and (13.34) into Eq. (13.35) together with the transmitting antenna gain G_T, we get

$$P_R = \left[\frac{\lambda}{4\pi d} \right]^2 P_T G_T G_R \qquad (13.36)$$

Equation (13.36) includes only the power loss from the spreading of the transmitted wave. If other losses, such as atmospheric absorption or ohmic losses of the waveguides leading to the antennas, are also present, Eq. (13.36) is modified as:

$$\frac{P_R}{P_T} = \left[\frac{\lambda}{4\pi d}\right]^2 \frac{G_T G_R}{L_0} \tag{13.37a}$$

$$\frac{P_R}{P_T} = \frac{G_T G_R}{L_p L_0} \tag{13.37b}$$

where:

$L_p = \left[\frac{4\pi d}{\lambda}\right]^2$ denotes the loss associated with propagation of electromagnetic waves from the transmitter to the receiver.

L_p depends on the carrier frequency and separation distance, d. This loss is always present. L_0 = loss factor for additional losses. When we express Eq. (13.37a) in terms of decibels, we get

$$P_R = 20\log\left[\frac{\lambda}{4\pi d}\right] + P_T + G_T + G_R - L_0 \tag{13.38}$$

The product $P_T G_T$ is called the Equivalent Isotropic Radiated Power (EIRP) and term $20\log\left[\frac{\lambda}{4\pi d}\right]$ is referred to as **free-space loss** (L_p) in dB.

13.4 RECEIVER NOISE

The movement of electrons through a device is not constant but undergoes random fluctuations called **thermal noise**. The fluctuations are proportional to the noise temperature of the device, the bandwidth being considered, and a constant. Passive devices (e.g., resistors) have noise temperatures equal to their physical temperature. Active devices (transistors, integrated circuits, vacuum tubes, and such) have noise temperatures that are higher than their physical temperature. The effects of thermal noise can be expressed as the average noise power generated internally to the receiver and is

$$P_{N,int} = kT_e B_w \tag{13.39}$$

where:
k is Boltzmann's constant and is $1.38 \times 10^{-23} \, J/°K$ (J = Joules of energy, K = temperature in degrees Kelvin),
B_w = bandwidth, and
T_e = effective noise temperature of the device and is given as

$$T_e = T_0(N_f - 1) \tag{13.40}$$

where:
$T_0 = 290°K$, and
N_f is the noise figure of the receiver.

The noise figure is expressed either in degrees Kelvin or in decibels, with 3 dB corresponding to $290°K$.

The noise figure is a property of the receiver. In a properly designed receiver, the noise figure is dominated by the initial stage of the receiver, but, in general, noise from all stages must be taken into account. Lower noise figures are obtained by careful design and selection of the first stages of the receiver. With modern semiconductors, noise figures of less than 1 dB per device are common.

The noise figure is a measure of the excess noise generated by a device. There is still a base level of noise generated by the physical temperature of the receiver. When extremely good noise performance is needed, lower noise levels are obtained by cooling the equipment to temperatures near absolute zero.

Even if the receiver were perfect and had a 0-dB noise figure, additional noise is caused by the antenna and the feed line from the antenna to the receiver. To include the effect of noise seen by the antenna, as well as the feed line loss between the antenna and the receiver, we add the temperature of the antenna, T_{ant}, and the equivalent temperature of the feed line loss, T_{fl}, to T_e to get:

$$P_N = k(T_{ant} + T_e + T_{fl})B_w = kTB_w \qquad (13.41)$$

where:
$T = T_{ant} + T_e + T_{fl}$

The antenna temperature is not a physical temperature; it represents an additional source of noise that depends on where the antenna is pointed and the frequency band of the received signal. For a receiving antenna on the ground, pointed at the sky between 10° and 90° with respect to horizontal, and a signal frequency between 1 and 20 GHz, $T_{ant} \approx 3°K$. This represents the noise of the "big bang" at the creation of the universe. If the antenna is pointed at the sun or the Milky Way, then the noise of the antenna increases. Moderate rainfall (≤ 10 mm/h) changes this value little, but severe rainstorms may increase it by 10° to 50° K because of scattering of the sky background noise into the antenna by raindrops.

When the antenna is pointed at the horizon or when a nondirectional antenna (such as on a car or handset) is used, the equivalent temperature of the antenna is the temperature of the ground–290°K.

By dividing Eq. (13.36) by Eq. (13.41), we obtain the received SNR at a receiver as:

$$SNR = \frac{P_R}{P_N} = \left[\frac{\lambda}{4\pi d}\right]^2 \frac{P_T G_T G_R}{L_0 kTB_w} \qquad (13.42)$$

E X A M P L E   1 3 – 2

Problem Statement
We consider a BS transmitting to an MS. The following parameters relate to this communication system:

- Distance between BS and MS: 8,000 m
- BS EIRP ($G_T = 8$ dB; $P_T = 1$ W): 8 dBW
- Transmitter frequency: 1.5 GHz($\lambda = 0.2$ m)
- MS receiver antenna gain: 0 dB
- Total system losses: 8 dB
- MS receiver noise figure, $N_f = 7$ dB (5)
- MS receiver antenna temperature = 290°K
- MS receiver bandwidth, $B_w = 0.2$ MHz
- Neglect any feed line loss between the antenna and the receiver

Calculate the received signal power at the MS receiver antenna and SNR of the received signal.

Solution

$$L_p : \text{Free Space Loss} = -20\log\left[\frac{0.2}{4\pi \times 8000}\right] = 114 \text{ dB}$$

$$P_R = -114 + 8 + 0 - 8 = -114 \text{ dBW} = -84 \text{ dBm}$$

$$T_e = 290(5-1) = 1160°\text{K}$$

$$P_N = 1.38 \times 10^{-23}(1160 + 290)(0.2 \times 10^6) = 4.0 \times 10^{-15} \text{ W} = -144 \text{ dBW}$$

$$SNR = \frac{P_R}{P_N} = -114 - (-144) = 30 \text{ dB}$$

13.5 THE PATH LOSS OVER A REFLECTING SURFACE

In outer space, the path between two antennas has no obstructions and no objects where reflections can occur. Thus the received signal is composed of only one component. When the two antennas are on the earth, however, there are multiple paths from the transmitter to the receiver. The effect of the multiple paths is to change the path loss between two points. The simplest case occurs when the antenna heights h_T and h_R are small compared with their separation, d, and the reflecting earth surface can be assumed to be flat. The received signal can then be represented by a scattered field, E_s, that can be approximated by combination of a direct wave and a reflected wave (refer to Figure 13.7).

$$E_s = [1 + a_r e^{j\Delta\theta}]E , \tag{13.43a}$$

$$E_s = [1 + a_r(\cos\Delta\theta + j\sin\Delta\theta)]E , \tag{13.43b}$$

where:
a_r = coefficient of reflection, and
$\Delta\theta$ = phase difference between the direct and reflected path.

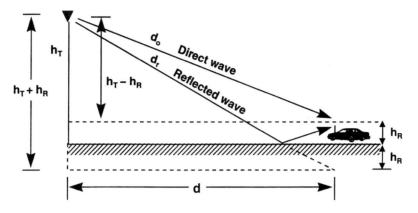

Fig. 13.7 A Simple Model for Path Loss with Reflection

$$\Delta\theta = \frac{2\pi}{\lambda}(d_r - d_o) = \left[\frac{2\pi}{\lambda}\right]\Delta d \tag{13.44}$$

where:
Δd = the difference between direct path (d_o) and reflected path (d_r)

In the mobile radio environment, $a_r = -1$ and $\Delta\theta$ is much less than one radian. Therefore, Eq. (13.43b) can be written as:

$$E_s \approx -Ej\Delta\theta \tag{13.45}$$

Since the received power level P_R is proportional to the square of the field strength, the power level at the antenna output located at a distance, d, from the transmitter, including the path loss, will be:

$$P_R = \left[\frac{\lambda}{4\pi d}\right]^2 \frac{P_T G_T G_R}{L_0}|j\Delta\theta|^2 \tag{13.46}$$

From Figure 13.7:

$$d_0 = \sqrt{(h_T - h_R)^2 + d^2}$$

and

$$d_r = \sqrt{(h_T + h_R)^2 + d^2}$$

$$\Delta d = d\left[1 + \left(\frac{1}{2}\right)\left(\frac{h_T + h_R}{d}\right)^2 + \dots\right] - d\left[1 + \left(\frac{1}{2}\right)\left(\frac{h_T - h_R}{d}\right)^2 + \dots\right]$$

When h_T and h_R << d, then

$$\Delta d \approx \frac{2h_T h_R}{d} \tag{13.47}$$

$$\therefore \Delta\theta \approx \left[\frac{4\pi}{\lambda d}\right] h_T h_R \tag{13.48}$$

Substituting for $\Delta\theta$ from Eq. (13.48) into Eq. (13.46) we get

$$P_R = \left[\frac{h_T h_R}{d^2}\right]^2 \frac{P_T G_T G_R}{L_0} \tag{13.49}$$

Expressing Eq. (13.49) in decibels, we get

$$P_R = 20\log\left[\frac{h_T h_R}{d^2}\right] + P_T + G_T + G_R - L_0 \tag{13.50}$$

It should be observed that Eq. (13.49) is independent of transmitting frequency.

E X A M P L E   1 3 – 3

Problem Statement

Using the data in Example 13-2 and with antenna height at the BS and MS to be 30 m, and 3 m, respectively, calculate the received signal power at the MS receiver antenna and the SNR of the received signal.

Solution

$$\text{Path Loss} = -20\log\left[\frac{30 \times 3}{8000^2}\right] = 117 \text{ dB}$$

$$P_R = -117 + 8 + 0 - 8 = -117 \text{ dBW} = -87 \text{ dBm}$$

$$P_N = -144 \text{ dBW (from Example 13-2)}$$

$$\therefore SNR = \frac{P_R}{P_N} = -117 - (-144) = 27 \text{ dB}$$

13.6 FADE MARGIN

As we discussed earlier, the local mean signal strength in a given area at a fixed radius R from a particular BS antenna is log-normally distributed. The local mean (i.e., the averaged signal strength) in dB is expressed by a normal random variable m with a mean $\bar{m}$ (measured in dBm) and standard deviation σ_m (dB). If S_{min} is the receiver sensitivity, we determine the fraction of the locations (at $d = R$) wherein an MS would experience a received signal above the receiver sensitivity. The receiver sensitivity is the value that provides an acceptable signal under Rayleigh fading conditions. The probability distribution function for log-normally distributed random variable is:

$$p(m) = \frac{1}{\sigma_m \sqrt{2\pi}} e^{-[(m - \overline{m})^2 / (2\sigma_m^2)]} \qquad (13.51)$$

$$P_{S_{min}}(R) = P[m \geq S_{min}] = \int_{S_{min}}^{\infty} p(m)dm = \frac{1}{2} - \frac{1}{2}erf\left(\frac{S_{min} - \overline{m}}{\sigma_m \sqrt{2}}\right) \qquad (13.52)$$

E X A M P L E   1 3 – 4

Problem Statement

If the mean received signal and receiver sensitivity are −100 dBm and −110 dBm, respectively, and the standard deviation is 10 dB, calculate the probability that the signal exceeds the receiver sensitivity.

Solution

$$P_{S_{min}}(R) = \frac{1}{2} - \frac{1}{2}erf\left(\frac{-110 + 100}{10\sqrt{2}}\right) = 0.5 + 0.5erf(0.707) = 0.84 = 84\%$$

Next, we determine the fraction of the coverage within an area in which the received signal strength from a radiating BS antenna exceeds S_{min}. We define the fraction of the useful service area F_u as that area, within an area for which the signal strength received by a mobile antenna exceeds S_{min}. If $P_{S_{min}}$ is the probability that the received signal exceeds S_{min} in an incremental area dA, then

$$F_u = \frac{1}{\pi R^2} \int P_{S_{min}} dA \qquad (13.53)$$

Using the power law for signal attenuation, we express mean signal strength $\overline{m}$ as

$$\overline{m} = \alpha - 10\gamma \log\frac{d}{R} \qquad (13.54)$$

where:
α accounts for the transmitter Equivalent Radiated Power (ERP), receiver antenna gain, feed line losses, and so on, and
γ = propagation path-loss exponent $2 \leq \gamma \leq 5$

Substituting Eq.(13.54) into (13.52), we get

$$P_{S_{min}} = \frac{1}{2} - \frac{1}{2}erf\left[\frac{S_{min} - \alpha + 10\gamma \log(d/R)}{\sigma_m \sqrt{2}}\right] \qquad (13.55)$$

Let $a = (S_{min} - \alpha)/(\sigma_m \sqrt{2})$ and $b = (10\gamma \log d/R)/(\sigma_m \sqrt{2})$

Substituting Eq.(13.55) into (13.53), we get

$$\therefore F_u = \frac{1}{2} - \frac{1}{R^2}\int_0^R x\{erf[a + b\log(x/R)]\}dx \qquad (13.56)$$

Let $t = a + b\log(x/R)$; then

$$\therefore F_u = \frac{1}{2} - \frac{2}{b}e^{(-2a)/b}\int_{-\infty}^a e^{(2t)/b}erf(t)dt \qquad (13.57)$$

or

$$F_u = \frac{1}{2}\left[1 - erf(a) + e^{(1-2ab)/b^2}\left(1 - erf\left[\frac{1-ab}{b}\right]\right)\right] \qquad (13.58)$$

If we choose α such that $\bar{m} = S_{min}$ at $d = R$, then a = 0 and

$$F_u = \frac{1}{2}\left\{1 + e^{1/b^2}[1 - erf(1/b)]\right\} \qquad (13.59)$$

Figure 13.8 shows the relation in terms of the parameter σ_m/γ.

13.7 PROPAGATION MODELS

Propagation models are used to determine how many cell sites are required to provide the coverage requirements for the network. Initial network design typically is engineered for coverage. Later on network growth is based on capacity. Some systems may need to start with wide area coverage and high capacity and, therefore, may start at a later stage of growth.

The coverage requirement is coupled with the traffic-loading require-ments, which rely on the propagation model chosen to determine the traffic distribution and the off-loading from an existing cell site to new cell sites as part of a capacity relief program. The propagation model helps to determine where the cell sites should be located to achieve an optimal position in the net-work. If the propagation model used is not effective in placing cell sites cor-rectly, the probability of incorrectly deploying a cell site into the network is high.

The performance of the network is affected by the propagation model chosen because it is used for interference predictions. As an example, if the propagation model is inaccurate by 6 dB, then S/I ratio could be 23 dB or 11 dB (assuming that S/I = 17 dB is the design requirement). Based on traffic-

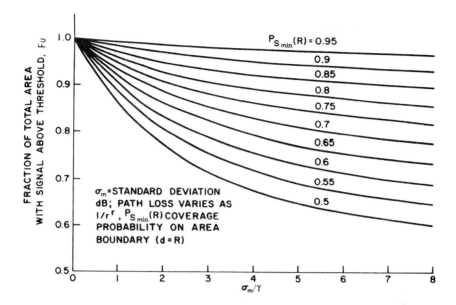

Fig. 13.8 Fraction of Total Area with Average Power above Threshold
(After: Jakes, W. C., Jr. [editor], *Microwave Mobile Communications*, John Wiley & Sons, New York, NY, 1974, p. 127.) ©1974, Lucent Technologies. Used with permission.

loading conditions, designing for a high S/I level could negatively affect financial feasibility. On the other hand, designing for a low S/I ratio would degrade the quality of service.

The propagation model is also used in other system performance aspects including handover optimization, power level adjustments, and antenna placements. Although no propagation model can account for all variations experienced in the real world, it is essential that one should use one or several models for determining the path losses in the network. Each of the propagation models being used in the industry has pros and cons. It is through a better understanding of the limitations of each of the models that a good RF engineering design can be achieved in a network. Also, calibrating the model to the actual propagation environment will be helpful in gaining confidence in the model.

13.7.1 Modeling for the Outside Environment

13.7.1.1 Analytical Model   The propagation loss between the BS and the MS in the outside environment has been extensively studied. The propagation loss is generally expressed by the following expression [6,9]:

$$P(R) = N(R_0, \sigma) + 10\gamma\log\frac{R}{R_0} + x_\sigma \qquad (13.60)$$

where:
$P(R)$ = loss at distance R relative to the loss at a reference distance R_0,
γ = path loss exponent,
σ = standard deviation, typically 8 dB, and
$N(R_0, \sigma)$ = path loss at reference distance R_0.
x_σ = log normal fading component

The second term on the right side of Eq. (13.60) represents a constant attenuation in the outside environment between the BS and the MS. Typically, γ approximately equals 4, although it may range between 2 (which equals the loss in free space) and 5. If γ is equal to 4, then the signal will be attenuated 40 dB if the distance increases 10 times with respect to the reference distance. The first term in Eq. (13.60) represents the variation in the loss above or below the average path loss. x_σ is a log-normal distribution with an average equal to zero and a standard deviation of approximately 8 dB. It has been found that this value is applicable for a wide range of radio environments, including urban and rural areas. An urban environment corresponds to a large metropolitan area with a large number of high-rise commercial buildings and other structures; a suburban environment consists of mostly residential buildings and few commercial buildings; an open environment is the rural area with few residential buildings.

13.7.1.2 Empirical Models   Several empirical models have been suggested and used to predict propagation path losses. We will discuss two widely used models–the Hata-Okumura model and the Walfisch-Ikegami Model–and will also present the models suggested for use in IMT-2000 specifications.

The Hata-Okumura Model [5]   Most of the propagation tools use a variation of Hata's model. Hata's model is an empirical relation derived from the technical report made by Okumura [8] so that the results could be used in computational tools. Okumura's report consists of a series of charts that have been used in radio communication modeling. The following are the expressions used in Hata's model to determine the mean loss L_{50}. Hata's model is applicable to urban, suburban, and open environment.

Urban Area

$$L_{50} = 69.55 + 26.16\log f_c - 13.82\log h_b - a(h_m) + (44.9 - 6.55\log h_b)\log R \;\; \text{dB} \quad (13.61)$$

where:
f_c = frequency (MHz),
L_{50} = mean path loss (dB),
h_b = BS antenna height (m),
$a(h_m)$ = correction factor for mobile antenna height (dB),
R = distance from BS (km).

The range of the parameters for which the Hata model is valid is:

$150 \le f_c \le 1{,}500$ MHz

$30 \le h_b \le 200$ m

$$1 \le h_m \le 10 \text{ m}$$

$$1 \le R \le 20 \text{ m}$$

$a(h_m)$ is computed as:

For a small or medium-sized city:

$$a(h_m) = (1.1\log f_c - 0.7)h_m - (1.56\log f_c - 0.8) \text{ dB} \qquad (13.62)$$

For a large city:

$$a(h_m) = 8.29(\log 1.54 h_m)^2 - 1.1 \text{ dB, } f_c \le 200 \text{ MHz} \qquad (13.63)$$

or

$$a(h_m) = 3.2(\log 11.75 h_m)^2 - 4.97 \text{ dB, } f_c \ge 400 \text{ MHz} \qquad (13.64)$$

Suburban Area

$$L_{50} = L_{50}(urban) - 2\left[\log\left(\frac{f_c}{28}\right)^2 - 5.4\right] \text{ dB} \qquad (13.65)$$

Open Area

$$L_{50} = L_{50}(urban) - 4.78(\log f_c)^2 + 18.33\log f_c - 40.94 \text{ dB} \qquad (13.66)$$

Hata's model does not account for any of the path-specific corrections used in Okumura's model.

Okumura's model [8] tends to average over some of the extreme situations and does not respond sufficiently quickly to rapid changes in the radio path profile. The distance-dependent behavior of Okumura's model is in agreement with the measured values. Okumura's measurements are valid only for the building types found in Tokyo. Experience with comparable measurements in the United States has shown that a typical U.S. suburban situation is often somewhere between Okumura's suburban and open areas. Okumura's suburban definition is more representative of residential metropolitan areas with large groups of row houses.

Okumura's model requires that considerable engineering judgment be used, particularly in the selection of the appropriate environmental factors. Data is needed in order to be able to predict the environmental factors from the physical properties of the buildings surrounding a mobile receiver. In addition to the appropriate environmental factors, path-specific corrections are required to convert Okumura's mean path-loss predictions to the predictions that apply to the specific path under study. Okumura's techniques for correction of irregular terrain and other path-specific features require engineering interpretations and are thus not readily adaptable for computer use.

The Walfisch-Ikegami Model [14]   This model (also known as the European Committee of Scientific and Technology [COST] 231 model) is used to estimate the path loss in an urban environment for cellular communication (Figure 13.9). The model is a combination of the empirical and deterministic models for estimating the path loss in an urban environment over the frequency range of 800–2000 MHz. This model is used primarily in Europe for GSM systems and in some propagation models in the United States. The model contains three elements: free-space loss, roof-to-street diffraction and scatter loss, and multiscreen (i.e., diffraction and scatter loss from other structures) loss. The expressions used in this model are

$$L_{50} = L_f + L_{rts} + L_{ms} \qquad (13.67)$$

or

$$L_{50} = L_f \text{ when } L_{rts} + L_{ms} \leq 0 \qquad (13.68)$$

where:
L_f = free-space loss,
L_{rts} = rooftop-to-street diffraction and scatter loss, and
L_{ms} = multiscreen loss.

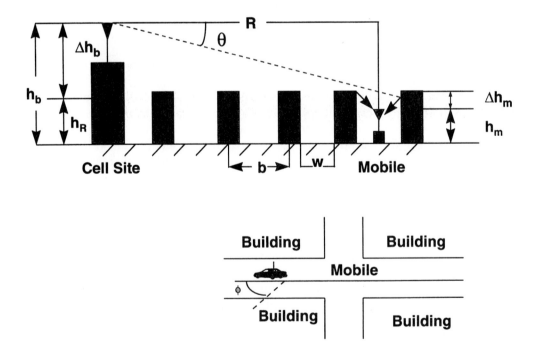

Fig. 13.9 The Walfisch-Ikegami Propagation Model

Free-space loss is given as:

$$L_f = 32.4 + 20\log R + 20\log f_c \text{ dB} \tag{13.69}$$

The rooftop-to-street diffraction and scatter loss is given as:

$$L_{rts} = -16.9 - 10\log W + 10\log f_c + 20\log\Delta h_m + L_0 \text{ dB} \tag{13.70}$$

where:
W = street width (m), and
$\Delta h_m = h_r - h_m$ (m).

$$L_0 = -9.646 \text{ dB } 0 \le \phi \le 35 \text{ degree}$$

$$L_0 = 2.5 + 0.075(\phi - 35) \text{ dB } 35 \le \phi \le 55 \text{ degree}$$

$$L_0 = 4 - 0.114(\phi - 55) \text{ dB } 55 \le \phi \le 90 \text{ degree}$$

where:
ϕ = incident angle relative to the street.

The multiscreen loss is given as:

$$L_{ms} = L_{bsh} + k_a + k_d\log R + k_f\log f_c - 9\log b \tag{13.71}$$

where:
b = distance between buildings along the radio path (m)
$L_{bsh} = -18\log 11 + \Delta h_b, h_b > h_r$
$L_{bsh} = 0, h_b < h_r$
$k_a = 54, h_b > h_r$
$k_a = 54 - 0.8h_b, R \ge 500 \text{ m}, h_b \le h_r$
$k_a = 54 - 1.6\Delta h_b R, R < 500 \text{ m}, h_b \le h_r$

Note: Both L_{bsh} and k_a increase the path loss with lower BS antenna heights.
$$k_d = 18, h_b < h_r$$

$$k_d = 18 - \frac{15\Delta h_b}{\Delta h_m}, h_b \ge h_r$$

$$k_f = 4 + 0.7\left(\frac{f_c}{925} - 1\right) \text{for a midsize city and suburban area with moderate}$$
tree density

$$k_f = 4 + 1.5\left(\frac{f_c}{925} - 1\right) \text{for a metropolitan center}$$

The range of parameters for which the model is valid is:

$$800 \le f_c \le 2000 \text{ (MHz)}$$
$$4 \le h_b \le 50 \text{ (m)}$$
$$1 \le h_m \le 3 \text{ (m)}$$
$$0.02 \le R \le 5 \text{ (km)}$$

The following default values can be used for the model:

b = 20 – 50 (m)

W = b/2

$\phi = 90°$

Roof = 3 m for pitched roof and 0 m for flat roof

h_r = 3 (number of floors) + roof

Using the following data, a comparison of the path loss using the Hata model and the Walfisch-Ikegami model is given in Table 13.1.

f_c = 880 MHz; h_m = 1.5 m; h_b = 30 m; roof = 0 m; h_r = 30 m; ϕ = 90 degrees; b = 30 m; and W = 15 m.

The path losses predicted by Hata's model are 13 to 16 dB lower that those predicted by the Walfisch-Ikegami model. Hata's model ignores effects from street width, street diffraction, and scatter losses, which the Walfisch-Ikegami model includes.

Correction Factor for Attenuation Due to Trees   Weissberger [15] has developed a modified exponential delay model that can be used where a radio path is blocked by dense, dry, in-leaf trees found in temperate climates. The additional path loss can be calculated from the following expression:

$$L_t = 1.33(f_c)^{0.284}(h_f)^{0.588} \text{ dB, for } 14 \leq h_f \leq 400 \text{ m} \tag{13.72}$$

$$= 0.45(f_c)^{0.284}h_f \text{ dB, for} 0 \leq h_f \leq 14 \text{ m} \tag{13.73}$$

where:
L_f = loss in dB,
f_c = frequency in GHz, and
h_f = tree height in meters.

Table 13.1 A Comparison of Path Loss Using Hata and Walfisch-Ikegami Models

Distance (km)	Path Loss (dB)	
	Hata's Model	**Walfisch-Ikegami Model**
1	126.16	139.45
2	136.77	150.89
3	142.97	157.58
4	147.37	162.33
5	150.79	166.01

The difference in path loss for trees with and without leaves has been found to be about 3 to 5 dB. For a frequency of 900 MHz, the above equations are reduced to

$$L_f = 1.291(h_f)^{0.588} \text{ dB, for } 14 \le h_f \le 400 \text{ m} \tag{13.74}$$

$$= 0.437 h_f \text{ dB, for } 0 \le h_f \le 14 \text{ m} \tag{13.75}$$

13.7.2 Models for Indoor Environment

Experimental studies have indicated that a portable receiver moving in a building experiences Rayleigh fading for obstructed propagation paths and Ricean fading for LOS paths, regardless of the type of building. Rayleigh fading is short-term fading resulting from signals traveling separate paths (multipath) that partially cancel each other. An LOS path is clear of building obstructions; in other words, there are no reflections of the signal. Ricean fading results from the combination of a strong LOS path and a ground path plus numerous weak reflected paths.

Quantification of propagation between floors is important for in-building wireless systems of multifloor buildings that need to share frequencies within the building. Frequencies are reused on different floors to avoid cochannel interference. The type of building material, aspect ratio of building sides, and types of windows have shown to impact the RF attenuation between floors. Measurements have indicated that the loss between floors does not increase linearly in dB with increasing separation distance. The greatest floor attenuation in dB occurs when the receiver and transmitter are separated by a single floor. The overall path loss increases at a smaller rate as the numbers of floors increase. Typical values of attenuation between floors is 15 dB for one floor of separation and an additional 6 to 10 dB per floor of separation up to four floors of separation. For five or more floors of separation, path loss increases by only a few dB for each additional floor (see Table 13.2).

The signal strength received inside a building due to an external transmitter is important for wireless systems that share frequencies with neighboring buildings or with an outdoor system. Experimental studies have shown that the signal strength received inside a building increases with height. On the lower floors of a building, the urban cluster induces greater attenuation and reduces the level of penetration. On higher floors, an LOS path may exist, thus causing a stronger incident signal at the exterior wall of the building. RF penetration is found to be a function of frequency as well as height within a building. Penetration loss decreases with increasing frequency. Measurements made in front of a window showed 6 dB less penetration loss on coverage than those measurements made in parts of the building without windows. Experimental studies also showed that building penetration loss decreased at a rate of about 2 dB per floor from ground level up to the 10th floor and then began to increase around the 10th floor. The increase in penetration loss at the higher floors was attributed to shadowing effects of adjacent buildings.

The mean path loss is a function of distance to the γ-th power [9].

$$L_{50}(R) = L(R_0) + 10 \times \gamma \log\left(\frac{R}{R_0}\right) \text{ dB} \qquad (13.76)$$

where:
$L_{50}(R)$ = mean path loss (dB),
$L(R_0)$ = path loss from transmitter to reference distance R_0 (dB),
γ = mean path-loss exponent,
R = distance from the transmitter (m), and
R_0 = reference distance from the transmitter (m).

We choose R_0 equal to 1 m and assume $L(R_0)$ due to free-space propagation from the transmitter to a 1-m reference distance. Next, we assume the antenna gain equals the system cable losses (obviously, this is not always true) and get a path loss, $L(R_0)$, of 31.7 dB at 914 MHz over a 1-m free-space path.

The path loss was found to be log-normally distributed about Eq. (13.76). The mean path loss exponent γ and standard deviation σ are the parameters that depend on building type, building wing, and number of floors between the transmitter and receiver. The path loss at a transmitter-receiver (T-R) separation of R meters can be given as:

$$L(R) = L_{50}(R) + X_\sigma \text{ dB} \qquad (13.77)$$

where:
$L(R)$ = path loss at a T-R separation distance R meters,
X_σ = zero mean log-normally distributed random variable with standard
              deviation σ dB

Table 13.2 gives a summary of the mean path-loss exponents and the standard deviation above or below the mean for different environments [9].

In a multifloor environment, Eq. (13.77) can be modified to emphasize that the mean path-loss exponent is a function of the number of floors between the transmitter and receiver. The value of γ (multifloor) is given in Table 13.2.

$$L_{50}(R) = L(R_0) + 10 \times \gamma(multifloor)\log\left(\frac{R}{R_0}\right) \qquad (13.78)$$

Another path-loss prediction model suggested in [9] uses the Floor Attenuation Factor (FAF). A constant FAF (in dB), which is a function of the number of floors and building type, was included in the mean path loss predicted by a path-loss model that uses the *same-floor* path-loss exponent for the building type.

$$L_{50}(R) = L(R_0) + 10 \times \gamma(same-floor)\log\left(\frac{R}{R_0}\right) + FAF \text{ dB} \qquad (13.79)$$

Table 13.2 Mean Path-Loss Exponents and Standard Deviations [9]

Type	γ	σ (dB)
• All Building		
All Locations	3.14	16.3
Same Floor	2.76	12.9
Through 1 Floor	4.19	5.1
Through 2 Floors	5.04	6.5
• Office Building 1		
Entire Building	3.54	12.8
Same Floor	3.27	11.2
West Wing 5th Floor	2.68	8.1
Central Wing 5th Floor	4.01	4.3
West Wing 4th Floor	3.18	4.4
• Grocery Store	1.81	5.2
• Retail Store	2.18	8.7
• Office Building 2		
Entire Building	4.33	13.3
Same Floor	3.25	5.2

where:
R is in meters and $L(R_0)$ = 31.7 dB at 914 MHz.

Table 13.3 provides the FAFs, and standard deviation (in dB) of the difference between the measured and predicted path loss. Values for the FAFs in Table 13.3 are an average (in dB) of the difference between the path loss observed at multifloor locations and the mean path loss predicted by the simple R^γ model (Eq. [13.76]), where γ is the *same-floor* exponent listed in Table 13.2 for the particular building structure and R is the shortest distance, measured in three dimensions, between the transmitter and receiver.

13.7.2.1 Soft Partition and Concrete Wall Attenuation Factor Model The path-loss effects of soft partitions (wall boards) and concrete walls (in dB) between the transmitter and receiver for the same floor was modeled in [9] and has been given as:

$$L_{50}(R) = 20\log\left(\frac{4\pi R}{\lambda}\right) + p \times AF(soft-partition) + q \times AF(Concrete-wall) \quad (13.80)$$

Table 13.3 Average Floor Attenuation Factor [9]

	FAF (dB)	σ
• Office Building 1		
Through 1 Floor	12.9	7.0
Through 2 Floors	18.7	2.8
Through 3 Floors	24.4	1.7
Through 4 Floors	27.0	1.5
• Office Building 2		
Through 1 Floor	16.2	2.9
Through 2 Floors	27.5	5.4
Through 3 Floors	31.6	7.2

where:
p = number of soft partitions between the transmitter and receiver,
q = number of concrete walls between the transmitter and receiver,
λ = wave length (m),
AF = 1.39 dB for each soft partition, and
AF = 2.38 for each concrete wall.

E X A M P L E   1 3 – 5

Use the two models (Eqs. [13.78] and [13.79]) to predict the mean path loss at a distance $R = 30$ m through three floors of an office building; assume the mean path-loss exponent for *same-floor* measurements in the building is $\gamma = 3.27$ (see Table 13.2), the mean path-loss exponent for three-floor measurements is $\gamma = 5.22$, and the average FAF is 24.4 dB (see Table 13.3).

From Eq (13.78):

$$L_{50}(30) = 31.7 + 10 \times 5.22 \log\left(\frac{30}{1}\right) = 108.8 \text{ dB}$$

From Eq. (13.79):

$$L_{50}(30) = 31.7 + 10 \times 3.27 \log\left(\frac{30}{1}\right) + 24.4 = 104.4 \text{ dB}$$

The results obtained by the two models are reasonably close.

13.7.3 IMT-2000 Models

For IMT-2000, the operating environments are identified by appropriate sub-sets consisting of indoor office environment, outdoor-to-indoor and pedestrian environment, and vehicular environment. For narrowband technologies, delay spread is characterized by its rms value alone. For wideband technologies, the number, strength, and relative time delay of the many signal components become important. In addition, for some technologies (e.g., those using power control), the path-loss models include the coupling between all cochannel propagation links to provide accurate predictions. Also, in some cases, the shadow-fading temporal variations of environment must be modeled. The key parameters of the propagation models are

Delay spread, its structure, and its statistical variation

Geometrical path-loss rule $R^{-\gamma}$ (e.g., $2 \leq \gamma \leq 5$)

Shadow fading

Multipath fading characteristics (e.g., Doppler spectrum, Rician vs. Rayleigh) for envelope of channels

Operating radio frequency

13.7.3.1 Indoor Office Environment   This environment is characterized by small cells and low transmit powers, with both BSs and pedestrian users located indoors. The rms delay spread ranges from around 35 ns to 460 ns. The path-loss rule varies due to scatter and attenuation by walls, floors, and metallic structures such as partitions and filing cabinets. These objects also produce shadowing effects. Log-normal shadowing with a standard deviation of 12 dB can be expected. Fading characteristics range from Rician to Rayleigh, with Doppler frequency offsets set by walking speeds. The path-loss model for this environment is:

$$L_p = 37 + 30\log R + 18.3 \cdot n^{[(n+2)/(n+1) - 0.46]} \text{ dB} \qquad (13.81)$$

where:
R = separation between transmitter and receiver (m), and
n = number of floors in the path.

13.7.3.2 Outdoor-to-Indoor and Pedestrian Environment   This environment is characterized by small cells and low transmit power. BSs with low antenna heights are located outdoors; pedestrian users are located on streets and inside buildings and residences. Coverage into buildings in high-power systems is included in the vehicular environment. The rms delay spread varies from 100 to 1800 ns. A geometrical path-loss rule R^{-4} is applicable. If the path is a LOS on a canyon-like street, the path loss follows a R^{-2} rule, where there is Fresnel zone clearance. For the region with longer Fresnel zone clearance, a path-loss rule of R^{-4} is appropriate, but a range of up to R^{-6} may be encountered due to trees and other obstructions along the path. Log-normal shadow fading with a standard deviation of 10 dB is reasonable for outdoors and 12 dB for indoors. Average building penetration loss of 18 dB with a standard deviation of 10 dB is appropriate. Rayleigh and/or Rician fading rates are generally set by walking speeds, but faster fading due to reflections from moving vehicles may occur some of the time. The following path-loss model has been suggested for this environment:

$$L_p = 40\log R + 30\log f_c + 49 \text{ dB} \tag{13.82}$$

This model is valid for a non-LOS (NLOS) case only and describes the worst-case propagation. Log-normal shadow fading with a standard deviation equal to 10 dB is assumed. The average building penetration loss is 18 dB with a standard deviation of 10 dB.

13.7.3.3 Vehicular Environment   This environment consists of larger cells and higher transmit power. The rms delay spread from 4 microseconds to about 12 microseconds may occur on elevated roads in hilly or mountainous terrain. A geometrical path-loss rule of R^{-4} and log-normal shadow fading with a standard deviation of 10 dB are used in the urban and suburban areas. Building penetration loss averages 18 dB with a 10 dB standard deviation.

In rural areas with flat terrain, the path loss is lower than that of urban and suburban areas. In mountainous terrain, if path blockages are avoided by selecting BS locations, the path-loss rule is closer to R^{-2}. Rayleigh fading rates are set by vehicle speeds. Lower fading rates are appropriate for applications using stationary terminals. The following path-loss model is used in this environment:

$$L_p = 40(1 - 4 \times 10^{-2}\Delta h_b)\log R - 18\log(\Delta h_b) + 21\log f_c + 80 \text{ dB} \tag{13.83}$$

where:
Δh_b = BS antenna height measured from average rooftop level (m).

13.8 DELAY SPREAD

The radio signal follows different paths because of multipath reflection. Each path has a different path length, so the time of arrival for each path is different. The effect, which smears or spreads out the signal, is called delay spread. As an example, if an impulse is transmitted by the transmitter, by the time this impulse is received at the receiver, it is no longer an impulse but rather a pulse that is spread (refer to Figure 13.10). In a digital system, the delay spread causes intersymbol interference, thereby limiting the maximum symbol rate of a digital multipath channel.

The mean delay spread τ_d is:

$$\tau_d = \frac{\int_0^\infty tD(t)dt}{\int_0^\infty D(t)dt} \tag{13.84}$$

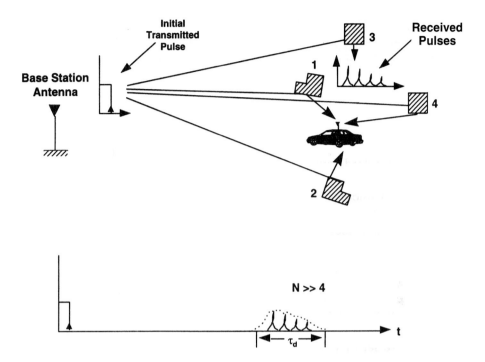

Fig. 13.10 Delay Spread Phenomenon

where:
$D(t)$ is the delay probability density function, and

$\int_0^\infty D(t)dt = 1.$

Some representative delay functions are:

Exponential

$$D(t) = \frac{1}{\tau_d}e^{\frac{t}{\tau_d}}$$

Uniform

$$D(t) = \frac{\tau_d}{2}, \ 0 \le t \le 2\tau_d$$

$$D(t) = 0, \text{ elsewhere}$$

The measured data suggest that the mean delay spreads are different in different environments (refer to Table 13.4).

Table 13.4 Measured Data for Delay Spread

Type of Environment	Delay Spread τ_d (μs)
Open Area	< 0.2
Suburban Area	0.5
Urban Area	3.0

13.8.1 Delay Spread Values in IMT-2000 Model

A majority of the time, rms delay spreads are relatively small, but occasionally there are worst-case multipath characteristics that lead to much larger rms delay spreads. Measurements in outdoor environments show that rms delay spread can vary over an order of magnitude within the same environment. Delay spreads can have a major impact on the system performance. To accurately evaluate the relative performance of radio transmission technologies, it is important to model the variability of delay spread as well as the worst-case locations where delay spread is relatively large. Three multipath channels are defined by IMT-2000 for each environment. Channel A represents the low-delay-spread case that occurs frequently; channel B corresponds to the median-delay-spread case that also occurs frequently; and channel C is the high-delay-spread case that occurs only rarely. Table 13.5 provides the rms values of delay spread for each channel and for each environment.

13.9 COHERENCE BANDWIDTH

The coherence bandwidth B_c is a statistical measure of the range of frequencies over which the channel passes all spectral components with approximately equal gain and linear phase. The coherence bandwidth represents a

Table 13.5 The rms Delay Spread (IMT-2000)

Environment	Channel A		Channel B		Channel C	
	τ_{rms} (ns)	% Occurrence	τ_{rms} (ns)	% Occurrence	τ_{rms} (ns)	% Occurrence
Indoor Office	35	50	100	45	460	5
Outdoor-to-indoor and pedestrian environment	100	40	750	55	1,800	5
Vehicular (high antenna)	400	40	4,000	55	12,000	5

frequency range for which either the amplitudes or phases of two received signals have a high degree of correlation. A signal's spectral components in that frequency range are affected by the channel in a similar manner, as, for example, exhibiting fading or not. As an approximation

$$B_c \approx 1/\tau_{dmax} \qquad (13.85)$$

where:
τ_{dmax} = maximum delay spread.

The maximum delay spread, τ_{dmax}, is not necessarily the best indicator of how any given system will perform on a channel because different channels with the same value of τ_{dmax} may exhibit very different profiles of signal intensity over the delay span. A more useful measurement is often expressed in terms of the rms delay spread, τ_{drms}. An exact relationship between coherence bandwidth and delay spread does not exist. Several approximations have been proposed.

Consider two fading signal envelopes at two frequencies f_1 and f_2, respectively, where

$$\Delta f = |f_1 - f_2| \qquad (13.86)$$

If the two frequencies are further apart than the coherence bandwidth, B_c, the signals will fade independently. This concept is useful for diversity reception.

We measure the correlation between the fading of the signals on the two frequencies by computing the correlation function, $R_T(\Delta f)$.

If $|R_T(\Delta f)| = 0.5$, then the coherence bandwidth for two fading amplitudes of two received signals is

$$\Delta f > B_c = \frac{1}{2\pi\tau_{drms}} \qquad (13.87)$$

If we measure the coherence bandwidth as the frequency interval over which the channel's complex frequency transfer function has a correlation of at least 0.9, then the coherence bandwidth is approximately given as:

$$B_c \approx \frac{1}{50\tau_{drms}} \qquad (13.88)$$

For mobile radio, a generally accepted useful model for the urban environment is an array of radially uniformly spaced scatters, all with equal magnitude reflection coefficients but independent, randomly occurring reflection phase angles. The model is referred to as the dense-scatters channel model. Using such a model, the coherence bandwidth has been defined for a bandwidth over which the channel's complex frequency transfer function has a correlation of at least 0.5 [12]; in other words

$$B_c \approx \frac{0.276}{\tau_{drms}} \tag{13.89}$$

A more popular approximation of B_c corresponding to a bandwidth with a correlation of at least 0.5 is [12]

$$B_c \approx \frac{1}{5\tau_{drms}} \tag{13.90}$$

A channel is called a frequency-selective channel if $B_c < 1/T_s = B_w$, where the symbol rate $1/T_s$ is nominally taken to be equal to the signal bandwidth B_w. Frequency-selective fading distortion occurs whenever a signal's spectral components are not all affected equally by the channel.

Frequency-nonselective or flat-fading occurs whenever $B_c > B_w$. Hence, all of the signal's spectral components are affected by the channel in a similar manner. Flat fading does not introduce channel-induced ISI distortion, but performance degradation can still be expected due to a loss in SNR whenever the signal is fading. In order to avoid channel-induced ISI distortion, the channel is required to be flat fading by ensuring that

$$B_c > B_w = 1/T_s \tag{13.91}$$

Thus, the channel coherence bandwidth sets an upper limit on the transmission rate that can be used without incorporating an equalizer in the receiver.

The GSM symbol rate (or bit rate, since modulation is binary) is 271 ksymbols/s and bandwidth is $B_w = 200$ kHz. Let the typical rms delay spread in an urban environment be $\tau_{drms} = 2$ µs; then, using Eq. (13.90), the coherence bandwidth $B_c \approx 100$ kHz. It is therefore apparent that, since $B_c < B_w$, the GSM receiver must use some form of mitigation to overcome frequency-selective distortion. To accomplish this goal, the Viterbi equalizer [13] is typically used.

13.10 DOPPLER SPREAD

The width of the Doppler power spectrum is called **spectral broadening** or **Doppler spread**, denoted f_d and sometimes called the **fading bandwidth** of the channel. The Doppler shift of each arriving path is generally different from that of another path. The effect on the received signal is seen as a Doppler spreading or spectral broadening of the transmitted signal frequency rather than as a shift. Doppler spread and coherence time T_0 are reciprocally related as:

$$T_0 = \frac{1}{f_d} \tag{13.92}$$

f_d is the typical fading rate of the channel. When T_0 is defined as the time duration over which the channel's response to a sinusoid has a correlation greater than 0.5, then

$$T_0 \approx \frac{9}{16\pi f_d} \qquad (13.93)$$

A popular rule of thumb is to define T_0 as the geometric mean of Eqs. (13.92) and (13.93)

$$T_0 = \sqrt{\frac{9}{16\pi f_d^2}} = \frac{0.423}{f_d} \qquad (13.94)$$

The time required to traverse a distance $\lambda/2$ (equal to fade interval) when traveling at a constant velocity v is

$$T_0{}' = \frac{\lambda/2}{v} = \frac{0.5}{(v/\lambda)} = \frac{0.5}{f_d} \approx T_0 \qquad (13.95)$$

With frequency equal to 900 MHz and velocity equal to 120 km/h, the coherence time is about 5 ms and Doppler spread is approximately 100 Hz. For a voice-grade channel with a typical transmission rate of 10,000 symbols/s, the fading rate is considerably less than the symbol rate. Under such conditions, the channel would manifest slow-fading effects. A channel is referred to as fast fading if the symbol rate $1/T_s$ is less than the fading rate $1/T_0$; in other words, the fast fading is characterized by [13]

$$B_w < f_d \text{ or} \qquad (13.96a)$$

$$T_s > T_0 \qquad (13.96b)$$

A channel is referred to as slow fading if the signaling rate is greater than the fading rate. Thus, to avoid signal distortion caused by fast fading, the channel must be made to exhibit slow fading by ensuring that the signaling rate exceeds the channel fading rate

$$B_w > f_d \text{ or} \qquad (13.97a)$$

$$T_s < T_0 \qquad (13.97b)$$

The channel fading rate f_d sets a lower limit on the signaling rate that can be used without suffering fast-fading distortion. A better way to state the requirement for mitigating the effects of fast fading would be that we desire $B_w \gg f_d$ (or $T_s \ll T_0$). If this condition is not satisfied, the random frequency due to varying Doppler shifts will limit the system performance significantly. The Doppler effects yield an irreducible error rate that cannot be overcome by simply increasing the S/I ratio. This irreducible error rate is most pronounced for any modulation that involves switching the carrier phase. For voice-grade applications with error rates of 10^{-3} to 10^{-4}, a large value of Doppler shift is considered to be on the order of $0.01\,B_w$. Thus, to avoid fast-fading distortion and Doppler-induced irreducible error rate, the signaling rate must exceed the fading rate by a factor of 100 to 200. The exact factor depends on the signal modulation, receiver design, and required bit error rate.

The GSM system is required to provide mitigation for distortion due to signal dispersions of about 15–20 μs. Since the bit duration in GSM is 3.69 μs, the Viterbi equalizer has a memory of 4- to 6-bit intervals. The objective of the Viterbi equalizer is to find, for each L_0-bit interval in the message, the most likely L_0-bit sequence out of 2^{L_0} possible sequences that might have been transmitted (where L_0 is the observation interval for the window). Determining the most likely L_0-bit sequence requires that 2^{L_0} meaningful reference waveforms be created by modifying (or disturbing) the 2^{L_0} ideal waveform in the same way the channel has disturbed the transmitted messages. Therefore, the 2^{L_0} reference waveforms are convolved with the windowed estimate of channel impulse response, $h_w\,(t)$, in order to derive the disturbed or channel-corrected reference waveform. Next, the channel-corrected reference waveforms are compared against the received-data waveforms to yield metric calculations. However, before comparison, the received-data waveforms are convolved with known windowed autocorrelation function, transforming them in a manner comparable to that applied to the reference waveform. The filtered message signal is compared to all possible 2^{L_0} channel-corrected reference signals, and matrices are computed as required by the Viterbi-decoding algorithm (VDA) [2]. The VDA provides the estimated maximum likelihood of the transmitted sequence.

13.11 ISI

In practical radio systems, the presence of a transmitter band-pass filter is essential to save spectrum as much as possible. However, such a band-limited channel could degrade the transmission performance due to ISI. Therefore, we should reduce the signal bandwidth as much as possible without introducing any ISI.

In a time-dispersive medium, the transmission rate R_b for a digital transmission is limited by delay spread. If a low-bit-error-rate performance is required, then

$$R_b < \frac{1}{2\tau_d} \tag{13.98}$$

In a real situation, R_b is determined based upon the required bit error rate.

13.12 SUMMARY

In this chapter we discussed the short-term and long-term fading characteristics of a radio signal, using a numerical example for calculating the fading rate and the average duration of a short-term fade. There are two empirical propagation models used to calculate the path losses in an outdoor environment and two other empirical models for calculating path losses in an indoor environment. The chapter covered the suggested propagation models in IMT-2000. We introduced the concepts of the delay spread, coherence bandwidth, and Doppler spread and discussed the techniques used in GSM to provide mitigation for distortion due to fast fading and frequency-selective fading.

13.13 PROBLEMS

1. Calculate the level-crossing rate at a level of −10 dB and average duration of fade for a cellular system at 1800 MHz and a vehicle speed of 30 km/h. Assume the free-space speed of propagation for electromagnetic waves = 3×10^8 m/s. Neglect the effects of the motion of the scatters. Compare the results obtained using the approximate expressions.

2. Calculate the mean path loss at 1800 MHz frequency in a large urban environment at distances of 1, 2, 3, 4, and 5 km from the BS using the Hata-Okumura and COST 231 models. Plot the path loss vs. distance curves. Assume BS antenna height = 150 m and mobile antenna height = 1 m. Assume any other data. Comment on the results.

3. Using the COST 231 model, calculate the maximum cell radius that can be used in a large metropolitan area at 1.9-GHz frequency with a maximum allowable path loss of 134.66 dB. Assume the following data:

   - Body Loss: 2 dB
   - Indoor Penetration Loss: 10 dB
   - Mobile Antenna Height (h_m) = 2 m
   - BS Antenna Height (h_b) = 40 m
   - Street Orientation = 90°
   - b = 30 m
   - $W = b/2$
   - $h_r = 40$
   - $k_a = 54$

4. The measured rms delay spread in a GSM system is 5 ms; what is the coherence bandwidth (B_c) with a correlation of 0.5? Comment on the result in the context of the GSM system which has channel spacing of 200 kHz.

13.14 REFERENCES

1. Clarke, R. H., "A Statistical Theory of Mobile Radio Reception," *Bell System Technical Journal* 47, July–August 1968, pp. 957–1000.

2. Forney, G. D., "The Viterbi Algorithm," *Proceedings of IEEE* 61 (3), March 1978, pp. 268–78.

3. Garg, V. K., and Wilkes, J. E., *Wireless and Personal Communications Systems*, Prentice Hall, Upper Saddle River, NJ, 1996.

4. Hanzo, L., and Stefanov, J., "The Pan-European Digital Cellular Mobile Radio System—Known as GSM," Chap. 8 in *Mobile Radio Communications*, Edited by R. Steele, Prentech Press, London, 1992.

5. Hata, M., "Empirical Formula for Propagation Loss in Land Mobile Radio Services," *IEEE Transactions on Vehicular Technology* 29 (3), 1980.

6. Jakes, W. C., ed., *Microwave Mobile Communications*, John Wiley, New York, NY, 1974.

7. Lee, William C. Y., *Mobile Communications*, John Wiley, New York, NY, 1989.

8. Okumura, Y., et al., "Field Strength and its Variability in VHF and UHF Land Mobile Radio Service," *Review Electronic Communication Lab* 16 (9–10), 1968.

9. Rappaport, T. S., *Wireless Communications*, Prentice Hall, Upper Saddle River, NJ, 1996.

10. Sampei, S., *Applications of Digital Wireless Technologies to Global Wireless Communications*, Prentice Hall, Upper Saddle River, NJ, 1997.

11. Seidel, S. Y., and Rappaport, T. S., "914 MHz Path Loss Prediction Models for Indoor Wireless Communications in Multi-floor Buildings," *IEEE Trans., Antenna & Propagation*, 40 (2), February 1992.

12. Sklar, B., "Rayleigh Fading Channels in Mobile Digital Communication Systems Part I: Characterization," *IEEE Communication Magazine* 35 (9), September 1997, pp. 136–46.

13. Sklar, B., "Rayleigh Fading Channels in Mobile Digital Communication Systems Part II: Mitigation," *IEEE Communication Magazine* 35 (9), September 1997, pp. 148–55.

14. Walfisch, J., and Bertoni, H. L., "A Theoretical Model of UHF Propagation in Urban Environment," *IEEE Trans., Antennas & Propagation.*, in press.

15. Weissberger, M. A., "An Initial Critical Summary of Models for Predicting the Attenuation of Radio Waves by Rees," ESD-TR-81-101, Electromagnet Compact, Analysis Center, Annapolis, MD, July 1982.

CHAPTER 14

Planning and Design of a GSM Wireless Network

14.1 Introduction

In this chapter, we present the teletraffic models required in cellular/PCS network planning and design, followed by an overview of the methods used for subscriber location management in the GSM system. We discuss important aspects for planning a wireless network and focus on the steps in designing a wireless system, including service requirements, constraints for hardware implementation, propagation path loss, and system requirements. The chapter will also cover spectral efficiency of a wireless system, receiver sensitivity and link budget, selection of a modulation scheme, design of a TDMA frame, and the relationship between delay spread and symbol rate. We include a design example of the GSM system that illustrates a systematic procedure for determining the system's equipment needs.

14.2 Teletraffic Models

We need teletraffic models [1] for network planning and design to help us compare network architectures, allocate network resources, and evaluate protocols performance, among other things. Traffic models have been developed and used for wireline networks, but wireline models are not applicable to cellular/ PCS networks because they predict the aggregate traffic moving through telephone switches and do not account for subscriber mobility or caller distributions in a service area. Therefore, wireline network models need modifications to be applicable to the cellular/PCS network.

A cellular/PCS network operator requires appropriate strategies to select the optimum equipment for the best network design, to engineer the network efficiently, and to offer high-quality service to mobile users. The modeling framework for a cellular/PCS network can be divided into the call model, mobility model, and topology model.

14.2.1 Call Model

The call model generates call traffic for each individual user. It has two parts—call traffic model and caller distribution model. A **call traffic model** describes the usage of offered services and includes procedures for call handling and operation of supplementary services. A **caller distribution model** describes the behavior of individual callers. It accounts for real-life behaviors such as users calling a group of people more frequently.

14.2.2 Mobility Model

A mobility model describes the occurrence of procedures such as location update and handover. Several approaches have been used to develop a mobility model. These include

☞ **Fluid Model.** The fluid model assumes traffic flow to be like the flow of a fluid [2]. It is used to model macroscopic movement behavior. The model formulates that the amount of traffic flowing out of a region is proportional to the population density of the region, the average velocity of movement, and the length of the region boundary. For a region, the average number of crossings per unit time is given as:

$$\lambda = \frac{\upsilon \rho L}{\pi} \qquad (14.1)$$

where:
λ = number of crossings per unit time,
ρ = average population density in the region,
υ = average movement velocity in the region, and
L = perimeter of the region.

One of the limitations of the fluid model is that it describes aggregate traffic and is difficult to apply to situations where individual movement patterns are desired. The fluid flow model is more applicable to regions with a large population because it uses average population density and average movement velocity of the region.

☞ **Markovian Model.** Also known as the *random walk model*, the Markovian model describes individual movement. In this model, a mobile subscriber either remains within the region or moves to an adjacent region according to a transition probability distribution. One of the limitations of this model is that there is no concept of trips.

☞ **Gravity Model.** Various variations of the gravity model have been used extensively in transportation research to model human movement behavior. Gravity models have been applied to regions of varying sizes, from city models to national and international models. The main difficulty with applying the gravity model is that many parameters are required for calculations; it is therefore hard to model geography with many regions.

14.2.3 Topology Model

A topology model describes the topology of the network area. The model is required to specify the reference points for user information and traffic signaling. It provides parameters to be used in the optimization process of network.

14.3 MOBILITY IN CELLULAR/PCS NETWORKS [2]

Mobility can be categorized as **radio mobility**, which mainly concerns the handover process, and **network mobility**, which mainly deals with location management (location updating and paging).

Location management schemes are essentially based on users' mobility and incoming call rate characteristics. The location update (LU) procedure allows the system to keep track of the mobile user's location to direct the incoming call. Location registration is also used to bring the user's service profile near his or her location and allows the network to provide users with the services rapidly.

The paging process transmits paging messages to all those cells where the mobile terminal can be located. Thus, if the LU cost is high, the paging cost will be low (paging will be performed over a small area). On the other hand, if the LU cost is low (user location knowledge is fuzzy), the paging cost will be high (paging will be performed over a wider area).

Location management involves two methods—the periodic LU method and the LU on location area (LA) border crossing. The VLR stores the LA identifier, and the HLR keeps the VLR identifier.

In the periodic LU method, an MS periodically sends its identity to the network. The drawback of this scheme is its resource consumption. For example, if the user does not move from an LA for several hours, resources are consumed unnecessarily.

In the LU-on-LA-crossing scheme, each BS periodically broadcasts the identity of its LA. The MS is required to regularly listen to network broadcast information and store the current LA identity. If the received number differs from the one stored in the MS, a LU procedure is automatically triggered by the MS. The advantage of this scheme is that it requires LUs only when the MS actually moves. A highly movable mobile user generates a large number of LUs; a low-mobility user triggers only a few.

A hybrid location management scheme combines the two schemes. The MS generates its LUs each time it detects an LA crossing. However, if no communication has occurred between the MS and the network for a specified amount of time, the MS generates a LU. This periodic LU typically allows the system to recover user location data in case of a database failure.

GSM deals with three types of LU procedures: intra-VLR LU, inter-VLR LU using TMSI, and inter-VLR LU using IMSI. Also, a fourth type—the IMSI-attached procedure—is triggered when the mobile is powered on in the LA where it was powered off. To minimize the location management cost (i.e., LU plus paging traffic and processing), the LA must be designed carefully by using a proper subscriber mobility model.

14.3.1 Application of a Fluid Flow Model

In the following pages, we present an application of the simple mobility model based on a fluid flow assumption for a DCS-1800 and PCS-1900 network in an urban environment with small cells and high user density. We investigate the impact of LUs on RF resource occupancy at the network level and determine the number of the transactions processed by MSC/VLR. The transaction is defined here as the messages received or transmitted by the MSC/VLR. We make the following assumptions:

- ☞ Cells are hexagonal.
- ☞ Maximum blocking probability of an SDCCH is 1 percent.
- ☞ MSs are uniformly distributed on the cell area.
- ☞ Movements of the MSs are uncorrelated; the directions of their movements are uniformly distributed on $[0, 2\pi]$.

The optimum number of cells (N_{copt}) per LA is given as:

$$N_{copt} = \sqrt{\left(\frac{C_{page}}{C_{LU}}\right) \cdot \frac{v}{\pi R}} \tag{14.2}$$

where:
C_{page} = cost of paging (in terms of the number of paging messages required to find an MS)
C_{LU} = cost of LUs (in terms of the number of LU messages required to update the location of the MS)
R = cell radius
v = mean mobile velocity in LA

Using the Eq. (14.1) number of LUs, $\lambda_{LU}^{(j)}$ in an LA perimeter of the j-th cell per hour will be:

$$\lambda_{LU}^{(j)} = \frac{vL\rho}{\pi} \tag{14.3}$$

where:

$L = 6R\left[\frac{1}{3} + \frac{1}{2\sqrt{3N_c - 3}}\right]$, length (km) of cell exposed perimeter,

ρ = density of MS in the cell (MSs per km^2),
N_c = number of cells per location area (LA), and
R = cell radius.

A location update uses an SDCCH. The SDCCH allocated to an MS consists of 4 time slots (for the SDCCH blocks) every 51 TDMA multiframes. If we consider a channel as a single time slot in a TDMA frame, then a channel can accommodate 8 SDCCHs; in other words, an LU consumes (during signaling exchange) 1/8 TCH/F channel (i.e., 8 MSs can alternatively share the same time slot every 51 TDMA multiframes).

The SDCCH/SACCH resource occupancy in the j-th cell due to MS LUs is given as:

$$T_{rLU}^{(j)} = \lambda_{LU}^{(j)} \cdot \left[\sum_{i=1}^{3} (p_{LU}^{(i,j)} \cdot t_{LU}^{(i)})\right] \tag{14.4}$$

where:

$p_{LU}^{(i,j)}$ = percentage of LUs in the i-th case for the j-th cell,

$t_{LU}^{(i)}$ = average duration of one LU in the i-th case (i = 1: intra-VLR; i = 2: inter-VLR with TMSI; and i = 3 inter-VLR with IMSI),

$T_{rLU}^{(j)}$ = resource occupancy in the j-th cell due to MS LUs, and

$\lambda_{LU}^{(j)}$ = number of transactions processed by MSC/VLR for one LU in the i-th case.

The total number of transactions due to LUs generated in $N_{LA} \cdot N_p$ LA perimeter cells (which we number from 1 to $N_{LA} \cdot N_p$) and processed per hour by the MSC/VLR is given as:

$$TN_{LU}^{(j)} = \lambda_{LU}^{(j)} \cdot \left[\sum_{j=1}^{N_{LA} \cdot N_p} \left(\sum_{i=1}^{3} p_{LU}^{(i,j)} \cdot t_{LU}^{(i)}\right)\right] \tag{14.5}$$

where:
N_{LA} = number of LAs managed by MSC/VLR,

$N_p = 6\sqrt{\frac{N_c}{3}} - 3$ = number of cells located on the perimeter of an LA, and

$TN_{LU}{}^{(j)}$ = total number of transactions due to LUs.

E X A M P L E　　　　　1 4 – 1

Problem Statement

Use the following data for DCS-1800 and evaluate the impact of LUs on the radio resource and calculate MSC/VLR transaction load.

- Density of MS in the cell = 10,000 MS/km²
- Cell radius = 500 m
- Average moving velocity of an MS = 10 km/hour
- Number of cells per LA = 10
- Number of LAs per MSC/VLR = 5
- Number of transactions and duration of the transactions to MSC/VLR per LU for different LUs are given in Table 14.1.

We consider two cases:

Case 1: an optimistic situation in which generated LUs in a cell are only intra-VLR LUs

Case 2: a pessimistic situation where generated LUs in a cell are inter-VLR LUs.

Solution

Case 1: We consider the j-th cell located at the border of two LAs related to the same VLR. In this case only, the intra-VLR LUs are processed in the cell.

$$L = 6R\left[\frac{1}{3} + \frac{1}{2\sqrt{3N_c} - 3}\right] = 6 \times 500 \cdot \left(\frac{1}{3} + \frac{1}{2\sqrt{3 \times 10} - 3}\right) = 1.377 \text{ km}$$

$$\lambda_{LU}^{(j)} = \frac{vL\rho}{\pi} = \frac{10 \times 1.377 \times 10,000}{\pi} = 43,826 \text{ LUs per hour}$$

$$T_{rLU}^{(j)} = \lambda_{LU}^{(j)} \cdot \left[\sum_{i=1}^{3} (p_{LU}^{(i,j)} \cdot t_{LU}^{(i)})\right] = \frac{43,826 \times (600/1000)}{3600} = 7.3 \text{ Erlangs}$$

Table 14.1 Number and Duration of Transactions for Different LUs

Transaction Type	No. of Transactions /LU	Duration of a Transaction
Intra-VLR LU	2	600 ms
Inter-VLR LU with TMSI	14	3500 ms
Inter-VLR LU with IMSI	16	4000 ms

This requires 14 channels at 1% blocking (see Erlang B table, appendix A) or 14/8 = 1.75 TCH (about 1/4 of an RF channel).

Case II: Consider the j-th cell located at the border of two LAs related to two different VLRs. In this case, only inter-VLR LUs will be processed in the cell. We assume 80% of LUs are with TMSI and 20% of LUs are with IMSI.

$$T_{rLU}^{(j)} = \frac{43{,}826}{3600} \cdot [0.8 \times 3.5 + 0.2 \times 4.0] = 43.83 \text{ Erlangs}$$

This requires 57 channels at 1% blocking (see Erlang B table, appendix A) or 57/8 = 7 TCH (about 1 RF channel).

MSC/VLR Transaction Load

We assume that one LA is in the center of the area and the remaining four LAs are on the border of the area. We also assume that, in the perimeter cells at the border LAs, only intra-VLR LUs are generated. For the other half of the perimeter cells at the border LAs, only inter-VLR LUs are generated. The number of cells where intra-VLR LUs are generated will be:

$$N_p + \frac{4}{2}N_p = 3N_p = 18\sqrt{\frac{N_c}{3}} - 9 = 18\sqrt{\frac{10}{3}} - 9 = 23.86 \approx 24$$

Similarly, the number of cells where inter-VLR LUs take place will be:

$$\frac{4}{2}N_p = 2N_p = 12\sqrt{\frac{N_c}{3}} - 6 = 12\sqrt{\frac{10}{3}} - 6 = 15.9 \approx 16$$

$$TN_{LU}{}^{(j)} = \lambda_{LU}^{(j)} \cdot \left[\sum_{j=1}^{N_{LA} \cdot N_p} \left(\sum_{i=1}^{3} p_{LU}^{(i,j)} \cdot t_{LU}^{(i)} \right) \right]$$

$$= 43{,}826 \cdot [2 \times 24 + 16(0.8 \times 14 + 0.2 \times 16)] = 12.2 \times 10^6 \text{ transactions at peak hour}$$

Transactions at Peak Hour

The results show that, under heavy traffic conditions, the impact of LUs can be significant. In terms of radio channels used, we conclude that between 1/4 to 1 GSM RF carrier can be used for LA boundary crossings. Although this burden cannot directly cause call blocking on the radio interface (in a DCS-1800 network with a 4 x 3 reuse cluster, the average number of RF carriers per cell with 3 operators is about 10), it is nevertheless not a negligible impact on TCH consumption. In terms of processing at the MSC/VLR side, with a processing load of about 12.2 $\times 10^6$ transactions per hour, it is clear that blocking can rapidly occur with the given scenario. The MSC/VLR resource dedicated to LU processing cannot be used to provide services. A major concern of operators, whose objectives are to provide users with rich and sophisticated services, is requiring more processing resources at the MSC/VLR side.

14.4 PLANNING OF A WIRELESS NETWORK

The planning of a wireless network is a multidiscipline task in which competing requirements are balanced. Network plans change throughout the life of a wireless network. During the initial start-up phase, *coverage* is the big issue and *traffic demand* is minimal. Too small a network initially will require expansion later; in the first few years, human resources need to be focused on the rollout of the network, not on reengineering the existing network.

To determine the size of a wireless network, we need to know about

☞ Network topology; BTS-BSC-MSC
☞ Link capacity; BTS-BSC, BSC-MSC
☞ BSC sizing
☞ MSC sizing

Cell and frequency planning in a wireless network involves

☞ Initial traffic estimation
☞ Traffic growth plan
☞ Initial design selection
☞ Initial cell plan
☞ Network expansion plan
☞ Selection of reuse pattern
☞ Development of initial theoretical plan
☞ Computer simulation
☞ Localized performance measurement
☞ Finalized network design

To design a traffic plan for wireless network deals, we must have information about

☞ The number of users
☞ Users' behavior, i.e., heavy users or light users
☞ Busy-hour traffic as a percentage of the total traffic
☞ Users' distribution over the service area
☞ Division of the service area into zones of different traffic density

Growth of a wireless network involves

☞ Time to roll out the network
☞ Requirements of additional cells
☞ Requirements for microwave link sites
☞ Cost to rework network vs. initial deployment

Basic traffic engineering of a wireless network is based on coverage. In this case, a maximum cell radius based on the required in-building coverage criterion is used. Often a maximum cell radius of less than 10 km is selected to determine the required number of cell sites in the coverage area. For some in-building coverage cell radius may be reduced to 5 to 7 km or even less to achieve good in-building coverage.

To design for capacity, market forecasts of average subscriber call minutes per month are used to determine the number of busy hours per day. This information is then used to calculate the **average busy-hour traffic** per subscriber. Knowing the spectrum availability, the frequency reuse plan is established. The traffic capacity per cell is determined based on a Grade of Service (GoS) criterion. Market forecasts for total number of subscribers for the given geographic zone are used to establish the number of BSs in the zone. The geographic area for each zone is quantified, and the number of BSs and cell sizes for each zone is calculated. During the first few years the network design is coverage driven, and in later years it is traffic-demand driven. To limit expanding cell sites in the first few years when human resources are focused on network roll-out, the initial installation is based on a three- or four-year design.

E X A M P L E   1 4 – 2

Problem Statement

Using the following data for a GSM system, calculate:

1. Average busy-hour traffic per subscriber
2. Traffic capacity per cell
3. Required number of BSs per zone and the hexagonal cell radius for the zone.

- Subscriber usage per month = 120 minutes
- Days per month = 24
- Busy hours per day = 5
- Allocated spectrum = 5 MHz
- Frequency reuse plan = 4/12
- RF channel width = 200 kHz, full rate
- Capacity of a BTS = 32 Erlangs
- Subscribers in the zone = 60,000
- Area of the zone = 500 km^2

Solution

$$\text{Erlangs per subscriber} = \frac{120}{24 \times 5 \times 60} = 0.0167$$

$$\text{Number of RF carriers} = \frac{5000}{200} = 25$$

$$\text{RF carrier per sector} = \frac{25}{4 \times 3} \cong 2$$

TCHs per sector $= 2 \times 8 = 16$

Traffic capacity of a sector at 2% GoS (using Erlang B table in appendix A) = 9.82 Erlangs

Traffic per BTS $= 9.82 \times 3 \cong 29.5 < 32$ Erlangs (this is OK)

$$\text{Maximum subscribers per BTS} = \frac{29.5}{0.0167} \cong 1,766$$

$$\text{Number of BTS in a zone} = \frac{60,000}{1,766} \cong 34$$

$$\text{Average hexagonal cell radius} = \sqrt{\frac{500}{34 \times 2.6}} = 2.38 \text{ km}$$

14.5 RADIO DESIGN FOR A CELLULAR/PCS NETWORK [5]

Many factors need to be considered in the design of a cellular/PCS network. For example, the extent of radio coverage for indoor locations, the quality of service for different environments, efficient use of the spectrum, and the evolution of the network are some of the key factors that need to be carefully evaluated by all prospective service providers. Often these factors are further complicated by the constraints imposed by the operating environments and regulatory issues. A system designer must carefully balance all trade-offs to ensure that the network is robust, future proof, and of high service quality.

14.5.1 Radio Link Design

For a wireless communications system, the first important step is to design the radio link. This is required to determine the BS density in different environments as well as the corresponding radio coverage. For a wireless network to provide good QoS in indoor and outdoor environments, flexibility and resilience should be incorporated into the design. The transmit power of handsets will be the determining factor for a GSM system with balanced up/downlink power.

Although the mobile antenna gain does not affect the balancing of the link budget, it is an important factor in the design of the power budget for handset coverage. From a user point of view, a cellular/PCS network should imply that there is little restriction on making or receiving calls within a building or traveling in a vehicle using handsets. A system should be designed to allow the antenna of a handset to be placed in nonoptimal positions. In addition, the antenna may not even be extended when calls are being made or received. In normal system designs, it is assumed that the gain of a mobile antenna is 0 dBi (dBi refers to the gain relative to an isotropic antenna). However, allowing for the handset antennas to be placed in suboptimal positions, a more conservative gain of −3 dBi should be used. In reality, antenna gain could be as low as −6 to −8 dBi because of the positioning of an antenna in an arbitrary position or with the antenna retracted into the handset housing, depending on specific handsets and their corresponding housing designs.

14.5.2 Coverage Planning

The most important design objective of a cellular/PCS network is to provide a near-ubiquitous radio coverage. One of the most important considerations in the radio coverage planning process is the propagation model. The accuracy of the prediction by a particular model depends on its ability to account for the detailed terrain, vegetation, and buildings. This accuracy is of vital importance in determining path loss and, hence, cell sizes and infrastructure requirements of a cellular/PCS network. An overestimation will lead to an inefficient use of the network resources, whereas an underestimation will result in poor radio coverage. Propagation models generally tend to oversimplify real-life propagation conditions and may be grossly inaccurate in complex metropolitan urban environments. The empirical propagation models provide only general guidelines; they are too simplistic for accurate network design. Accurate field measurements should be made to provide information on radio coverage in an urban environment. Measured data can be used either directly in the planning process to access the feasibility of the individual cell site or to calibrate the coefficients of the empirical propagation model to achieve better characterization of a specific environment.

Radio propagation in an urban environment is subject to shadowing. To ensure that the signal level in 90 percent of the cell area is equal to or above the specified threshold, a shadow fading margin, which is dependent on the standard deviation of the signal level, must be included in the link budget. For a typical urban environment, a shadow fading margin of 8 to 9 dB should be used based on the assumption that path loss follows an inverse exponent law. Path loss is inversely proportional to the distance of separation raised to a power between 2 and 5. The value of the power is dependent upon propagation characteristics.

Another critical factor that affects radio coverage is penetration loss for both buildings and vehicles. If radio coverage for the outer portion of a building is sufficient, then an assumed penetration loss of 10 to 15 dB should be adequate. However, if calls are expected to be received and originated within the inner core of the building, a penetration loss of about 20 to 30 dB should be used. Similarly, for in-vehicle coverage, penetration loss is equally important. A car could experience a penetration loss of 3 to 6 dB, whereas vans and buses have even larger variations. The penetration loss at the front of a van should be no more than that experienced in a car, but the loss at the back of a van could be as high as 10 to 12 dB, depending on the amount of window space. Thus, for design purposes, a high penetration loss should be assumed to ensure a good service quality. For an urban environment, because building penetration loss is the dominant factor, in-vehicle penetration will generally be sufficient.

14.6 DESIGN OF A WIRELESS SYSTEM

The design of a wireless system must satisfy

- ☞ Service requirements
- ☞ Constraints for hardware implementation
- ☞ Propagation path loss
- ☞ System requirements

14.6.1 Service Requirements

Currently the basic services of wireless systems include voice transmission and low-rate data transmission less than or equal to 14.4 kbps. In the future, multimedia services with variable-rate transmission are envisioned. The ITU-R Study Group 8 (SG 8) is developing third-generation wireless communications services of about 2 Mbps to support voice as well as data services.

One of the important considerations in the design of a TDMA system is the processing delay due to TDMA framing. Therefore, a TDMA frame length must be determined from the acceptable maximum processing delay. We also have to consider coverage area requirement, the bit rate for SACCH, and the required BER performance for each type of information.

14.6.2 Constraints for Hardware Implementation

The constraints for hardware implementation are important considerations in the design of mobile terminals. To support a high transmission rate, we must evaluate the DSP load required for antifrequency-selective fading techniques. When the DSP load is too high for a handheld mobile terminal, we need some DSP load reduction methods including the reduction of symbol rate, use of higher modulation level, increase of the maximum number of carriers assigned to a terminal, or modification of the fading compensation techniques.

Another important factor in the hardware design is the required linearity of the power amplifier. A requirement of higher linearity decreases the battery life due to lower power efficiency unless an appropriate nonlinearity compensation method is used. A higher spectrally efficient modulation scheme requires higher linearity. Thus, the required linearity of the power amplifier should be determined from a trade-off between the battery life and the required spectral efficiency.

The noise factor of the receiver amplifier is also an important factor because a lower noise figure tends to increase the terminal cost.

14.6.3 Propagation Path Loss

An accurate estimate of propagation path loss is essential when designing terminals and establishing cell location and configuration. Propagation path loss due to distance and shadow fading is required to determine coverage area and cell configuration. Multipath fading, which rapidly changes the received signal level, affects terminal design including the selection of antifading compensation techniques.

14.6.4 System Requirements

Bandwidth allocation is based on each country's regulations. These regulations dictate the number of service providers that support similar services using the same frequency band as well as geographical locations and specifications of different services that use the same or adjacent channels.

14.7 SPECTRAL EFFICIENCY OF A WIRELESS SYSTEM [6]

An efficient use of the frequency spectrum is the most important criterion in the design of a mobile communications system. Several methods are employed to improve spectral efficiency including reduction of the channel width, information compression, variable bit-rate control, and improved channel assignment algorithms. Spectral efficiency of a mobile communications system also depends on the choice of a multiple access scheme and modulation technique. Spectral efficiency, η_T, of a mobile communications system is given as:

$$\eta_T = \frac{n_c \cdot T_c}{A_c \cdot B_w} \tag{14.6}$$

where:
n_c = number of voice channels in each cell,
T_c = traffic per voice channel (Erlangs),
A_c = area of each cell (m^2), and
B_w = system bandwidth (MHz).

If B_c is the channel spacing and N is the reuse factor, then

$$n_c = \frac{B_w}{B_c \cdot N} \tag{14.7}$$

We substitute Eq. (14.7) into Eq. (14.8) to get

$$\eta_T = \eta_s \cdot \eta_c \cdot \eta_t = \frac{1}{NA_c} \cdot \frac{1}{B_c} \cdot T_c \qquad (14.8)$$

where η_s is the spectral efficiency with respect to space and is given as

$$\eta_s = \frac{1}{N \cdot A_c} \ (1/m^2).$$

η_c is the spectral efficiency with respect to frequency and is given as

$$\eta_c = \frac{1}{B_c} \ (channel/Hz);$$

and η_t is the spectral efficiency with respect to time and is given as

$$\eta_t = T_c \ (Erlangs/channel).$$

In this discussion we have assumed voice transmission services. In case of nonvoice transmission, spectral efficiency is often expressed in terms of bps/Hz/cell instead of Erlangs/Hz/m^2.

For nonvoice transmission, we can rewrite Eq. (14.6) as:

$$\eta_T = \eta_s \cdot \eta_c = \frac{1}{N} \cdot \frac{\log_2 M \cdot R_s}{B_c} \ (bps/Hz/cell) \qquad (14.9)$$

where:

$$\eta_c = \frac{\log_2 M \cdot R_s}{B_c} \ (bps/Hz),$$

$$\eta_s = \frac{1}{N} \ (1/cell),$$

R_s = symbol rate, and
M = number of bits per symbol.

η_t is determined by channel spacing and is independent of η_s and η_c. η_s and η_c are determined by modulation and access technology. They are dependent on each other. For example, using a modulation scheme with a higher modulation level, we increase η_c but degrade η_s because robustness to cochannel interference is decreased with an increasing modulation level due to its shorter minimum signal distance. Therefore, we should maximize the product of $\eta_c \cdot \eta_s$ when we optimize parameters of the modulation and access techniques.

14.8 RECEIVER SENSITIVITY AND LINK BUDGET

We are required to calculate the minimum received signal power to satisfy a certain BER. This is referred to as receiver sensitivity, or S_{min}. For voice transmission, a BER of 10^{-3} is usually used. We should also know under which conditions—AWGN or fading—the BER is evaluated. When the holding time for a radio channel is relatively long when compared to the fading duration as in the case of voice transmission, we evaluate the BER under fading conditions. In this case the fade margin is determined from the statistics of the large-scale signal variation based on the log-normal distribution. On the other hand, when the holding time of a radio channel is relatively short as in the case of data transmission, we evaluate the BER under static conditions. In this case, fade margin is determined from the combined statistics of both large-scale and small-scale signal variations based on the log-normal and Rayleigh distributions, respectively. In this section we focus only on voice transmission. At the receiver, noise is generated at the front-end RF amplifier (see Figure 14.1).

The thermal noise power spectral density at the receiver RF amplifier is given as:

$$N_T = N_0 + N_f = 10 \log kT + N_f \quad \text{dB} \tag{14.10}$$

where:
k is the Boltzmann's constant and is assumed to be 1.38×10^{-20} mW/Hz/°K,
T is the temperature in °K,
with $T = 290$°K, N_0 is about -174 dBm/Hz, and
N_f is the noise figure of the receiver amplifier.

Next we obtain the relationship between the minimum signal power, S_{min}, and the required $(S/N_T)_{req}$ or $(E_b/N_T)reqd$.

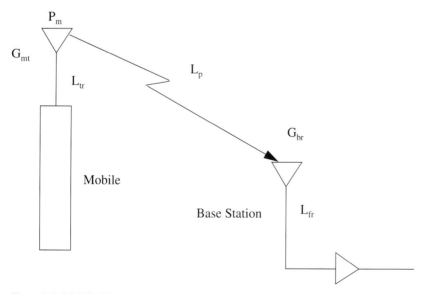

Fig. 14.1 Mobile Transmitter and BS Receiver

$$S_{min} = (S/N_T)_{reqd} + N_0 B_c N_f = (E_b/N_T)_{reqd}(R_b/B_c) + N_0 B_c N_f \quad (14.11)$$

or

$$S_{min} = (E_b/N_T)_{reqd} + 10\log R_b + N_T \text{ dB} \quad (14.12)$$

where E_b is the bit energy,
R_b is the information rate,
B_c is the channel width, and
N_T is the total noise density.

If P_m is the average transmitter power of the MS over the burst duration for a single TCH, the EIRP is given as:

$$P_{EIRP} = P_m - L_{tr} + G_{mt} \quad (14.13)$$

where:
L_{tr} = power loss due to the transmitter cable, and
G_{mt} = mobile transmitter antenna gain relative to an isotropic antenna.

The received signal strength will be

$$P_{rec} = P_{EIRP} - L_p + G_{br} - L_{fr} \quad (14.14)$$

Substituting Eq. (14.13) into Eq. (14.14) we get

$$P_{rec} = P_m - L_p + (G_{br} + G_{mt}) - (L_{fr} + L_{tr}) \quad (14.15)$$

where:
L_p is the propagation path loss due to distance between the transmitter
       and receiver,
G_{br} is the BS receiver antenna gain, and
L_{fr} is the power loss due to cable and connectors in the BS receiver.

The maximum allowable path loss L_{pmax} will be:

$$L_{pmax} = (P_m - S_{min}) + (G_{br} + G_{mt}) - (L_{tr} + L_{fr} + f_m) \quad (14.16)$$

where:
f_m is the fade margin to reduce outage probability due to shadowing.

Using Eq. (14.16) we can determine the coverage area of the system. The key factors in Eq. (14.16) are the relationship between the path loss and distance and the estimate of the fade margin f_m.

E X A M P L E   1 4 – 3

Problem Statement
Calculate the minimum signal power required for the acceptable quality of voice at the BS receiver of a GSM system. Assume receiver noise figure N_f to be 5 dB, receiver temperature equal to 290°K (room temperature), and Boltzman's constant equal to 1.38×10^{-20} Joules/°K.

What is the maximum allowable path loss? The effective isotropic radiated power of the MS is 1 W (30 dBm), transmitter cable loss is 0 dB, receiver cable loss amounts to 2.5 dB, and transmitter and receiver antenna gain are 0 dBi and 12 dBi, respectively. Assume a fade margin of 10 dB and required E_b/N_T of 13.5 dB.

Solution

- BS noise floor

$$N_T = 10\log KT + N_f = 10\log(1.38 \times 10^{-20} \times 290) + 5 = -174 + 5 = -169 \text{ dB/Hz}$$

- Required S/N_T

$$(S/N_T)_{reqd} = (E_b/N_T)_{reqd} + 10\log(R_b/B_c) = 13.5 + 10\log(271/200) = 14.8 \text{ dB}$$

- Minimum signal power required

$$S_{min} = (E_b/N_T)_{reqd} + 10\log R_b + N_T = 13.5 + 10\log 271 \times 10^3 - 169 = -101.2 \text{ dBm}$$

- Maximum allowable path loss

$$L_{pmax} = 30 - (-101.2) + (12 + 0) - (2.5 + 0 + 10) = 130.7 \text{ dB}$$

14.9 SELECTION OF MODULATION SCHEME

In the TDMA system, several factors must be considered in selecting a modulation scheme. These factors are

- ☞ Modulation level, M
- ☞ Symbol rate, R_s
- ☞ Roll-off factor of transmitter and receiver root Nyquist filter, α
- ☞ Transmitter power requirement, P_t
- ☞ Maximum delay spread, τ_{dmax}
- ☞ Channel coding and interleaving schemes
- ☞ Antifading method
- ☞ Terminal cost and size
- ☞ Battery life
- ☞ Robustness of terminal

The symbol rate R_s depends on various factors. For a high transmission rate, we must consider the maximum delay spread τ_{dmax} to be compensated. The number of computations increases linearly with the square of the symbol rate R_s. This results in a substantial increase in the DSP load. The DSP load can be reduced either by increasing the modulation level M or increasing the assigned carriers to a terminal. Both these approaches demand higher linearity of the transmitter amplifier. This may result in reducing battery life. A trade-off is generally made between the linearity of the RF amplifier, battery life, and DSP load.

For a linear modulation scheme such as $\pi/4$-QPSK, a root Nyquist filter is considered to be an optimum filter for the transmitter and receiver. The choice of roll-off factor α is important. A smaller value of α is desirable for reducing adjacent-channel interference [3]. A value of $\alpha = 0.5$ has been used for the TDMA systems in North America and Japan.

Channel coding is important to improve receiver sensitivity and spectral efficiency. It is chosen to maximize spectral efficiency at an appropriate BER depending on the type of transmitted information. The coding scheme for a voice encoder is based on BER from 10^{-3} to 10^{-2}, whereas for nonvoice transmission a BER of 10^{-6} to 10^{-5} is used.

14.10 Design of TDMA Frame [5]

In a TDMA system, variable-rate transmission is achieved by changing the number of assigned time slots in the frame to a user. The minimum bit rate R_{bmin} is supported by assigning one time slot in the frame to a user, and the maximum bit rate R_{bmax} is provided by assigning all the time slots in the frame to a user.

The required number of time slots N_{slot} in a TDMA frame can be given by:

$$N_{slot} = \frac{R_{bmax}}{R_{bmin}} \cdot \frac{1}{N_{ca}} \tag{14.17}$$

where N_{ca} is the number of carriers assigned to user.

The length of the frame T_f is determined from the maximum processing delay τ_{pmax}.

$$T_f \leq \frac{1}{2}\tau_{pmax} \tag{14.18}$$

If the bit rate for the SACCH is a_1 times the minimum information bit rate, the required symbol rate R_s can be given as:

$$R_s = \frac{R_{bmin} \cdot (1 + a_1) \cdot N_{slot}}{\eta_f \cdot R_c \log_2 M} \tag{14.19}$$

where:
η_f = TDMA frame efficiency,
R_c = coding rate of the channel encoder,
M = modulation level,
N_{slot} = number of time slots in the TDMA frame,
R_{bmin} = minimum bit rate supported,
and $a_1 \cdot R_{bmin}$ bit rate for SACCH.

In the TDMA frame, the guard time is used to prevent slot collision due to a difference in terminal locations. With a guard time of G, we can support a cell radius R of $0.5Gc$, where c is the velocity of light. Guard time may reduce frame efficiency as well as spectral efficiency. Alignment of the uplink transmission timing controlled by the BS is an effective method to reduce the guard time.

14.11 RELATIONSHIP BETWEEN DELAY SPREAD AND SYMBOL RATE

Table 14.2 provides the relationship between delay spread in terms of symbols and corresponding delay time for various symbol rates. When the compensated delay spread is limited to 6 symbols, the maximum delay spread is 3 μs for 2 Msymbols/s, 6 μs for 1 Msymbols/s, 12 μs for 500 ksymbols/s, and 24 μs for 250 ksymbols/s. For cellular systems with a microcell of radius 100–500 m, we can expect maximum delay spread of less than 1 μs. Thus, we can use a symbol rate of 2 Msymbols/s. For a cell radius of about 1 km with a BS antenna that is not too high, we can use 1 Msymbols/s in most of the cases because the maximum delay spread in such a case is expected to be less than several μs. On the other hand, for a large cell with a radius of about 10 km, we must reduce the symbol rate to 500 ksymbols/s or less because of the large delay spread.

Table 14.2 Relationship between Compensated Maximum Delay Spread in Terms of Symbols & Corresponding Delay Time

Compensated Max. Delay Spread (symbols)	Compensated Max. Delay Spread (μs) Symbol Rate			
	2 M	1 M	500 k	250 k
3	1.5	3.0	6.0	12.0
4	2.0	4.0	8.0	16.0
5	2.5	5.0	10.0	20.0
6	3.0	6.0	12.0	24.0
7	3.5	7.0	14.0	28.0
8	4.0	8.0	16.0	32.0
9	4.5	9.0	18.0	36.0
10	5.0	10.0	20.0	40.0

EXAMPLE 1 4 – 4

Problem Statement

Design a cellular system using the GMSK modulation scheme. The maximum transmitted power of the mobile is 1 W (30 dBm). The following data are given:

- Carrier frequency = 900 MHz
- Log-normal fading with standard deviation = 8 dB, providing a shadow margin of 10.5 dB with 90% cell coverage

- Mobile antenna gain = 1 dBi
- BS antenna gain = 16 dBi
- Cable and connector loss at mobile = 1 dB
- Cable and connector loss at BS = 1 dB
- Noise figure of receiver amplifier at the BS = 4 dB
- Path loss according to Hata's model

The BER performance vs. E_b/N_T for the GMSK and QPSK modulation is given in Table 14.3.

The maximum delay spread equivalent to 6-symbol duration is assumed and selection combining is used as the diversity combining technique.

Solution

We first calculate receiver sensitivity with GMSK modulation by assuming a symbol rate of 2 Msymbols/s and BER 10^{-2}.

$$S_{min} = -174 + 4 + 10\log 2 \times 10^6 + 14 = -93 \text{ dBm}$$

Similarly, with QPSK modulation the receiver sensitivity at 2 Msymbols/s and a BER 10^{-2} will be:

$$S_{min} = -174 + 4 + \log 2 \times 10^6 + 12 = -95 \text{ dBm}$$

We made similar calculations for BER = 10^{-2} and BER = 10^{-3} for the two modulation schemes, each at symbol rates of 2.0, 1.0, 0.5, and 0.25 Msymbols/s. The results are summarized in Table 14.4.

Next we use Hata's model to calculate the path loss:

$$L_p = 69.55 + 26.16\log f_c - 13.83\log h_t - a(h_r) + [44.9 - 6.55\log h_t] \cdot \log d \text{ dB}$$

where:
f_c = carrier frequency, $150 \le f_c \le 1500$ MHz,
h_t = BS antenna height, $30 \le h_t \le 200$ m,
$a(h_r)$ = correction factor for the mobile antenna height,
h_r = mobile antenna height (assume 3 m), and
d = distance from BS, $1 \le d \le 20$ km.

For a large city $a(h_r) = 3.2(\log 11.75 h_r)^2 - 4.97$ dB , $f_c \ge 400$ MHz

Table 14.3 BER Performance for GMSK and QPSK

Modulation	BER	
	10^{-2}	10^{-3}
GMSK	14.0 dB	18.0 dB
QPSK	12.0 dB	15.6 dB

Table 14.4 Receiver Sensitivity with Parameters of Required BER and Symbol Rates

		Receiver Sensitivity	
Required BER	Symbol Rate (Msymbols/s)	GMSK (dBm)	QPSK (dBm)
BER = 10^{-2}	2.0	−93.0	−95.0
	1.0	−96.0	−98.0
	0.50	−99.0	−101.0
	0.25	−102.0	−104.0
BER = 10^{-3}	2.0	−89.0	−91.4
	1.0	−92.0	−94.4
	0.50	−95.0	−97.4
	0.25	−98.0	−100.4

We assume a cell radius of 10 km and a BS antenna height = 160 m. With a mobile antenna height of 3 m, the correction factor is:

$$a(h_r) = 3.2(\log 11.75 \times 3)^2 - 4.97 = 2.69 dB$$

$$L_p = 69.55 + 26.16\log(900) - 13.83\log 160 - 2.69 + [44.9 - 6.55\log 160] \cdot \log 10 = 144.1 \text{ dB}$$

The required transmitted power for a GMSK MS at BER = 10^{-2} with 0.25 Msymbols/s will be:

$$P_t = S_{min} - (G_t + G_r) + (L_{ft} + L_{fr} + f_m) + L_p$$

$$P_t = -102 - (16 + 1) + (1 + 1 + 10.5) + 144.1 = 37.62 \text{ dBm} > 30 \text{ dBm}$$

Since the required transmitted power is more than the maximum power of the MS, we reduce the cell radius to 5 km. The path loss with a 5-km cell radius will be:

$$L_p = 69.55 + 26.16\log 900 - 13.83\log 160 - 2.69 + [44.9 - 6.55\log 160] \cdot \log 5 = 134.95 \text{ dB}$$

The required power for the MS at BER = 10^{-2} and a symbol rate of 0.25 Msymbols/s will be:

$$P_t = -102 - (16 + 1) + (1 + 1 + 10.5) + 134.95 = 28.45 \text{ dBm} < 30 \text{ dBm}$$

The required power for the MS with a QPSK at BER = 10^{-2} and a symbol rate of 0.25 Msymbols/s will be:

$$P_t = -104 - (16 + 1) + (1 + 1 + 10.5) + 134.95 = 26.45 \text{ dBm} < 30 \text{ dBm}$$

The results are summarized in Table 14.5.

Table 14.5 Summary of Results for GMSK and QPSK

Parameters	GMSK	QPSK
BS Antenna	Height = 160 m Gain = 16 dBi	Height = 160 m Gain = 16 dBi
MS Antenna	Height = 3 m Gain = 1 dBi	Height = 3m Gain = 1 dBi
Transmitter Power (MS)	1 W (30 dBm)	1 W (30 dBm)
Receiver Noise Figure (BS)	4.0 dB	4.0 dB
Symbol Rate	0.25 Msymbols/s	0.25 Msymbols/s
Max. delay spread to be compensated for	24 μs (6 symbols)	24 μs (6 symbols)
Coverage	90% of cell area	90% of cell area
Required Transmitter Power (MS)	28.45 dBm	26.45 dBm

E X A M P L E   1 4 – 5

Problem Statement

Repeat the example 14.4 for the PCS system using QPSK modulation with a carrier frequency of 1.8 GHz; BS antenna height of 50 m and gain of 12 dBi; MS antenna height of 0.2 m and gain of 0 dBi; and the maximum transmitted power of the MS is equal to 0.40 W (26 dBm). All other information is the same as in example 14.4.

We assume a cell radius of 1 km and calculate the path loss using Hata's model.

Solution

$$a(h_r) = 3.2(\log 11.75 \times 0.2)^2 - 4.97 = -4.53 \text{ dB}$$

$$L_p = 69.55 + 26.16\log(1800) - 13.83\log(50) - (-4.53) + (44.9 - 6.65\log 50) \cdot \log 1 \quad \text{dB}$$

$$L_p = 69.55 + 85.16 - 23.50 + 4.53 = 135.74 \text{ dB}$$

Required transmitted power for QPSK at BER = 10^{-3} with 2 Msymbols/s will be:

$$P_t = S_{min} - (G_t + G_r) + (L_{ft} + L_{fr} + f_m) + L_p$$

$$P_t = -91.4 - (12 + 0) + (1 + 1 + 10.5) + 135.74 = 44.84 \text{ dBm} > 26 \text{ dBm}$$

Since the required transmitted power is more than the maximum transmitted power of the MS, we reduce the cell radius to 250 m. The path loss with d = 250 m will be 115.41 dB and the required power will be:

$$P_t = -91.4 - 12 + 12.5 + 115.41 = 24.51 dBm < 26 \text{ dBm (this is OK)}$$

System Parameters for 250-m PCS cell:

BS Antenna	Height = 50 m
	Gain = 12 dBi
MS Antenna	Height = 0.2 m
	Gain = 0 dBi
Required Transmitter Power (MS)	24.51 dBm
Transmitter Power (MS)	0.4 W (26 dBm)
Receiver Noise Figure (BS)	4.0 dB
Symbol Rate	2.0 Msymbols/s
Maximum delay spread to be compensated for	3 µs (6 symbols)

E X A M P L E   1 4 – 6

Problem Statement

Design a TDMA frame for a cellular system to support variable bit rates from 8 kbps to 128 kbps. A user can be assigned multiple carriers (not more than 2). Assume GMSK modulation, a coding rate of R_c = one-half, frame efficiency of 75%, and the symbol rate of the SACCH—a_1 = 0.1 R_s. The cell radius is limited to 5 km and maximum processing delay to 90 ms. The velocity of light is c = 3 x 10^8 mps.

Solution

We first calculate the required number of time slots with N_{ca} = 1 and N_{ca} = 2.

$$N_{slot} = \frac{R_{bmax}}{R_{bmin}} \cdot \frac{1}{N_{ca}} = \frac{128}{8} \cdot \frac{1}{N_{ca}} = \frac{16}{N_{ca}}$$

For N_{ca} = 1, N_{slot} = 16 and for N_{ca} = 2, N_{slot} = 8.

Next we calculate the required symbol rate for each value of N_{slot}.

For N_{slot} = 16

$$R_s = \frac{R_{bmin} \cdot (1 + a_1) \cdot N_{slot}}{\eta_f \cdot R_c \log_2 M} = \frac{8 \cdot (1 + 0.1) \cdot 16}{0.75 \times 0.5} = 375 \text{ ksymbols/s}$$

Similarly, R_s = 187.5 ksymbols/s with N_{slot} = 8.

For the frame length $T_f \le \frac{90}{2} \le 45$ ms, we choose T_f = 40 ms. We then calculate the guard time as:

$$R = 0.5Gc \text{ or } G = \frac{2R}{c}$$

The results are summarized as

N_{ca}	1	2
N_{slot}	16	8

R_s	375 ksymbols/s	187.5 ksymbols/s
T_{slot}	40/16 = 250 µs	40/8 = 500 µs
G	33.34 µs	33.34 µs
G / T_{slot}	13.34%	6.67%

Since the ratio between the guard time and time slot should be less than or equal to 10 percent, we select the TDMA frame with 8 time slots, each of 500 µs, and symbol rate of 187.5 ksymbols/s. Each user will be assigned two carriers.

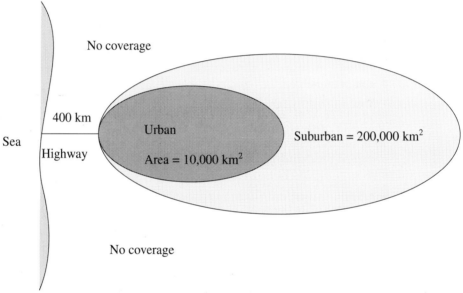

Fig. 14.2 Design of a GSM System

14.12 DESIGN EXAMPLE FOR A GSM SYSTEM [7]

Problem Statement

The government has allowed your company to set up a GSM mobile phone business. Your task is to determine how many pieces of equipment will be required (see Figure 14.2).

Your mobile phone company has been awarded the right to build a network to cover the city, suburban area, and highway connecting the city with the ocean. You do not have to purchase spectrum.

The GSM system data is

☞ RF channel bandwidth (*Channel*): 200 kHz

☞ Number of speech channels per RF channel (*SpeechperRF*): 8

☞ BTS antenna gain (G_{ms}): 20 dBi

☞ MS antenna gain (G_{bts}): 0 dBi

☞ Receive cable connector loss (L_{rm}): 2 dB

☞ Noise figure, N_f: 7

☞ GoS = 2%

☞ Maximum system temperature in degrees Kelvin (T): 310°

☞ Percentage of mobile call origination (*orig*): 60%

☞ Percentage of powered terminals (*powered*): 50%

☞ Number of VLR within an MSC (*VLRperMSC*): 1

☞ GSM SS7 link (*linkspeed*): 64 kbps

☞ S/I required = 14 dB

☞ Frequency reuse factor (N) = 4

☞ Growth rate per year (*growth*) = 0.03

☞ Rollout time in years (*ROT*) = 5

☞ Average traffic per subscriber (E) = 0.02 Erlang

☞ Market share (*Share*) = 0.4

☞ Percent area that needs to have an SNR of at least SNR_1 (14 dB) [*AREA$_1$*] = 0.9

☞ Percent area that needs to have an SNR of at least SNR_2 (5 dB) [*AREA$_2$*] = 0.95

☞ Maximum power (p_{max}) = 1 W (30 dBm)

☞ Boltzman's constant in Joules/°Kelvin (k) = 1.38 x 10^{-23}

Equipment

Components	Capacity	Maximum Connections
MSC	250,000 BHC	10 BSCs
BSC	2500 E	150 BTSs
HLR		300,000 Customer

Population Projection

Year	Urban (M)	Suburban (M)
0	10.0000	4.0000
1	10.3000	4.1200
2	10.6090	4.2436
3	10.9273	4.3709
4	11.2155	4.5021
5	11.5927	4.6371

SS7 Bytes

☞ Call origination (*SS7orig*): 670

   ✘ MSC to/from VLR (*MSCtoVLRorig*): 550
   ✘ MSC to/from PSTN (*MSCtoISDNorig*): 120

☞ Call termination (*SS7term*): 858

   ✘ MSC to/from VLR (*MSCtoVLRterm*): 612
   ✘ MSC to/from PSTN (*MSCtoISDNterm*): 120
   ✘ VLR to/from HLR (*VLRtoHLRterm*): 126

☞ Inter-MSC handover (*SS7interho*): 531

   ✘ MSC to/from VLR (*MSCtoVLRinterho*): 148
   ✘ MSC to/from MSC (*MSCtoMSCinterho*): 383

☞ Intra-MSC handover (*SS7intraho*): 148

   ✘ MSC to/from VLR (*MSCtoVLRintraho*): 148

☞ LU, visited VLR (*SS7visitlu*): 896

   ✘ MSC to/from VLR (*MSCtoVLRvisitlu*): 406
   ✘ VLR to/from VLR (*VLRtoVLRvisitlu*): 213
   ✘ New VLR to/from HLR (*newVLRtoHLRvisitlu*): 182
   ✘ Old VLR to/from HLR (*oldVLRtoHLRvisitlu*): 95

☞ LU, home VLR (*SS7homelu*): 461

   ✘ MSC to/from VLR (*MSCtoVLRhomelu*): 406
   ✘ VLR to/from HLR (*VLRtoHLRhomelu*): 55

For the **urban** area:

1. Population, $Pop_{urban} = 10 \times 10^6$
2. Market penetration, $Pene_{urban} = 0.25$
3. One-way bandwidth in MHz, $BW_{urban} = 8.33$
4. Average call holding time in seconds, $ht_{urban} = 120$
5. One-mile intercept path loss in dB/mile, $I_{o_{urban}} = -120$
6. Slope of the path-loss curve, $\gamma_{urban} = 3.63$
7. Total urban area in km², $Area_{urban} = 10,000$
8. Average speed of each subscriber in km/hour, $v_{urban} = 6$

For the **suburban** area:

1. Population, $Pop_{suburban} = 4 \times 10^6$
2. Market penetration, $Pene_{suburban} = 0.15$
3. One-way bandwidth in MHz, $BW_{suburban} = 4.166$
4. Average call holding time in seconds, $ht_{suburban} = 150$

5. One-mile intercept path loss in dB/mile, $Io_{suburban} = -109$
6. Slope of the path-loss curve, $\gamma_{suburban} = 3.84$
7. Total urban area in km^2, $Area_{suburban} = 200,000$
8. Average speed of each subscriber in km/hour, $v_{suburban} = 50$

For the **rural** area:

1. One-way bandwidth in MHz, $BW_{rural} = 4.166$
2. Average call holding time in seconds, $ht_{rural} = 150$
3. One-mile intercept path loss in dB/mile, $Io_{rural} = -103$
4. Slope of the path-loss curve, $\gamma_{rural} = 3.0$
5. Average speed of each subscriber in km/hour, $v_{rural} = 100$
6. Length of highway in km, $Length = 400$

Solution

First we determine the number of sectors required per cell in each environment to satisfy the signal-to-interference ratio (SIR) requirement.

$$\text{The reuse ratio } q = \sqrt{3N} = \sqrt{3 \times 4} = 3.464$$

The worst-case SIR in dB with an omnidirectional antenna in a hexagonal cell is given as (see chapter 4):

$$SIR_{omni_{envr}} = 10\log\left[\frac{1}{6(q-1)^{-\gamma_{env}}}\right]$$

For SIR = 14 dB with $\gamma_{urban} = 3.63$ and $\gamma_{suburban} = 3.84$, we get

Urban environment with omnidirectional antenna: SIR = 6.436 < 14 dB (no good)

Suburban environment with omnidirectional antenna: SIR = 7.258 <14 dB (no good)

Since the cell organization along the highway in the rural area is different (see Figure 14.2), the relationship for SIR in dB is also different with an omnidirectional antenna.

$$SIR_{omni_{rural}} = 10\log\left[\frac{1}{[2N+1]^{-\gamma_{rural}} + [2N-1]^{-\gamma_{rural}}}\right]$$

Rural environment with N = 4 and $\gamma_{rural} = 3$ using omnidirectional antenna: SIR = 23.678 > 14 dB (OK)

Since the required SIR for the urban and suburban environment is not satisfied, we try using a three-sector antenna. The worst-case SIR in dB with a three-sector antenna is given as (see chapter 4):

$$SIR_{3-sector_{envr}} = 10\log\left[\frac{1}{q^{-\gamma_{env}} + (q+0.7)^{-\gamma_{env}}}\right]$$

We evaluate SIR for an urban and suburban environment with a three-sector antenna:

Urban environment with three-sector antenna: SIR = 17.79 > 14 dB (OK)

Suburban environment with three-sector antenna: SIR = 18.98 > 14 dB (OK)

Thus, by using a three-sector antenna in the urban and suburban environment and an omnidirectional antenna in the rural environment, we satisfy the requirement of a minimum SIR of 14 dB.

Next we calculate the maximum cell radius for each environment. The traffic density in an environment at the end of a five-year period is given as:

$$TrafficDensity_{env} = \frac{E \cdot Pop_{env} \cdot Pene_{env} \cdot (1 + growth)^{ROT} \cdot Share}{Area_{env}}$$

For urban environment: $E = 0.02$ Erlang/subscriber, $Pop_{env} = 10 \times 10^6$, $Pene_{env} = 0.25$, $growth = 0.03$, $ROT = 5$, $Share = 0.4$, and $Area = 10,000$ km^2.

$$TrafficDensity_{urban} = \frac{0.02 \times 10 \cdot 10^6 \times 0.25 \times (1+0.03)^5 \times 0.4}{10,000} = 2.319 \text{ Erlangs/km}^2$$

For suburban environment: $E = 0.02$ Erlang/subscriber, $Pop_{env} = 4 \times 10^6$, $Pene_{env} = 0.15$, $growth = 0.03$, $ROT = 5$, $Share = 0.4$, and $Area = 200,000$ km^2.

$$TrafficDensity_{suburban} = \frac{0.02 \times 4 \cdot 10^6 \times 0.15 \times (1+0.03)^5 \times 0.4}{200,000} = 0.028 \text{ Erlang/km}^2$$

The number of voice channels required in each environment is (assuming one RF channel in each environment is not used to carry traffic—this channel is used as a control channel):

$$VoiceChannel_{env} = \left(\frac{BW_{env}}{Channel_{env}} - 1\right) \times \frac{SpeechperRF}{N \cdot Sector_{env}}$$

For urban environment: $BW_{env} = 8.33 \times 10^6$ Hz, $Channel_{env} = 200$ kHz, $SpeechperRF = 8$ RF channels per carrier, reuse factor $N = 4$, and number of sectors per cell is 3.

$$\text{Number of RF Channels} = \frac{8.33 \times 10^6}{200 \times 10^3} - 1 \approx 40$$

$$VoiceChannel_{urban} = \frac{8}{4 \times 3} \times 40 = 26.67$$

For suburban environment: $BW_{env} = 4.16 \times 10^6$ Hz, $Channel_{env} = 200$ kHz, $SpeechperRF = 8$ RF channels per carrier, reuse factor $N = 4$, and number of sectors per cell is 3.

$$\text{Number of RF Channels} = \frac{4.16 \times 10^6}{200 \times 10^3} - 1 \approx 19$$

$$VoiceChannel_{suburban} = 19 \times \frac{8}{4 \times 3} = 12.67 \approx 12$$

For rural environment: $BW_{env} = 4.16 \times 10^6$ Hz, $Channel_{env} = 200$ kHz, $SpeechperRF = 8$ RF channels per carrier, reuse factor $N = 4$, and number of sectors per cell is 1.

$$\text{Number of RF Channels} = \frac{4.16 \times 10^6}{200 \times 10^3} - 1 \approx 19$$

$$VoiceChannel_{rural} = 19 \times \frac{8}{4 \times 1} = 38$$

We calculate the traffic per sector with GoS = 0.02 using the Erlang B traffic tables (see appendix A):

$$TrafficperSector_{urban} = 18.387 \text{ Erlangs (urban)}$$

$$TrafficperSector_{suburban} = 6.615 \text{ Erlangs (suburban)}$$

$$TrafficperSector_{rural} = 29.168 \text{ Erlangs (rural)}$$

The cell area (in km^2) in the urban and suburban environments is calculated as:

$$Cellarea_{urban} = \frac{TrafficperSector_{urban} \times 3}{TrafficDensity_{urban}} = \frac{3 \times 18.387}{2.319} = 23.787 \text{ km}^2$$

$$Cellarea_{suburban} = \frac{TrafficperSector_{suburban} \times 3}{TrafficDensity_{suburban}} = \frac{3 \times 6.615}{0.028} = 708.85 \text{ km}^2$$

The cell radius (hexagonal) in the urban and suburban environments will be:

$$CellRadius_{urban} = \sqrt{\frac{Cellarea_{urban}}{2.6}} = \sqrt{\frac{23.787}{2.6}} = 3.025 \approx 3.0 \text{ km}$$

$$CellRadius_{suburban} = \sqrt{\frac{Cellarea_{suburban}}{2.6}} = \sqrt{\frac{708.85}{2.6}} = 16.52 \approx 16.5 \text{ km}$$

It should be noted that the cell radius along the highway in the rural environment is not governed by traffic but is based on the maximum transmitted power in the rural environment.

Next we calculate the maximum distance from a BTS that an MS can roam based on the maximum transmitted power in a given environment. It must be greater than the cell radii, as determined earlier.

The thermal noise in dB is given as

$$Noise = 10 \cdot \log(k \cdot T \cdot Channel \cdot 10^{N_f/10}) = 10 \cdot \log(1.38 \times 10^{-23} \cdot 200 \times 10^3 \cdot 10^{7/10})$$

$$= -143.677 \text{ dB}$$

Next we determine the minimum received power in urban, suburban, and rural environments required to satisfy criteria for 90-percent and 95-percent coverage.

$$p_{rmin1_{env}} = SNR_1 + Noise - 10\log\left[1 - 10^{\frac{SNR_1 - SIR_{env}}{10}}\right] \text{ dB, and}$$

$$p_{rmin2_{env}} = SNR_2 + Noise - 10\log\left[1 - 10^{\frac{SNR_2 - SIR_{env}}{10}}\right] \text{ dB.}$$

☞ **Urban**

$SNR_1 = 14$ dB, $Noise = -143.677$ dB, $SIR_{envr} = 17.79$ dB, then $p_{rmin1_{urban}} = -127.327$ dB; and $SNR_2 = 5$ dB, then $p_{rmin2_{urban}} = -138.442$ dB

☞ **Suburban**

$SNR_1 = 14$ dB, $Noise = -143.677$ dB, $SIR_{envr} = 18.979$ dB, then $p_{rmin1_{suburban}} = -128.02$ dB; and $SNR_2 = 5$ dB, then $p_{rmin2_{suburban}} = -138.5$ dB

☞ **Rural**

$SNR_1 = 14$ dB, $Noise = -143.677$ dB, $SIR_{envr} = 23.678$ dB, then $p_{rmin1_{rural}} = -129.182$ dB; and $SNR_2 = 5$ dB, then $p_{rmin2_{rural}} = -129.618$ dB

The radius (in km), which covers 90 percent of a cell's area, is given by:

$$radius1_{envr} = \sqrt{\frac{cellarea_{envr} \cdot AREA_1}{\pi}}$$

We calculate $radius1_{urban}$ and $radius1_{suburban}$ as follows:

$$Cellarea_{urban} = 2.6 \times 3^2 = 23.4 \text{ km}^2$$

$$radius1_{urban} = \sqrt{\frac{23.4 \times 0.90}{\pi}} = 2.59 \text{ km}$$

$$Cellarea_{suburban} = 2.6 \times 16.5^2 = 707.85 \text{ km}^2$$

$$radius1_{suburban} = \sqrt{\frac{707.85 \times 0.90}{\pi}} = 14.24 \text{ km}$$

Similarly we calculate $radius2_{urban}$ and $radius2_{suburban}$ for $AREA_2 = 0.95$ as follows:

$$radius2_{urban} = \sqrt{\frac{23.4 \times 0.95}{\pi}} = 2.66 \text{ km}$$

$$radius2_{suburban} = \sqrt{\frac{707.85 \times 0.95}{\pi}} = 14.63 \text{ km}$$

Next, using the path-loss relationship, we calculate the maximum distance in km that an MS can be from the BTS with p_{max} being the maximum transmitted power.

$$d_{max1_{envr}} = \frac{\left[10^{\frac{I_{0_{envr}} + G_{ms} + G_{bts} - L_{rm} + p_{max} - p_{rmin1_{envr}}}{10}} \right]^{1/\gamma_{envr}}}{miperkm}$$

☞ **Urban**

$I_{0_{envr}} = -120$ dB, $G_{ms} = 0$ dB, $G_{bts} = 20$ dBi, $L_{rm} = 2$ dB, $p_{max} = 0$ dB, $p_{rmin1_{envr}} = -127.327$ dB, $\gamma_{urban} = 3.63$, and $miperkm = 0.6214$; then $d_{max1_{urban}} = 8.02$ km $> radius1_{urban} = 2.59$ km; and for $p_{rmin2_{envr}} = -138.442$ dB, $d_{max2_{urban}} = 16.238$ km $> radius2_{urban} = 2.66$ km.

These results indicate that there is enough signal power to satisfy the SNR criteria for 90 percent and 95 percent of the cell area in the urban environment.

☞ **Suburban**

$I_{0_{envr}}$ = −109 dB, G_{ms} = 0 dB, G_{bts} = 20 dBi, L_{rm} = 2, p_{max} = 0 dB, $p_{rmin1_{envr}}$ = −128.016 dB, $\gamma_{suburban}$ = 3.84, and $miperkm$ = 0.6214; then $d_{max1_{suburban}}$ = 14.8 km > radius1$_{suburban}$ = 14.24 km; and for $p_{rmin2_{envr}}$ = −138.5 dB, $d_{max2_{suburban}}$ = 27.76 km > radius2$_{suburban}$ = 14.63 km

These results indicate that there is enough signal power to satisfy the SNR criteria for 90 and 95 percent of the cell area in the suburban environment.

☞ **Rural**

$I_{0_{envr}}$ = −103 dB, G_{ms} = 0 dB, G_{bts} = 20 dBi, L_{rm} = 2 dB, p_{max} = 0 dB, $p_{rmin1_{envr}}$ = −129.182 dB, γ_{rural} = 3.0, and $miperkm$ = 0.6214; then $d_{max1_{rural}}$ = 47.78 km; and for $p_{rmin2_{envr}}$ = −129.618 dB, $d_{max2_{suburban}}$ = 49.42 km

Thus the cell radius in the rural area = $\sqrt{\dfrac{(47.78)^2 \times \dfrac{\pi}{0.90}}{2.6}}$ = 55.37 km

$$Cellarea_{rural} = 2.6 \times 55.37^2 = 7.97 \times 10^3 \text{ km}^2$$

Having established the cell radius in each environment, we determine the number of cells.

☞ **Urban**

Number of cells required in urban area = $\dfrac{Area_{urban}}{Cellarea_{urban}} = \dfrac{10{,}000}{2.6 \times (3)^2} \approx 427$

☞ **Suburban**

Number of cells required in suburban area =

$\dfrac{Area_{suburban}}{Cellarea_{suburban}} = \dfrac{200{,}000}{2.6 \times (16.5)^2} \approx 283$

☞ **Rural**

Number of cells required to cover highway in rural area =

$\dfrac{Length}{2 \times Cell_{radius_{rural}}} = \dfrac{400}{2 \times 55.37} = 3.611 \approx 4$

Assuming that a BSC serves only one type of environment, the required numbers of BSCs in each environment based on traffic will be:

☞ Number of BTSs /BSC in urban environment = $\dfrac{2500}{3 \times 18.387}$ = 45.32 ≈ 45

☞ Number of BSCs required in urban environment = $\dfrac{427}{45}$ = 9.49 ≈ 10

☞ Number of BTSs/BSC in suburban environment =

$\dfrac{2500}{3 \times 6.615}$ = 125.98 ≈ 125

☞ Number of BSCs required in suburban environment = $\dfrac{283}{125}$ = 2.26 ≈ 3

☞ Number of BSCs required in rural area = 1

Next we determine the number of MSCs required in each environment based on maximum traffic. We assume that an MSC serves only one type of environment.

$$MSC_{cap}E_{envr} = \frac{MSC_{cap} \cdot ht_{envr}}{sec\,perhr}$$

Number of MSCs in a given environment =

$$\frac{Cellarea_{envr} \cdot TrafficDensity_{envr} \cdot Cellnum_{envr}}{MSC_{cap}E_{envr}}$$

☞ **Urban**
$Cellarea_{urban}$ = 23.4 km², $TrafficDensity_{urban}$ = 2.319 E/km², $Cellnum_{urban}$ = 427,

$$MSC_{cap}E_{urban} = \frac{MSC_{cap} \cdot ht_{urban}}{sec\,perhr} = \frac{250000 \cdot 120}{3600} = 8333.3$$

$$\text{Number of MSCs required} = \frac{23.4 \times 2.319 \times 427}{8333.3} = 2.78 \approx 3$$

☞ **Suburban**
$Cellarea_{suburban}$ = 707.85 km², $TrafficDensity_{suburban}$ = 0.028 E/km², $Cellnum_{suburban}$ = 283,

$$MSC_{cap}E_{suburban} = \frac{MSC_{cap} \cdot ht_{suburban}}{sec\,perhr} = \frac{250000 \cdot 150}{3600} = 10416.67$$

$$\text{Number of MSCs required} = \frac{707.85 \times 0.028 \times 283}{10416.67} = 0.538 \approx 1.0$$

☞ **Rural area along highway**
Use 1 MSC

$$TrafficDensity_{rural} = \frac{29.168}{7.97 \times 10^2} = 0.00366 \text{ E/km}^2$$

We assume that an HLR can support multiple MSCs.

Capacity of an HLR: $HLR_{cap} = 3 \times 10^5$ subscribers

$$Subscriber_{envr} = \frac{Cellnum_{envr} \cdot TrafficdDensity_{envr} \cdot Cellarea_{envr}}{E/(subscriber)}$$

$$Subscriber_{urban} = \frac{427 \times 2.319 \times 23.4}{0.02} = 1.156 \times 10^6$$

$$Subscriber_{suburban} = \frac{283 \times 0.028 \times 707.85}{0.02} = 2.804 \times 10^5$$

$$Subscriber_{rural} = \frac{4 \times 7.97 \times 10^3 \times 0.00366}{0.02} = 5.833 \times 10^3$$

$$\text{Number of HLR} = \frac{(1.156 \times 10^6 + 2.804 \times 10^5 + 5.833 \times 10^3)}{3 \times 10^5} = 4.8 \approx 5$$

Next we calculate the number of handovers per hour per MSC.

$$BTSperMSC_{envr} = \frac{Cellnum_{envr}}{MSCnum_{envr}}$$

$$BTSperMSC_{urban} = \frac{Cellnum_{urban}}{MSCnum_{urban}} = \frac{427}{3} \approx 142$$

$$BTSperMSC_{suburban} = \frac{Cellnum_{suburban}}{MSCnum_{suburban}} = \frac{283}{1} = 283$$

$$BTSperMSC_{rural} = \frac{Cellnum_{rural}}{MSCnum_{rural}} = \frac{4}{1} = 4$$

$$MSCarea_{envr} = BTSperMSC_{envr} \cdot Cellarea_{envr}$$

$$MSCarea_{urban} = BTSperMSC_{urban} \cdot Cellarea_{urban} = 142 \times 23.4 = 3.23 \times 10^3 \text{ km}^2$$

$$MSCarea_{suburban} = BTSperMSC_{suburban} \cdot Cellarea_{suburban}$$
$$= 283 \times 707.85 = 2 \times 10^5 \text{ km}^2$$

$$MSCarea_{rural} = BTSperMSC_{rural} \cdot Cellarea_{rural} = 4 \times 7.97 \times 10^3 = 3.19 \times 10^4 \text{ km}^2$$

Assuming that each MSC serves a hexagonal area in either the urban or suburban area, we calculate the perimeter of each MSC area in km.

$$MSCperimeter_{envr} = 6 \cdot \sqrt{\frac{MSCarea_{envr}}{2.6}}$$

$$MSCperimeter_{urban} = 6 \cdot \sqrt{\frac{3.323 \times 10^3}{2.6}} = 214.5 \, \text{km}$$

$$MSCperimeter_{suburban} = 6 \cdot \sqrt{\frac{2 \times 10^5}{2.6}} = 1.665 \times 10^3 \, \text{km}$$

The rate of an MS crossing out of or into an MSC area per hour is given as:

$$Crossingrate_{envr} = \frac{\dfrac{Subscribers_{envr}}{MSCarea_{envr}} \times v_{envr} \times MSC_{perimeter}}{\pi}$$

$$Crossingrate_{urban} = \frac{\dfrac{1.156 \times 10^6}{3.323 \times 10^3} \times 6 \times 214.5}{\pi} = 1.425 \times 10^5$$

$$Crossingrate_{suburban} = \frac{\dfrac{0.2804 \times 10^6}{2 \times 10^5} \times 50 \times 1.665 \times 10^3}{\pi} = 3.709 \times 10^4$$

$$Crossingrate_{rural} = \frac{v_{rual} \cdot Subscribers_{rural}}{\left(\dfrac{length}{MSCnum_{rural}}\right) \times 2} = \frac{100 \times 5.833 \times 10^3}{400 \times 2} = 729.2$$

Total handover (HO) per hour per MSC in each environment will be:

$$HOtotal_{envr} = BTSperMSC_{envr} \times \frac{Crossinggrate_{envr}}{\sqrt{BTSperMSC_{envr}}} \times E$$

$$HOtotal_{urban} = \frac{142 \times 1.425 \times 10^5 \times 0.02}{\sqrt{142}} = 3.396 \times 10^4$$

$$HOtotal_{suburban} = \frac{283 \times 3.709 \times 10^4 \times 0.02}{\sqrt{283}} = 1.248 \times 10^4$$

$$HOtotal_{rural} = Crossingrate_{rural} \times BTSperMSC_{rural} \times E = 729.2 \times 4 \times 0.02 = 58.336$$

The total number of inter-MSC handovers will be:

$$HOinter_{envr} = Crossingrate_{envr} \times E$$

$$HOinter_{urban} = 1.425 \times 10^5 \times 0.02 = 2.85 \times 10^3$$

$$HOinter_{suburban} = 3.709 \times 10^4 \times 0.02 = 741.8$$

$$HOinter_{rural} = 729.2 \times 0.02 = 14.58$$

The total number of intra-MSC handovers will be:

$$HOintra_{envr} = HOtotal_{envr} - HOinter_{envr}$$

$$HOintra_{urban} = 3.396 \times 10^4 - 2.85 \times 10^3 = 3.111 \times 10^4$$

$$HOintra_{suburban} = 1.248 \times 10^4 - 741.8 = 1.174 \times 10^4$$

$$HOintra_{rural} = 58.336 - 14.58 = 43.756$$

Next we calculate LUs per hour per MSC.

$$LUtotal_{envr} = VLRperMSC \times Crossingrate_{envr} \cdot (powered - E)$$

$$LUtotal_{urban} = 1 \times 1.425 \times 10^5 \cdot (0.5 - 0.02) = 6.84 \times 10^4$$

$$LUtotal_{suburban} = 1 \times 3.709 \times 10^4 \cdot (0.5 - 0.02) = 1.78 \times 10^4$$

$$LUtotal_{rural} = 1 \times 729.2 \cdot (0.5 - 0.02) = 350$$

LUs for the home VLR per hour per MSC will be:

$$LUhome_{envr} = \frac{HOintra_{envr}}{HOtotal_{envr}} \times LUtotal_{envr}$$

$$LUhome_{urban} = \frac{3.111 \times 10^4}{3.394 \times 10^4} \times 6.84 \times 10^4 = 6.27 \times 10^4$$

$$LUhome_{suburban} = \frac{1.174 \times 10^4}{1.248 \times 10^4} \times 1.78 \times 10^4 = 1.674 \times 10^4$$

$$LUhome_{rural} = \frac{43.752}{58.336} \times 350 = 262.5$$

LUs for the visited VLR per hour per MSC will be:

$$LUvisited_{envr} = \frac{HOinter_{envr}}{HOtotal_{envr}} \times LUtotal_{envr}$$

$$LUvisited_{urban} = \frac{2.85 \times 10^3}{3.39 \times 10^4} \times 6.84 \times 10^4 = 5.74 \times 10^3$$

$$LUvisited_{suburban} = \frac{741.5}{1.248 \times 10^4} \times 1.78 \times 10^4 = 1.058 \times 10^3$$

$$LUvisited_{rural} = \frac{14.584}{58.336} \times 350 = 87.5$$

Next we determine the SS7 traffic (in bytes per hour) for each MSC.

$$SS7origrate_{envr} = Subscribers_{envr} \cdot \left(\frac{E \cdot sec\,per\,hr}{ht_{envr}}\right) \cdot orig \cdot SS7orig$$

$$SS7termrate_{envr} = Subscribers_{envr} \cdot \left(\frac{E \cdot sec\,per\,hr}{ht_{envr}}\right) \cdot (1 - orig) \cdot SS7orig$$

$$SS7intrahorate_{envr} = SS7intraho \cdot HOintra_{envr}$$

$$SS7interhorate_{envr} = SS7interho \cdot HOinter_{envr}$$

$$SS7visitlurate_{envr} = SS7visitlu \cdot LUvisited_{envr}$$

$$SS7homelurate_{envr} = SS7homelu \cdot LUhome_{envr}$$

$$SS7rate_{envr} = SS7origrate_{envr} + SS7termrate_{envr} + SS7intrahorate_{envr}$$

$$+ SS7interhorate_{envr} + SS7visitlurate_{envr} + SS7homelurate_{envr}$$

Next we convert to kbps from bytes per hour

$$SS7kbps_{envr} = \frac{SS7rate_{envr} \cdot bitperbytes}{sec\,per\,hr \cdot 1000}$$

☞ **Urban**

$$SS7origrate_{urban} = 1.156 \times 10^6 \cdot \left(\frac{0.02 \times 3600}{120}\right) \cdot 0.6 \times 670 = 278.83 \times 10^6$$

$$SS7termrate_{urban} = 1.156 \times 10^6 \cdot \left(\frac{0.02 \times 3600}{120}\right) \cdot (1 - 0.6) \times 858 = 238.04 \times 10^6$$

$$SS7intrahorate_{urban} = 148 \times 3.111 \times 10^4 = 4.604 \times 10^6$$

$$SS7interhorate_{urban} = 531 \times 2.85 \times 10^3 = 1.513 \times 10^6$$

$$SS7visitlurate_{urban} = 896 \times 5.74 \times 10^3 = 5.143 \times 10^6$$

$$SS7homelurate_{urban} = 461 \times 6.27 \times 10^4 = 2.89 \times 10^6$$

$$SS7rate_{urban} = 531.02 \times 10^6 \text{ bytes/hour/MSC}$$

$$SS7kbps_{urban} = \frac{531.02 \times 10^6 \times 8}{3600 \times 1000} = 1.18 \times 10^3$$

☞ **Suburban and rural.** Similarly, we calculate the SS7 rate for the suburban and rural environments.

$$SS7kbps_{suburban} = 245.483$$

$$SS7kbps_{rural} = 5.112$$

Next we determine the SS7 links between MSC, VLR, and HLR.

$$MSCtoVLR_{envr} = MSCtoVLRorig \cdot Subscribers_{envr} \cdot \left(\frac{E \cdot sec\,perhr}{ht_{envr}} \right) \cdot orig$$

$$+ \, MSCtoVLRterm \cdot Subscribers_{envr} \cdot \left(\frac{E \cdot sec\,perhr}{ht_{envr}} \right) \cdot (1 - orig)$$

$$+ \, MSCtoVLRinterho \cdot HOinter_{envr} + MSCtoVLRintraho \cdot HOintra_{envr}$$

$$+ \, MSCtoVLRvisitlu \cdot LUvisited_{envr} + MSCtoVLRhomelu \cdot LUhome_{envr}$$

$$MSCtoVLRkbps_{envr} = \frac{MSCtoVLR_{envr} \cdot bitsperbyte}{sec\,perhr \cdot 1000}$$

$$MSCtoVLRlinks_{envr} = \frac{MSCtoVLRkbps_{envr}}{linkspeed}$$

☞ **Urban**

$$MSCtoVLR_{urban} = 550 \cdot (1.156 \times 10^6) \cdot \left(\frac{0.02 \times 3600}{120} \right) \cdot 0.6 \, +$$

$$612 \cdot (1.156 \times 10^6) \cdot \left(\frac{0.02 \times 3600}{120}\right) \cdot (1 - 0.6) \ +$$

$$148 \times (2.85 \times 10^3) + 148 \times (3.111 \times 10^4) \ +$$

$$406 \times (5.74 \times 10^3) + 406 \times (6.27 \times 10^4) \ = 431.4 \times 10^6$$

$$MSCtoVLRlinks_{urban} \ = \ \frac{(431.4 \times 10^6) \times 8}{3600 \times 1000 \times 64} \ = \ 14.97 \approx 15$$

☞ **Suburban and rural.** We obtain the SS7 links between MSC and VLR for the suburban and rural environment the same way.

$$MSCtoVLRlinks_{suburban} = 3$$

$$MSCtoVLRlinks_{rural} = 1$$

VLR-to-HLR Link

$$VLRtoHLR_{envr} \ = \ VLRtoHLRterm \cdot Subscribers_{envr} \cdot \left(\frac{E \cdot sec\,perhr}{ht_{envr}}\right) \cdot (1 - orig)$$

$$+ \ (new\,VLRtoHLRvisitlu + old\,VLRtoHLRvisitlu) \cdot LUvisited_{envr}$$

$$+ \ VLRtoHLRhomelu \cdot LUhome_{envr}$$

$$VLRtoHLRlinks_{envr} \ = \ \frac{VLRtoHLR_{envr} \cdot bitsperbyte}{sec\,perhr \cdot 1000 \cdot linkspeed}$$

☞ **Urban**

$$VLRtoHLR_{urban} \ = \ 126 \cdot (1.156 \times 10^6) \cdot \left(\frac{0.02 \times 3600}{120}\right) \cdot (1 - 0.6) \ +$$

$$(182 + 95) \cdot (5.74 \times 10^3) + 55 \cdot (6.27 \times 10^4) \ = \ 39.998 \times 10^6$$

$$VLRtoHLRlinks_{urban} \ = \ \frac{(39.998 \times 10^6) \times 8}{3600 \times 1000 \times 64} \ = \ 1.39 \approx 2$$

☞ **Suburban and rural.** Calculated the same way, the VLR-to-HLR links both equal 1.

☞ **MSC-to-MSC Link**

$$MSCtoMSC_{envr} \ = \ MSCtoMSCinterho \cdot HOinter_{envr}$$

$$MSCtoMSClinks_{envr} = \frac{MSCtoMSC_{envr} \cdot bitsperbyte}{sec\,perhr \times 1000 \times 64}$$

☞ **Urban**

$$MSCtoMSC_{urban} = 383 \cdot (2.85 \times 10^3) = 1.092 \times 10^6$$

$$MSCtoMSClinks_{urban} = \frac{(1.092 \times 10^6) \times 8}{3600 \times 1000 \times 64} = 0.038 \approx 1$$

☞ **Suburban and rural.** The MSC-to-MSC links both equal 1.

☞ **VLR-to-VLR Link**

$$VLRtoVLR_{envr} = VLRtoVLRvisitlu \cdot LUvisited_{envr}$$

$$VLRtoVLRlinks_{envr} = \frac{VLRtoVLR_{envr} \cdot bitsperbyte}{sec\,perhr \times 64000}$$

☞ **Urban**

$$VLRtoVLR_{urban} = 213 \cdot (5.74 \times 10^3) = 1.223 \times 10^6$$

$$VLRtoVLRlinks_{urban} = \frac{(1.223 \times 10^6) \times 8}{3600 \times 64000} = 0.043 \approx 1$$

☞ **Suburban and rural.** VLR-to-VLR link is 1.

☞ **MSC-to-PSTN Switch**

$$MSCtoISDN_{envr} = MSCtoISDNorig \cdot Subscribers_{envr} \cdot \left(\frac{E \cdot sec\,perhr}{ht_{envr}}\right) \cdot orig +$$

$$MSCtoISDNterm \cdot Subscribers_{envr} \cdot \left(\frac{E \cdot sec\,perhr}{ht_{envr}}\right) \cdot (1-orig)$$

$$MSCtoPSTNlinks_{envr} = \frac{MSCtoISDN_{envr} \cdot bitsperbytes}{sec\,perhr \times 64000}$$

☞ **Urban**

$$MSCtoISDN_{urban} = 120 \cdot (1.156 \times 10^6) \cdot \left(\frac{0.02 \times 3600}{120}\right) \cdot 0.6 +$$

$$126 \cdot (1.156 \times 10^6) \cdot \left(\frac{0.02 \times 3600}{120}\right) \cdot (1-0.6) = 84.9 \times 10^6$$

$$MSCtoPSTNlinks_{urban} = \frac{(84.9 \times 10^6) \times 8}{3600 \times 64000} = 2.95 \approx 3$$

☞ **Suburban and rural**. MSC-to-PSTN links equal 1.

14.13 SUMMARY

In this chapter, we discussed teletraffic models including the call, mobility, and topology models. The mobility model based on the fluid flow model has been used extensively because of its simplicity. One of the limitations of this model is that it describes aggregate traffic and is difficult to apply to situations where individual movement patterns are required.

We found that, under heavy traffic conditions, the impact of LUs can be important. First, in terms of radio channels used, we observed that between one-fourth to one GSM RF carrier is used for LA boundary crossings. Although this may not cause call blocking on the radio interface, it has a non-negligible impact on TCH consumption. Second, in terms of processing at the MSC/VLR site, it was found that blocking can be reached rapidly under certain scenarios. The MSC/VLR resource dedicated to LU processing cannot be used to provide services.

The planning and design of a wireless network is a multidiscipline task that needs balancing of competing requirements. Several factors need to be considered including the extent of radio coverage, QoS for different environments, the extent of coverage for indoor locations, efficient use of the spectrum, and the evolution of the network. In the initial start-up phase, coverage plays a primary role and traffic demand is minimal. A system designer must carefully balance all trade-offs to ensure that the network is robust, future proof, and of high quality.

14.14 REFERENCES

1. Lam, D., Cox, D. C., and Widom, J., "Teletraffic Modeling for Personal Communications Services," *IEEE Communications Magazine* 35 (2), February 1997.

2. Tabbane, S., "Location Management Methods for Third-Generation Mobile Systems," *IEEE Communications Magazine* 35 (2), August 1997.

3. Wirth, P. E., "The Role of Teletraffic Modeling in the New Communications Paradigms," *IEEE Communications Magazine* 35 (2), August 1997.

4. Morales-Andres, G., Villen-Altarmirano, M., "An Approach to Modeling Subscriber Radio Network," *Proc. Telecom Forum '87*, Geneva, Switzerland, November 1987.

5. Sampei, S., *Applications of Digital Wireless Technologies to Global Wireless Communications*, Prentice Hall, Upper Saddle River, NJ, 1997.

6. Garg, V. K., and Wilkes, J. E., *Wireless and Personal Communications Systems,* Prentice Hall, Upper Saddle River, NJ, 1996.

7. Wilborn, T., "Design of a GSM Cellular System," Design Project for Wireless and Personal Communications Class at University of Illinois, Urbana, April 1997.

CHAPTER 15

Management of GSM Networks

15.1 INTRODUCTION

Wireless systems, like other telecommunications networks, require large numbers of translations and other parameters to define their physical and logical configurations. Data changes are frequent for many reasons including system growth, changes in patterns of subscriber density, "quasidynamic" frequency channel assignments, trunk assignments, and on-going fine tuning of the system to improve QoS.

In the past each new network element of a wireless system had a corresponding operation support system to provide management capabilities. Each of these management systems had a different user interface, employed a different computing platform, and typically managed one type of network element. The total was the sum of all these independent, resource-consuming partial solutions. They made the network management task inefficient, complex, time consuming, and expensive to administer. As wireless service providers moved into a mixed vendors' environment where network elements manufactured by several vendors are used, they could no longer afford different management systems for each network element. Therefore, the standards group in the Alliance for Telecommunications Industry Solutions (ATIS) committee T1M1, ITU, ETSI, TIA committee TR 45, and Network Management

Forum (NMF) defined the interfaces and protocols for management systems under the umbrella of TMN.

The management of the GSM PLMN is based on TMN principles and methodology as defined in CCITT M.3010 [1] and CCITT M.3020 [2]. TMN principles provide a standardized management using the definition of a TMN information architecture including a management model and the management information exchange.

In section 15.2, we present the traditional approaches to network management (NM). In section 15.3, we briefly introduce the TMN and encourage those of you who are interested to refer to the appropriate CCITT documents. We provide management requirements for wireless networks in section 15.4. In section 15.5, we focus on the platform-centered NM approaches and present two widely used NM approaches. We discuss the NM interfaces and functionality in section 15.6. Section 15.7 covers the management of a GSM PLMN.

15.2 TRADITIONAL APPROACHES TO NM

Traditional NM practices include a wide array of procedures, processes, and tools for configuration, fault, performance, security, accounting, and other management functions. The management functions rely on a "master-slave" relationship between operation systems (OSs) and network elements (NEs). Typically, NEs have basic operations functionality with little or no ability to control activities or make decisions beyond call processing and information transmission tasks. The OSs perform the majority of Operations, Administration, Maintenance, and Planning (OAM&P) tasks including processing raw data generated by individual NEs, making decisions and instructing each individual NE to perform specific actions. The master-slave relationship contributes to operating inefficiencies in several ways. There is little sharing of logical resources such as data because NEs and OSs are designed independently. Also, each vendor's equipment has unique configuration and fault management interfaces with specific performance requirements. NM systems characterize each NE and vendor's interface on an individual basis. This adds considerable time and complexity in introducing new services or technologies. Other factors have also compounded this complexity. NM systems are often designed to optimize the task of an individual service provider's organization or work group at a particular point in time for a particular technology. This type of development is often undertaken independently by individual organizations without paying much attention to system-level interworking. Multiple copies of data, each tied to specific systems or job functions and to specific equipment vintages or implementations, are used throughout the network, creating a major data synchronization problem. Thus, it becomes difficult for a service provider to evolve services, network technologies, and NM processes in a cost-effective, timely, and competitive fashion in response to the rapidly changing environment of wireless communications.

15.3 TMN

TMN principles in M.3010 [1] specify the architecture of TMN including the identification of the different types of nodes and the interfaces between them. Table 15.1 lists the various physical components of the TMN.

TMN uses the concept of functional architecture which is defined in terms of functional blocks and reference points. Functional blocks are the logical entities that can be implemented in a variety of physical configurations. Table 15.1 shows the functional blocks, which are mandatory for each TMN node. TMN reference points represent exchange of information between two functional blocks. Table 15.2 gives the correspondence between TMN interfaces and reference points. Interfaces are denoted in uppercase; reference points are designated in lowercase.

15.3.1 TMN Layers

TMN divides management functionality into logical layers to represent a hierarchy of system functionality using Logical Layer Architecture (LLA) concepts. Each higher layer provides a higher level of system functional abstraction. Five TMN layers in M.3010 are

1. Network Element Layer (NEL)
2. Element Management Layer (EML)
3. Network Management Layer (NML)

Table 15.1 TMN Nodes and Their Mandatory Functional Blocks

TMN Nodes	Mandatory Functional Blocks
OS	OSF
Mediation Device (MD)	MF
Q-Adaptor (QA)	QAF
Workstation (WS)	WSF
NE	NEF

Table 15.2 TMN Interfaces and Reference Points

TMN Interfaces	TMN Reference Points
Q_3	q_3
Q_x	q_x
F	f
X	x

4. Service Management Layer (SML)

5. Business Management Layer (BML)

For resource management, only the first three layers are pertinent. The role of the three resource management layers is discussed below.

NML.   The NML provides a management view of the network that is under one administrative domain. It has responsibility for end-to-end management of a network. Through the EML, it manages subnetworks or NEs based on the view presented by the EML. It can provide detailed views of the portions, or segments, of the connections within its domain. The basic responsibility of the NML is to provide networkwide functions, for example, network traffic monitoring, network protection routing, and fault correlation and analysis from multiple NEs/nodes. NML provides applications and functionality in four principal areas:

☞ Control and coordination of network's view of all NEs within its domain

☞ Provision, cessation, or modification of network capabilities for the support of service to customers

☞ Maintenance of network capabilities

☞ Maintenance of statistical log and other data about the network and interaction with SML on performance, usage, availability, etc.

EML.   The EML manages each NE on an individual or group basis and supports an abstraction of the functions provided by the NEL. NEs of a similar type are managed at this layer. Aggregating nodes into subnetworks for wireless systems is advantageous in some cases in order to manage equipment from the same manufacturer. The EML filters message traffic going to the NML (e.g., NE alarm correlation). It is responsible for operation on one or several NEs/nodes—for example, remote operation and maintenance, remote hardware and software management, and local fault handling. The principal areas supported by the EML are

☞ Control and coordination of a subset of the network on an individual-network-function basis

☞ Control and coordination of a subset of NEs on a collective basis

☞ Maintenance of statistical log and other data about NEs

NEL.   The NEL performs basic management functions for wireless equipment, such as detecting faults and counting errors. It provides the agent service functions in support of a management request from the EML.

Classifying functions into NEL, EML, or NML does not imply any particular physical implementation; it simply provides a logical organization for the management functions. In the same way, an interface between two TMN layers is a "logical interface," which may or may not be physically implemented.

For example, two layers, such as EML and NEL, may reside in the same physical device (see Figure 15.1).

15.3.2 TMN Nodes

Figure 15.2 is a simplified example of a physical TMN architecture. The OS provides the supervisory or control systems in TMN. OSs can be interconnected and thus can form management hierarchies.

Mediation Devices may provide storage, adaptation, filtering, thresholding, or condensing operations on data received from subtending equipment. They are probably the most vague component of TMN.

The QA is used to connect a TMN system to a non-TMN system. QAs may be used for integrating existing networks into TMN.

The NE is the only node that actually resides in the management network. The primary job of an NE is to handle traffic and not management. It is, however, the ultimate origin or destination of management supervision and control.

The WS is an interface between the user and TMN. It provides the presentation function to the user. It should be noted that the WS as a TMN node is not the same as the workstation of the computer world.

The nodes communicate through the Data Communications Network (DCN), which is the transportation means used in the TMN.

Control and coordination of a subset of NEs is conducted on a collective basis.

15.3.3 TMN Interfaces

The manner in which management systems interact is governed by the interfaces. Standardized interfaces allow the nodes to interwork as long as the pro-

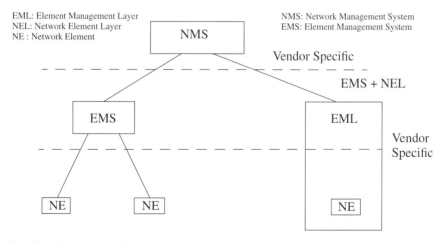

Fig. 15.1 Example of Layering

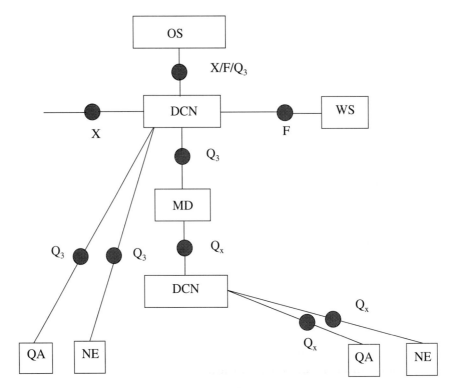

Fig. 15.2 A Simplified TMN Physical Architecture

tocols are specified to a level that allows applications to interact. The Q_3 interface connects an OS and an NE, or an OS and a QA, or an OS and an MD, or two OSs that belong to the same TMN. Q_x is like a Q_3 but with less functionality. It is intended to be used when cost or efficiency issues prevent a fully functional Q_3 interface. The X interface is used to communicate between OSs that belong to different TMNs, or between a TMN OS and a non-TMN OS that supports a TMN-like interface. The F interface is used for communicating between the WS and the other nodes.

15.3.4 TMN Management Services

CCITT divides management into five broad management functional areas to provide a framework for the determination of management service applications. The management functional areas are

☞ Performance management

☞ Fault management

☞ Configuration management

☞ Accounting management

☞ Security management

The management of customer access services is defined in CCITT M.3200 [3] which specifies a service related to the configuration and fault management functional areas for customer access equipment. CCITT M.3400 [4] specifies several service components for each of the relevant management functional areas.

Performance Management.   Performance management is concerned with

☞ Measurement data collection and storage

☞ Simple statistics showing summaries and cumulative counts

☞ Support of customized displays and reports for network performance information

Fault Management.   Fault management deals with

☞ GSM 12 series compliance with respect to alarm type, error ID, probable cause, severity, and descriptive information

☞ Alarm collection and storage

☞ Alarm acknowledgment

☞ Alarm filtering and reclassification

☞ Alarm monitoring

☞ Alarm forwarding to paging and e-mail

☞ Scheduled loop testing

Configuration Management.   The basic functions of configuration management are

☞ Capabilities to send files with many prespecified configuration commands to NEs

☞ Objects creation and deletion

☞ Logical interconnection schemes to show various NEs, subsystems, and functional blocks at each site

☞ Software management

☞ New features/upgrades

Security Management.   Security management—the handling of access authorizations—is of real significance. A series of measures are generally used to prevent unauthorized access. This is achieved using both user- and device-related access authorization mechanisms included in the software and smart cards with coded user identifiers. Some of the tasks performed in security management include

☞ Password authentication of OMC and NMS users

☞ Logging of OMC/NMS attempts

☞ Automatic log-off

Accounting Management.   Accounting management deals with the collection of billing information for the services provided by a network. Billing is a critical process as it delivers all the revenue to the service providers. In GSM, there is a relationship between each PLMN and other networks to which it interconnects. These networks may be national or international carriers and may even be fixed wireline. Each PLMN has to negotiate with other networks with which they interconnect as to the basis of sharing revenue. In all cases an accounting relationship is established between PLMNs and other networks.

15.4 MANAGEMENT REQUIREMENTS FOR WIRELESS NETWORKS

In a wireless network where both users and terminals are mobile rather than fixed, mobility management plays an important role. In addition, radio resources must be managed that do not exist in a wireline network. Furthermore, the wireless network can be a cellular system where radio resources and switching resources are owned and managed by one company. A second newer model being used for a PCS network has a wireline company performing the standard switching functions and a wireless company providing the radio-specific functions. With these new modes of operation, NM must include standard wireline requirements and new requirements specific to a cellular/PCS system.

15.4.1 Management of Radio Resources

The cellular/PCS network allows terminals to connect via radio links. These links must be managed independently of the ownership of the access network and the switching network. The cellular/PCS network may consist of multiple service providers or one common service provider. An environment with multiple service providers will require interoperable management interfaces.

15.4.2 Personal Mobility Management

Another important aspect of the cellular/PCS network is the management of personal mobility. Users are no longer in a fixed location—they may be anywhere in the world. They may be using wireline or wireless terminals to place and receive calls. In some cases the network will determine the user's location automatically; in other cases the user may need to report his or her location to the network. This will increase the load on management systems as subscribers manage various decision parameters about their mobility; for example, they may request different services based on the time of day or terminal busy conditions.

15.4.3 Terminal Mobility

The primary focus of a cellular/PCS network is to deliver service via wireless terminals. These terminals may appear anywhere in the worldwide wireless network. Single or multiple users may register on a wireless terminal. The terminal management functions may be integrated with the user management function or separate from it. Different service providers may choose to operate their systems in a variety of modes. The management functions must support all modes.

15.4.4 Service Mobility Management

Service mobility is the ability to use vertical features (e.g., call hold, call forwarding) in a transparent manner from remote locations or while in motion. For example, the user should have access to the messaging service anywhere, anytime. The ability to specify an event is provided via a user interface that is flexible enough to support a number of input formats and media. Users can specify addressing in a simple, consistent manner no matter where they are.

15.5 PLATFORM-CENTERED MANAGEMENT

In a platform-centered management of a heterogeneous network, management applications are centralized in platforms, separated from the managed data and control functions included in NEs. Platform-centered management systems require

☞ Standards to unify management-agent information access across multi-vendor NEs. These include standardization of managed information database structures at the agent as well as standardization of the access management protocol.

☞ Standards to unify the meaning of managed information across multi-vendor systems.

☞ Standards to unify platform-processing environments across multi-vendor platforms.

Simple Network Management Protocol (SNMP) [5] and OSI Common Management Information Protocol (CMIP) [6] address the first category of standards while complementary efforts are being pursued by other organizations such as NMF to address the last two standards. In this section, we concentrate on only the first category of standards—those that govern managed information organization and access protocol.

15.5.1 SNMP

The SNMP has been designed to be an easily implemented basic management tool that could be used to meet short-term network management needs mainly for computer networks. The SNMP standards provide a framework for the definition of management information and a protocol for the exchange of that information between managers and agents. A manager is a software module having the responsibility for managing part or all of the configuration on behalf of NM applications and users. An agent is also a software module in a managed NE with responsibility for maintaining local management information and delivering that information to a manager via SNMP. A management information exchange can be initiated by the manager by polling, or by the agent by using a trap. SNMP accommodates the management of those NEs that do not implement SNMP software by means of proxies. A proxy is an SNMP agent that maintains information on behalf of one or more non-SNMP devices.

SNMP uses a simple model for the structure of managed information (SMI), involving six application-defined data types and three generic ones. Temporal behaviors are described in terms of counters and gauges. An error *counter* represents the cumulative errors since device booting. A *gauge* is used to model management variables such as queue length. An error counter is not very useful in detecting rapid changes in error rates. Changes in error rates are reflected in the second derivative of an error counter. The manager must sample the counter frequently to estimate its second derivative, leading to unrealistic polling rates. SNMP extensions (RMON) improve handling this difficulty. The values of the managed variables are recorded in an agent's management information base (MIB) where they can be polled by the manager.

Typical problems of SNMP involve difficulty capturing the range of complex network behaviors that may not be feasible within a narrow model of managed information. The counters and gauges of SNMP are too restrictive to support event detection and correlation. The objects in SNMP are simple variables with a few basic characteristics, such as data type and read-only or read-write attributes. There is a limit to how far SNMP can be extended by simply defining new and more elaborate MIBs. RMON perhaps represents as far as one may go in trying to enhance the functionality of SNMP by adding to the semantics of MIB. As SNMP is applied to larger and more sophisticated networks, such as cellular/PCS networks, its deficiencies, which are mainly in the area of security and functionality, become more evident. For example polling does not scale well, and the weakness in agent-to-manager communication is compounded by the fact that traps are unreliable and cannot be used as a substitute for polling. Security issues are addressed in SNMPv2; however, they have not yet been widely accepted in the industry.

The SNMP agent is uniquely identified by an IP address and User Datagram Protocol (UDP) port number. Thus, only a single agent is accessible at a given IP. This agent can maintain only a single MIB. Therefore, within a single IP address, only a single MIB instance may exist. This unique binding of an MIB to an IP address can limit the complexity of data that an agent can offer. When a system requires multiple MIBs to manage its different components, to access via a single agent these MIBs need to be unified under a single MIB tree. Situations may occur in the stack of SNMP/UDP/IP, leading to great complexity in the organization of management information.

15.5.2 OSI Systems Management

As mentioned earlier, one of the criteria imposed on TMN is that it accommodate the management of diverse technologies. This requires that the TMN interfaces must be both general and flexible. Also, the requirement for consistency has motivated the use of standardized protocol suites. OSI systems management technology was selected as the basis for the TMN interfaces.

The overall framework for OSI management is aimed at satisfying NM requirements in five function areas specified by TMN, referred to as System Management Function Areas (SMFA). OSI includes a number of general-purpose system management functions to support the SMFAs. Each function area can be implemented as an application that relies on some subset of the system management functions. An example of SMFA is the event management that enables event forwarding and filtering.

A managed object (MO) is the conceptual view of a physical or logical resource that needs to be monitored and controlled in order to avoid failures and performance degradation in a network. MOs with the same properties are *instances* of an MO class. The MIB is the conceptual repository of the MO instances. An MO class is defined by attributes, the management operations, the behavior, and the notifications.

- ☞ **Attributes** are data elements and values that characterize the MO class.
- ☞ **Management operations** are operations that can be applied to MO instances.
- ☞ **Behavior** exhibited by an MO instance is based on the resource the MO class represents.
- ☞ **Notifications** are messages that MO instances emit spontaneously.

OSI management defines two roles—the manager and agent. The *manager* is the specific entity in the managing system that exerts the control, the coordination, and the monitoring. The manager issues requests to perform operations against the agent. It also receives the notifications emitted by MOs and sent by the agent.

The *agent* is the specific entity in the managed system to which the control, coordination, and monitoring are directed. The agent receives and executes the requests sent by the manager and sends the notifications to the manager.

Table 15.3 lists the capabilities of the manager and the agent.

Manager and agent may communicate using a seven-layer OSI protocol stack. A key element of the protocol suite is the Common Management Service Element (CMISE). CMISE is one of the building blocks used at the application layer and consists of a service definition—the Common Management Information Service (CMIS)—and a protocol specification—the CMIP. CMIP is used for the basic exchange of management information which is defined using the Guideline for Definition of Managed Objects (GDMO) and Abstract Syntax Notation one (ASN.1). CMIP provides a comprehensive set of features such as scooping, filtering, and multiple links replies to facilitate the exchange of management information between the manager and agent. These features offer distinct advantages to an OSI manager as compared to an SNMP manager by enabling the former to retrieve bulk information as well as selected information of interest from the agent.

All messages exchanged between the manager and the agent have a basic form of either requesting something of one or more objects or an object informing another system of some event. The requests may be as simple as returning the value of a parameter or as complicated as asking the NE to reconfigure itself.

The agent receiving the message is responsible for carrying out the request(s). It maps the request(s) on the MO(s) into request(s) on real resources. However, the mechanisms used for the mapping are implementation specific and not standardized.

Using these concepts, the resources are modeled so that the manager and the agent have a common view. This specification of object-oriented information is called information modeling. OSI management uses an extended object-oriented model of managed information as compared to SNMP. OSI management organizes managed object instances in a Management Information Tree (MIT), which is updated to reflect changes in the resources being managed. The manager issues CMIP commands to act upon object instances present in the MIT.

The OSI systems management standards provide power and flexibility in defining interface standards. The TMN interface standards consist of generic

Table 15.3 Capabilities of a Manager and an Agent

Manager Role	Agent Role
Create managed object	Respond to manager requests
Delete managed object	Report instance creation
Manipulate relationships (with relationship attributes)	Report instance deletion
Inspect attributes	Report relationship changes
Manipulate attributes	Report attribute changes

standards and technology-dependent standards. The generic standards are intended to be applicable across all telecommunications technologies and services. The objects specified in the technology-specific standards are often imported from the generic standards or are subclasses of generic objects.

Inheritance (also referred to as subclassifying) is the procedure of specifying a new object class based upon a previously defined object class. The new object class has all the characteristics of the base object class (superclass) with some new characteristics. This method of deriving technology-specific object classes from base generic object classes ensures a level of similarity between different technology-dependent information models.

Allomorphism—a capability that may be used to manage telecommunications technologies in a generic manner—is the procedure of specifying a subclass that masquerades as a superclass. One use of this is to allow a technology-specific object to be treated as a more generic object, enabling it to be managed as a generic object. A related use of allomorphism is to provide a generic set of management capabilities in certain situations, while providing vendor-specific enhancements in other situations. There has been insufficient use of TMN standards to determine if allomorphism is truly a useful concept.

Inheritance and allomorphism, along with the concept of generic and technology-specific standards, are the mechanisms that provide the generality and consistency desirable in TMN interfaces.

The OSI information model enables the definition of a more complex and flexible managed object set as compared to SNMP. Distinguishing features of OSI management are inheritance; ability to create and delete objects, thereby leading to a dynamic MIT; and flexibility to define specific procedures that can be invoked by a manager by issuing an M-action command. With the OSI approach it is no problem to handle a large number of objects at one time. Objects in excess of 50,000 can easily be processed.

The protocol suite at the Q interface is the OSI protocol suite. In using the OSI system management protocol suite, TMN has gained, among other things, reliable and robust communications capabilities and a wealth of application-layer building blocks. The application-layer building blocks are Association Control Service Element (ACSE) and CMISE. The ACSE provides a means for establishing associations and negotiating application protocol capabilities. Various other benefits are

☞ A semiformal specification technique (templates) to define the information model, including object classes, attributes, actions, and notifications

☞ The use of object-oriented techniques such as inheritance and allomorphism

☞ Naming rules to facilitate the structure of objects in a database

☞ A large set of objects already defined for routing alarms (event forwarding discriminators), logging data (logs), generating reports (scanners), etc.

☞ A data-specification language (ASN.1) for definition of data structures in an abstract (machine-independent) notation

☞ A method for encoding and decoding application-layer data, independent of machine-specific representation

15.6 NM INTERFACES AND FUNCTIONALITY

In Figure 15.3 we show a scheme to connect OMC and NMS. This arrangement can be applied to all digital wireless communications technologies currently available—CDMA, GSM, IS-136 TDMA, and Japan PDC.

The NMS provides NML functionality and is responsible for network-wide tasks such as network traffic monitoring, network surveillance, network congestion control, traffic routing, regional network health monitoring, alarm correlation, and so on. The NMS should support Q_3 interfaces between OMCs. It should allow connection between BMSs through a Q_3 interface. The NMS functionality must be in compliance with industry standards such as NMF,

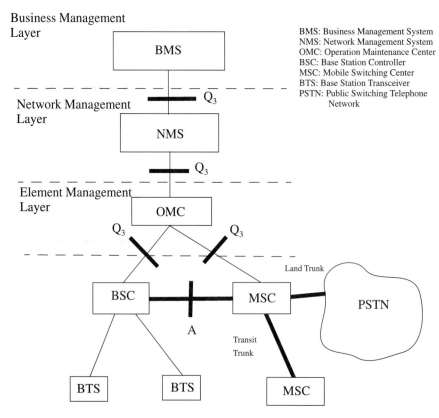

Fig. 15.3 NM Architecture and Interfaces

ETSI, and ITU-R. The design of the NMS must be flexible enough to incorporate future generation systems.

The OMC provides EML functionality. Its design must be in compliance with GSM Series 12 standards. As far as possible the OMC's design should be independent of air interface technology. The interface between the OMC and the BSC should be Q_3. A Q_3 interface should also be provided between the OMC and the MSC.

15.6.1 NMS Functionality

The use of a suitable NMS is a decisive factor in the technical and commercial success of the wireless network operator. A well-organized network management system must be able to react rapidly to changing requirements on the wireless communication network. Applications of the NMS routinely include collecting billing data from MSCs and handling intricate special situations such as network expansion and fault management of the entire network.

The OMC can be considered a "regional manager" for the network since it controls only parameters of the regional network. Since there are several regional network operators in a GSM PLMN, there should be several independent OMCs. The NMS should be regarded as an "entire network manager" for the system hardware and software. The controlling parameters in this case should be associated with the overall system.

In Figure 15.4 OMC_1 controls region #1, whereas OMC_2 controls region #2. Both the OMCs provide necessary information to the NMS for their control, decision, and intervention, if necessary. The NMS provides a view of the entire system and is responsible for management of the network.

The NMS resides at the top of the hierarchy and provides global network management as shown in Figure 15.4. The NMS receives its information from the network equipment via OMCs and filters the information. The NMS focuses on issues requiring systemwide coordination and coordinates issues regarding interconnects to other networks, such as PSTN. In summary, NM for the GSM PLMN consists of the following:

☞ Equipment for operation of NEs in the field (i.e., BTS, BSC, MSC, etc.)

☞ OMC as the central device for OA&M of BSSs and Mobile Switching Subsystems (MSSs) in individual regional networks within the GSM PLMN

☞ NMS for additional global technical management of the GSM PLMN and further administrative and commercial control functions at the total network level

The OMC monitors and controls various elements of its BSSs and MSSs and feeds the summary view to the NMS. Since the NMS knows the status of the entire network, it can direct OMC operators to change their approach for correcting a local regional problem because the NMS knows to what extent the problem in a particular region affects other regional networks. OMC operators are simply ignorant about this situation. In summary, the NMS must

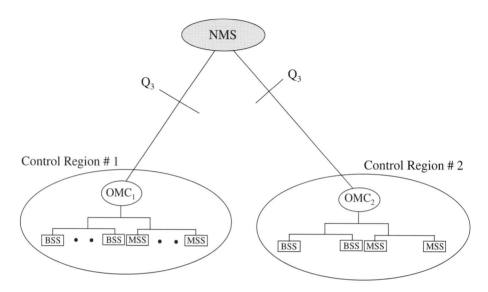

Fig. 15.4 Functionality of NMS and OMC

☞ Provide integrated operation of the entire network
☞ Monitor the network for high-level alarms
☞ Present status of all regional networks within a GSM PLMN
☞ Provide networkwide traffic management
☞ Monitor the status of automatic controls applied to the network equipment in response to overload
☞ Monitor QoS of each regional network
☞ Support network planning activities
☞ Take regional responsibility when OMC is not equipped with craftsperson(s)

15.6.2 OMC Functionality

The OMC is a centralized facility that supports day-to-day operations and management of a cellular/PCS network and that provides databases for long-run network engineering and planning tools. The OMC manages certain BSSs and MSSs of the GSM PLMN to provide regionalized management. From the NM point of view, an OMC operator can look into any part of the system closely. In case of a failure of some part of the regional system, the operator decides what has really gone wrong with the part of the system. The OMC operator also reconfigures parts of the system by placing the troubled elements of the system out of service. The OMC is concerned with the areas of fault and alarm management, configuration and operations management, performance management, and security management.

15.7 MANAGEMENT OF GSM NETWORK

15.7.1 TMN Applications

Figure 15.5 shows how TMN architecture is used in the GSM network. The **mscFunction** and **vlrFunction** are included in a single network element. The **bssFunction** is treated as a special case in that whichever network element contains this function also contains a mediation function and supports a TMN Q interface to an NE supporting BTS functionality.

In the area of fault and configuration management, committee work is limited to the BSS NE. Other NEs, such as HLR and VLR, are mostly related to accounting management. The MSC has not been dealt with because it is, to a large extent, not a GSM element and therefore mostly outside the scope of the work of the SMG technical committee.

In the area of security management, committee work is limited to the management of the security features that are defined for GSM systems rather than to the specification of requirements on the security of the NEs, data, and interfaces.

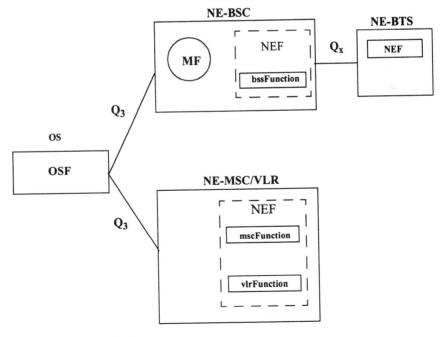

Fig. 15.5 GSM TMN Architecture

15.7.2 GSM Information Model

The GSM information model contains three types of objects. The first type represents purely functional resources; the second type represents functional resources that are generally related to equipment; the third type represents equipment resources directly. Most defined objects support attribute value change and create/delete notifications. Those with state deal with state change notifications. Objects related to equipment support user labeling attributes that allow name and location to be specified. These objects also support the reporting of equipment alarms. All together there are more than 100 managed object classes (MOCs) with about 500 attributes in a GSM PLMN model.

A management system is set up to monitor only the standardized functional objects, but it can still receive alarms from failed equipment. Only one object is defined to directly represent equipment resources. This general object supports user labeling, state control, and alarm notification capabilities.This was done to provide a standard equipment object that might be used to represent an arbitrary piece of equipment. Difference among equipment resources is achieved through the type, name, and location attributes and through containment.

Table 15.4 provides the defined service components and some general model information.

15.7.3 GSM Containment Tree

Figure 15.6 shows the containment tree of the high-level information model. The **network** object class is defined in CCITT M.3100 and imported into the PLMN model without any change. It represents an identifier for the telecommunications network being modeled. Subnetworks may be modeled by the nesting "network."

The **plmnNetwork** object class is defined in GSM 12.00. It is a subclass from a network and adds the specific details to characterize a PLMN, such as the network identity and lists of supported services.

The **managedElement** object class is defined in CCITT M.3100 to represent a component of a network that comprises functional and/or physical aspects, such as equipment and software. In the GSM high-level object model, functional NEs and the related equipment (if any) may be combined in various ways and are contained in a *managedElement* instance.The *managedElement* is the entity that supports a Q_3 interface for the purpose of being managed. For an example, a small *plmnNetwork* instance may contain three *managedElement* instances, one containing an *mscFunction* instance and a *vlrFunction* instance and the other two containing *bssFunction* instances. The GSM-specific NE functions provide a widely recognized grouping of defined functionality and thus provide an attachment point (containment) for the model of resources that are element-specific. The set of these functional element object classes is defined in GSM 12.00 and is detailed on pages 348–9.

Table 15.4 General Management Model Information for GSM

Management Area	Details	Specifications
Performance	Number of objects are specified that are purely of functional nature. Each object contains one or more measurement attributes with behavior specification as to exact measurement triggers and/or start/stop condition.	GSM 12.04 requirements to support the capabilities to conduct measurements, collect data, and forward data. GSM specifications also suggest to use CCITT-defined objects, including *log, eventForwardingDiscriminator*, and *simple scanner*.
Security	Service components are defined for system security control, security monitoring, and security alarm reporting. The model uses a few objects that can be instantiated under the functional elements. Attributes are defined to control encryption, encryption algorithm selection, when to use authentication, and the use of IMEI checks. Attributes to count various events related to determining the load on security features and to identifying attempted security breaches are also included.	GSM 12.03 covers only security that is unique to GSM it does not address generic security issues of data transfer or NE management access.
Fault	Supports alarm surveillance for the service component; service functions are defined for alarm reporting, routing, logging, and related functions. The service is defined only for the BSS NE. Model also supports the *eventForwardingDiscriminator* object in X.721 and other CCITT-defined objects and notification. The *equipmentRelatedAlarm Package* contains X.721 defined *environmental alarms* and *equipment alarms*.	GSM 12.20 provides parameters to allow the notification to include either the identification of an equipment object instance or to include the related equipment information directly (i.e., equipment name, location, type, etc.).
Configuration	Contains provisioning service component and the NE status and control service component. The service is defined only for BSS. Most of the object classes are on the functional side of the model with a single defined equipment object and three software/database objects. None of the objects on functional side are derived from M.3100 *equipment* MOC. The GSM-specific classes,	GSM 12.20-defined *related GSMEquipment* attribute is used to identify the equipment object instances that support the functionality. These same object classes are also provided with *equipmentRelatedAlarm*. Package contains notifications to allow the reporting of equipment alarms via instances of these object classes. The *gsmEquipment* object class is

Table 15.4 General Management Model Information for GSM (Continued)

Management Area	Details	Specifications
Configuration (continued)	*replaceableSoftwareUnit* and *executableSoftwareUnit* provide additional attributes to identify related files and to identify software patch containment relationship. The *operatingSoftwareUnit* provides attributes for administrative control, as well as capabilities to identify automatically restartable software units. Name bindings are defined for this object class to the set of functional objects that could report equipment alarms.	derived from M.3100 *equipment* with the addition of an *equipmentType* attribute to alleviate the need for additional subclassifying. The *relatedGSMFunctionalObjects* attribute is used to identify functional objects for the purpose of reporting equipment alarms through these objects.
Accounting	Provides services for customer administration, management of mobile equipment data, and tariff and charging administration. For customer administration functions "create/modify/delete" MSISDN in the HLR are supported. For mobile equipment data "create/interrogate/delete" equipment functions in the EIR are supported. For tariff and charging "create/set/get/delete" charging origin; "create/set/get/delete" charging destination; and "create/set/get/delete tariff class functions are supported. The model contains a collection of more than 50 purely functional object classes. The objects are mostly data record type with more than 200 attributes to record subscriber, equipment, service, tariff, and call information.	GSM 12.02 and GSM 12.05. The model also supports the use of CCITT object classes for *logging* and *eventForwarding*. This part of the model also specifies the use of file transfer capabilities to move the large amount of data.

☞ **bssFunction.** Derived from the CCITT X.721 *top* MOC to model containment for those aspects of BSS management that are common to this functional element.

☞ **hlrFunction.** Derived from the CCITT X.721 *top* MOC to provide state attributes and notifications for attribute value change, state change, and create/delete.

☞ **vlrFunction.** Derived from the CCITT X.721 *top* MOC to include attribute for VLR numbering and an ISDN number. It also provides state attributes and notifications for attribute value change, state change, and create/delete.

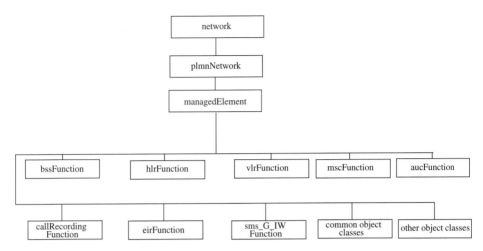

Fig. 15.6 GSM Containment Tree

☞ **mscFunction.** Derived from the CCITT X.721 *top* MOC to include attributes for MSC numbering and an ISDN number. It also provides state attributes and notifications for attribute value change, state change, and create/delete.

☞ **aucFunction.** Derived from the CCITT X.721 *top* MOC to provide state attributes and notifications for attribute value change, state change, and create/delete.

☞ **eirFunction.** Derived from the CCITT X.721 *top* MOC to include attributes for EIR numbering and an ISDN number. It also provides state attributes and notifications for attribute value change, state change, and create/delete.

In addition to the object classes defined to support functional element-specific modeled resources, several other GSM PLMN high-level object classes have been defined to address those functions that are not specific to the defined functional architectural areas. These are

☞ **callRecordingFunction:** This object class is derived from the CCITT X.721 *top* MOC to model resources for the generation, collection, and content of that data required to provide revenue from network usage. This function is described as a coherent whole, but includes aspects that may be implemented on MSCs and HLRs.

☞ **sms_G_IWFunction:** This object class is derived from the CCITT X.721 *top* MOC to model the manageable aspects of the resources of PLMN to provide and receive short messages from a message center. It also provides state attributes and notifications for attribute value change, state change, and create/delete.

Beyond those objects defined for containment groupings, a few other objects are defined for use throughout the object model. These objects are referred to as common managed objects and are defined to model resources necessary to support such items as bulk data transfer and equipment and software management. These are included in GSM 12.20 specifications. These object classes are

☞ **simpleFileTransferControl.** Derived from the CCITT X.721 *top* MOC to model resources for a method of bulk data transfer. It provides a set of actions and notifications to request and confirm the transfer of an identified file between the OS and the NE.

☞ **generalDataTransferControlFunction.** Derived from the CCITT X.721 *top* MOC to provide containment of various data transfer methods that may be implemented in a managed element.

☞ **gsmEquipment.** Derived from the CCITT M.3100 *equipment,* with the addition of an *equipmentType* attribute. The purpose is to alleviate the need for additional subclassifying. The *relatedGSMFunctionalObjects* attribute identifies functional objects for reporting equipment alarms through these objects.

☞ **replaceableSoftwareUnit.** Derived from the CCITT M.3100 *software,* with the addition of *relatedFiles* attribute to identify files that are downloaded or otherwise contained in the system to make up the software resource.

☞ **executableSoftwareUnit.** Derived from the CCITT M.3100 *software,* with the addition of a *relatedRSUs* attribute to identify the replaceable units that make up the software resource.

☞ **operatingSoftwareUnit.** Derived from the CCITT X.721 *top* MOC to provide attributes for administrative control as well as capabilities for identifying automatically restartable software units.

A comprehensive management information model for a GSM PLMN will require managed object classes that are defined in other standards or are defined by operators or manufacturers to be contained in this same hierarchy. These latter objects may be subclasses of those defined in the current set of specifications. These are grouped in Figure 15.6 as "other object classes."

15.7.4 Future Work Items

The current model does not include any objects for the management of telecommunications facilities that are used to interconnect the various NEs. GSM uses the modified version of the ITU SS7. Each defined NE needs to support management of this signaling system. The fault and configuration management currently address only the BSS. It should be expanded to cover the other main physical element in the GSM system, the MSC. There is currently nothing specified for fault isolation, fault recovery, and testing services.

15.8 Summary

In this chapter, we outlined the requirements needed to manage a wireless network. We discussed the SNMP- and OSI-based management schemes and presented the limitations of the SNMP in managing a wireless network. A brief discussion of the TMN architecture along with TMN layers, nodes, and interfaces was given. We also discussed the roles of five management functional areas specified in the TMN. The last section of the chapter was devoted to the management of a GSM PLMN. In this section, we presented the details of the information model and discussed the containment tree. We briefly described the managed object classes used in GSM. We concluded by pointing out the directions for future improvement of the GSM management model.

15.9 References

1. CCITT Rec. M.3010, "Principles for a Telecommunications Management Network," 1992.
2. GSM 12.00, "Objectives and Structure of GSM PLMN Management," 1994.
3. CCITT Rec. M.3400, "TMN Management Functions," 1992.
4. GSM 12.01, "Common Aspects of PLMN Network Management," 1994.
5. GSM 12.21, "Network Management Procedures and Messages on the A-bis Interface," 1994.
6. GSM 12.11, "Fault Management of the Base Station System," 1994.
7. GSM 12.06, "Network Configuration Management and Administration," 1994.
8. GSM 12.05, "Subscriber Related Call and Events Data," 1994.
9. GSM 12.03, "Security Management," 1994.
10. GSM 12.04, "Performance Management and Measurements for a GSM PLMN," 1994.
11. CCITT Rec. X.721, "Information Technology—Open Systems Interconnection—Structure of Management Information," 1992.
12. CCITT Rec. X. 738, "Information Technology—Open Systems Interconnection—Systems Management: Summarization Function," 1992.
13. CCITT Rec. X.739, "Information Technology—Open Systems Interconnection—System Management: Work Load Monitoring Function," 1992.
14. GSM 12.20, "BSS Management Information," 1994.
15. CCITT Rec. M.3200, "TMN Management Services: Overview," 1992.
16. CCITT Rec. Q.821, "Q_3 Interface for Alarm Surveillance," 1993.
17. CCITT Rec. M.3100, "Generic Network Information Model," 1992.
18. CCITT Rec. X.733, "Information Technology—Open Systems Interconnection—Systems Management: Alarm Reporting Function," 1992.
19. CCITT Rec. Q.731, "Information Technology—Open Systems Interconnection—Systems Management: State Management Function," 1993.
20. CCITT Rec. X.734, "Information Technology—Open Systems Interconnection—Systems Management: Event Report Management Function," 1993.

Low-Mobility Adjunct to GSM

16.1 INTRODUCTION

As cellular phone service becomes more popular, both operators and users want to extend the concept of wireless communications to all of their communications needs. In many countries, the wireline infrastructure is not available to support the demands for telephone service. Even in countries that have a well-developed wireline infrastructure, competition has resulted in new companies that want to offer telephone services. In each of these cases, the cost of a new wireline infrastructure is prohibitive. Therefore, a wireless solution is necessary. When one examines the necessary bandwidth and other requirements for offering traditional wireline telephone services over wireless systems, the solutions offered by traditional (high-tier) cellular/PCS services (i.e., CDMA, GSM, and TDMA) are often compromises. Therefore, operators that want to offer a range of services to pedestrians, wireless local loop, and low-speed vehicles often examine different radio technologies that are designed specifically for these markets.

Parts of this chapter are adapted from references [5] and [6], © 1997, Bellcore. Used with permission.

Table 16.1 compares the salient features of the high-tier and low-tier technologies. The advantages of a high-tier radio system are the large coverage area of the BSs and the high user velocities at which access can be supported. The trade-offs, however, are medium-quality voice and limited data service capabilities with high delays. The low-tier systems are at a disadvantage on coverage area size and user speeds. Their advantages include high-quality, low-delay voice and data capabilities.

This chapter describes the DECT (and its U.S. equivalent PWT) system and its suitability in the local loop to provide voice and data services.

The DECT is a radio interface standard developed in Europe mainly for indoor wireless applications and being promoted lately for wireless local loop applications as well. PWT is a DECT-based standard developed by TIA in the United States for the unlicensed PCS applications. PWT-E is the version that is suitable for licensed PCS applications. The goal for each of these systems is to provide a cordless phone that can be used anywhere in a region rather than just within a home or office. Once the BSs, also called radio ports (RPs), are in place, service can also replace wireline service by mounting fixed units on a home or apartment building and connecting to the house wiring.

This chapter provides detailed analyses for DECT/PWT in the following areas:

☞ System architecture
☞ Vertical services
☞ Data capabilities
☞ Capacity and spectrum analysis
☞ Operations
☞ Coverage

Table 16.1 A Comparison of High-Tier and Low-Tier Radio Technologies

	High Tier	Low Tier
BS	Large and expensive	Small and inexpensive
Coverage of a BS	2–10 km	0.5–1 km
Talk time for portables	1–4 hours	> 4 hours
Capacity	Low	High
Vehicular Service	> 110 km/h	< 60 km/h
Voice Quality	Poorer than wireline	Equivalent to wireline
Expected Usage	Low	High
Principal Application	Outdoor vehicle	Indoor & outdoor pedestrian and wireless local loop
Representative Systems	CDMA, GSM, TDMA	DECT/PWT, PACS, PHS

16.2 DECT

The DECT system is one of the major radio technologies developed for wireless local loop. This technology was originally developed for wireless PBXs and low mobility in Europe to replace the Cordless Telephone-2 (CT-2) standard. However, it was eventually enhanced in Europe and the United States to address Wireless Local Loop (WLL) aspects as well. The American version of DECT is called PWT for unlicensed use and PWT-E for licensed use.[*] For the remainder of this chapter, when we refer to DECT, we are also referring to PWT which only differs from DECT at the physical layer. Specific physical layer differences will be presented as needed.

DECT was originally designed as a radio system interface between a Fixed Part (FP) and a Portable Part (PP) (see Figure 16.1). We assume that the DECT radio system will be used in conjunction with a wireline network. The interface between a DECT radio system and supporting wireline network has been left up to vendors to implement. However, recently ETSI, the organization responsible for DECT standards, started working on these interfaces (discussed later in this chapter).

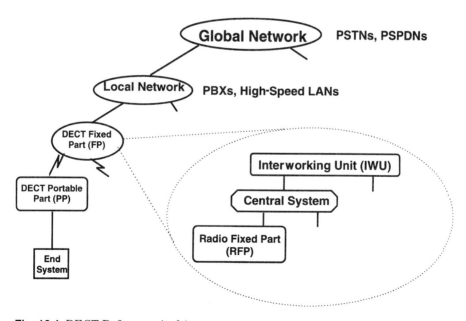

Fig. 16.1 DECT Reference Architecture

---

[*]For the United States, the protocol has been modified to support both the licensed and unlicensed PCS bands. In the unlicensed bands, the protocol supports the spectrum etiquette rules of the FCC.

The FP generically has three major functions:

1. **Radio Fixed Part (RFP)**—terminates the Common Interface (CI) air interface protocol
2. **Central system**—provides a cluster controller functionality managing a number of RFPs
3. **Interworking Unit (IWU)**—provides all the necessary function for the DECT radio system to interwork with the attached wireline network (e.g., PSTN, ISDN, PSPDN).

The DECT standards do not define these components or the interfaces between them. They are supplier proprietary. Some suppliers are using standard interfaces.

Note that there may be a local network (e.g., PBX, LAN) between the DECT radio system and the wireline network.

For the WLL application (Figure 16.2), the PP in the DECT reference model is a component attached to the building and typically called the Cordless Terminal Adapter (CTA). The CTA contains all the functionality of a PP and an additional interface to the inside wiring.

As shown in Figure 16.3, the DECT reference architecture may contain a controller/concentrator and an IWU. An NMS may also be part of the architecture.

European DECT uses GMSK with a bandwidth of 1.728 MHz. In the United States, GMSK did not meet the FCC spectrum emission requirements; therefore, the DECT protocol was modified to use $\pi/4$-QPSK which allows more efficient use of the spectrum. In the unlicensed PCS band (1910–1930 MHz), the FCC permits 10 MHz for voice and 10 MHz for data. In the licensed band, the entire band can be used for voice or data.

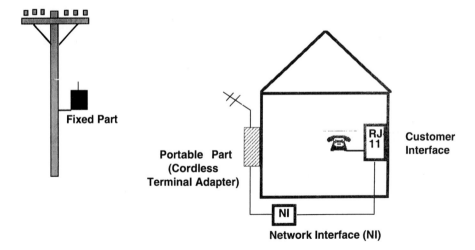

Fig. 16.2 DECT Reference Model for WLL Applications

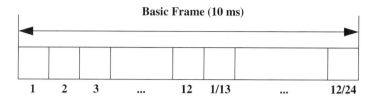

Fig. 16.3 Frame Structure of DECT

While other PCS technologies separate the band into a handset transmit band and a BS transmit band, DECT uses TDD with both the handset and BS transmitting on the same frequency (at different times).

DECT has 24 time slots in 10 ms. Twelve slots are defined for BS to handset transmission and 12 are defined for handset to BS transmission. The overall data rate for voice for handset/BS is 32 kbps using ADPCM which provides toll-quality voice. The transmission path between handset and BS uses a pair of time slots on the single RF channel.

While the standard is designed for low-mobility portable applications, it also permits fixed access units (or CTA as shown in Figure 16.4) for fixed WLL applications.

If the BS or handset requests data rates higher than 32 kbps, multiple time slots are used. If the requested data rate is lower than 32 kbps, then the transmission skips time slots (thus allowing other handsets to use that time slot at a similar lower data rate).

DECT uses DCA for assigning frequencies to the channels. The frequencies are dynamically selected right before their use. (Most cellular technologies assign channels to the base station at the time of installation.) The DCA mechanism provides an efficient use of the valuable radio spectrum.

The size of the cell covered by an RFP is rather small, less than 150 m for urban applications and 1–2 km for rural applications. For rural applications the coverage can be extended by using repeaters but capacity is lowered.

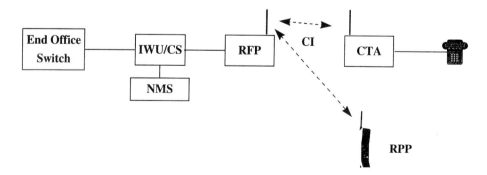

Fig. 16.4 A More Formal Reference Architecture for DECT

DECT is primarily designed to support pedestrian speed mobility. This speed is typically less than 10 km/h.

Table 16.2 provides a comparison of basic characteristics of DECT and the U.S. version of PWT.

16.2.1 A Typical Implementation of DECT

Ericsson's SuperCordless system is based on the U.S. standard PWT. It is primarily implemented for indoor wireless business applications. It is being marketed as a WLL product as well.

As shown in Figure 16.5, the fixed access unit consists of DECT PP/CTA functionality with interfaces to the fixed wireline phones. The BS provides the DECT RFP functionality. The IWU and central system functions of the DECT are provided by the radio exchange equipment.

Table 16.2 Comparison of Basic Characteristics of DECT and PWT

	DECT	PWT
Duplexing	TDD	
Channel Separation	1728 kHz	1250 kHz
Transmission Rate	1152 kbps	
Speech Codec	32 kbps ADPCM	
Modulation	GMSK	π/4-QPSK
Handset Peak Power Output	250 mW	50 mW
Frequency Planning	DCA	
Radius of Service	100–150 m	

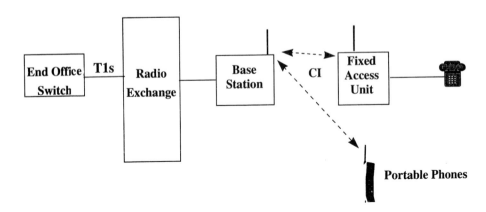

Fig. 16.5 A Typical Implementation of PWT-E Technology

The radio exchange can support 48 Erlangs of traffic and 128 radio ports. Each radio port can support 12 duplex users. For fixed services, standard traffic tables apply so that one radio port can support 5.9 Erlangs at 1-percent blocking. For portable services, 2 co-located radio ports support 1.2 Erlangs at a 1-percent GoS (that is, a combination of call blocking and handover blocking) [4]. A maximum of 10 radio carriers can be co-located and will supply 7.7 Erlangs for portable service and 122 Erlangs for fixed service. The range of a radio port is about 150 m in an urban environment supporting portables. In a rural environment supporting fixed wireless access units, the range is about 1.5–2 km (approximately 1–1.2 miles). With directional antennas, the range can be extended. No frequency planning is necessary; the radio port determines the correct frequency to use within the permitted frequency range.

16.2.2 Supplementary Services

DECT standards recommend the use of supplementary services defined for ISDN. The following supplementary services are explicitly defined in various standards and are the same as for GSM as described in chapter 9.

☞ Malicious Call Identification (MCID)

☞ Call Forwarding Busy (CFB)

☞ Call Forwarding Unconditional (CFU)

☞ User-to-User Signaling (UUS)

☞ Calling Line Identification Presentation (CLIP)

☞ Calling Line Identification Restriction(CLIR)

☞ Connected Line Identification Presentation (COLP)

☞ Connected Line Identification Restriction(COLR)

☞ Completion of Calls to Busy Subscribers (CCBS)

☞ FreePhone (FPH)

☞ Advice of Charge (AOC)

☞ Subaddressing (SUB)

☞ Terminal Portability (TP)

☞ Call Waiting (CW)

☞ Direct Dialing In (DDI)

☞ Multiple Subscriber Number (MSN)

☞ Closed User Group (CUG)

☞ Explicit Call Transfer (ECT)

☞ Single-Step Call Transfer (SCT)

☞ Call Forwarding No Reply (CFNR)

☞ Call Deflection (CD)

☞ Conference call add-on (CONF)

☞ Call Hold (CH)

☞ Three Party (3PTY)

Supplementary services in DECT are divided into two separate categories:

1. **Call-Related Supplementary Services (CRSS)**—explicitly associated with a single instance of CC entity. All CRSS information elements are contained in the associated CC messages (establish, information, release).

2. **Call-Independent Supplementary Services (CISS)**—refers to all CC instances (e.g., call forward on busy) or may refer to services that are unconnected to any CC instances. The CISS messages are invoked independent of any CC instance. Examples:

    *   Negotiation of account details
    *   Reverse charging
    *   Charge sharing
    *   Advice of charge
    *   Charge information (electronic receipt)

    DECT standards define several additional supplementary services:

☞ Queue management—used to register a PP in a queue for outgoing calls, e.g., in the case of a network congestion.

☞ Indication of subscriber number—this feature needs to be activated in order to receive the subscriber number.

☞ Control of echo control functions—used to connect or disconnect the FP echo control functions depending on the type of service and call routing information. Depending on the type of interface (e.g., 2-wire) and the type of local network (e.g., GSM) that connects the FP to the network, one of the four echo control functions can be used.

☞ Cost information—used to obtain cost information such as tariffing, charging, or charging pulses.

16.2.3 Data Capabilities

DECT provides data capabilities in a number of different areas to facilitate various data applications.

The Connection Oriented Message Service (COMS) includes the point-to-point connection-oriented packet service that supports only packet-mode calls.

The Connectionless Message Service (CLMS) includes the fixed- and variable-length message services. The fixed-length message service is used to support group paging and broadcast messages from FP to PP only. The variable-length message service operates in both directions.

The following are the optional user-plane data services defined for DECT.

☞ Transparent Unprotected Service (TRUP)—to transmit unprotected transparent data.

☞ Frame Relay Service (FREL)—reliable transmission of SDUs; performs checksum, segmentation, and reassembly.

☞ Frame Switching Service (FSWI)—for further study.

☞ Forward Error Correction Service (FEC)—for further study.

☞ Basic Rate Adaptation Service (BRAT)—provides for the transparent transport of synchronous continuous data rate at rates 64, 32, 16, and 8 kbps. This allows transparent interworking with ISDN B and D channels.

☞ Secondary Rate Adaptation Service (SRAT)—operates only in conjunction with BRAT service. It rate adapts data terminal equipment with V-series interfaces (async or sync from 50 bps to 56 kbps) to be interfaced to one of the input rates provided by the BRAT. It uses the procedures defined in V.110.

☞ Escape Service (ESC)—to allow an implementation-specific nonstandard service.

16.2.4 Capacity and Spectrum Analysis

Radio parameters were determined from reference [2]. This document is currently undergoing letter ballot to be a standard for PWT-E. Since the document is under ballot, it may change before adoption.

When determining the coverage of a system, both system capacity for handling traffic and radio coverage must be considered. Some areas of a system may need radio ports for capacity and some may need it for coverage.

For the licensed PCS bands, the channel spacing for PWT-E is 1.0 MHz. Thus, the system can have 15 RF carriers in both the transmit and receive bands (30 carriers total) for the 30-MHz PCS bands and 10 carriers for the 10-MHz PCS bands. When used in the unlicensed band, the carrier spacing is 1.25 MHz, allowing for 16 carriers in the unlicensed band. Since the system uses TDD, both radio port and handset (or fixed unit) transmit and receive in the same band. Thus, the PCS bands that are separated into BS and handset bands for other technologies are used by both units in both band segments. In this chapter, only the licensed band was considered.

DECT system traffic engineering depends on whether the service is for fixed WLL or for portable handsets (mobility application). When used for fixed WLL, standard traffic tables apply. Thus, Erlang B at 1-percent blocking can be used to determine the capacity of the system. When the system is used for mobility, special traffic engineering rules are applied. The PWT standard document [2] describes a table for engineering the system. The BSs are deployed in sufficient quantities so that a given handset can receive service from multiple BSs. Once a call is established, as the handset moves, a handover is necessary. The engineering rules apply a 10:1 weighting factor between a blocked handover and a blocked call. This is to ensure that, once a call is established, the handover will not be blocked. This method of traffic engineering reduces the capacity of the system compared to fixed usage.

In an urban environment, the low power and low sensitivity of the DECT BS limit the radio range; the BSs can be no further than 300 m apart. In rural areas, this number is higher. In this chapter, the 300-m number was used.

As described before, each BS has 12 duplex (transmit and receive) voice channels. Therefore, a maximum of 12 conversations can be handled on each BS.

The architecture of the system requires that all BSs be connected to a Radio Exchange (RE). Each RE can control a maximum of 128 BSs. If the traffic per BS is high, then traffic considerations will limit the number of BSs connected to an RE to less than 128. The BS can be a maximum of 6 km from the RE. The powering of the BS is with 18 volts direct current (DC) at about 0.5 amps. The BS interface is ISDN Basic Rate Interface (BRI). For our sample architecture for DECT, we assume that the RE can handle 48 Erlangs of traffic independent of the number BSs connected to it. If the total traffic on each BS is high, then fewer than 128 BSs will saturate the RE.

For BSs, the traffic engineering depends on whether the BSs are used for fixed or mobility applications. For fixed WLL applications, then standard traffic engineering tables apply. Thus one BS has 12 channels and handles 5.9 Erlangs of traffic at 1-percent blocking. For two BSs (near to each other) there are 24 channels with 15.3 Erlangs of traffic. Some typical numbers are supplied in Table 16.3. For additional traffic table information, consult the standard Erlang B traffic tables in appendix A or use the software supplied with this book.

Table 16.3 Traffic Analysis for Ericsson SuperCordless System

Number of Radio Ports	Traffic in Erlangs at 1% Blocking (No Mobility)	Traffic in Erlangs at 1% Blocking (with Mobility)	Capacity of RE in Erlangs
1	5.9	0.27	48
2	15.3	1.9	48
3	25.5	3.2	48
4	36.1	4.1	48
5	47	4.9	48
6	48	5.4	48
7	48	5.7	48
8	48	5.8	48
9–10	48	6	48

16.2.5 Operations

This discussion illustrates DECT operations flows within a TMN context as shown in Figure 16.6.

The service activation flow follows a relatively traditional sequence from initial customer contact through activation of the subscription at the switch. There is no requirement, however, for a formatted Universal Service Order (USO) although a service request event in some form must initiate activation and billing. The only event that could be construed as specific to WLL is the need to dispatch for installation of the DECT fixed access unit at the subscriber's premises. If inventory of the fixed access unit is desired, the data would be stored together with RE and BS data. There is no activation required at the RE or BSs. Note that, as mentioned previously, embedded systems require loop assignment as part of the service activation process. In the event that WLL is deployed in an environment with these embedded systems, workarounds must be instituted.

The service assurance flow supports maintenance of the radio equipment. The radio EM passes subnetwork performance and fault messages to the radio NM. This information can be filtered up to a cross-domain surveillance system if desired.

Customer-initiated trouble papers that require loop testing will require enhancement or workaround to embedded maintenance systems as previously mentioned. Maintenance of the switch is unaffected by the presence of WLL in the network and is not addressed.

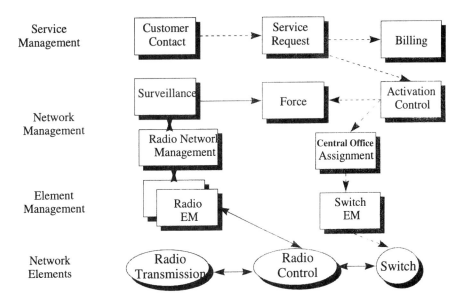

Fig. 16.6 DECT Operations Flows within a TMN Context Architecture

Network capacity provisioning for the radio equipment is driven by performance management activities and confined to the radio EM and NM.

As discussed previously, the DECT reference model does not address operations, and thus there are no defined standards for DECT operations systems or interfaces. In the absence of any standard, DECT vendors are obligated to develop operations products that are unique to their own product implementation decisions. For this reason DECT operations products of one supplier would not support another DECT supplier's radio equipment.

16.3 SUMMARY

In this chapter we examined DECT and PWT, the low-mobility wireless system that is an adjunct to GSM. The system provides low-cost and high-voice quality services to pedestrians and WLL subscribers. Table 16.2 provided a comparison of several key system parameters for these two technologies.

16.4 REFERENCES

1. ETS 300 175, "DECT Common Interface."
2. TR41.6/96-03-007, "Personal Wireless Telecommunications—Enhanced (PWT-E) Interoperability Standard (PWT-E)," July 1996.
3. Garg, V. K., and Wilkes, J. E., *Wireless and Personal Communications Systems,* Prentice Hall, Englewood Cliffs, NJ, 1996.
4. Yu, C. C., Morton, D., Stumpf, C., Ulema, M., White, R. G., Wilkes, J. E., "Low Tier Wireless Local Loop Radio Systems, Part 1 Introduction," *IEEE Communications Magazine,* March 1997.
5. Yu, C. C., Morton, D., Stumpf, C., Ulema, M., White, R. G., Wilkes, J. E., "Low Tier Wireless Local Loop Radio Systems, Part 2 Comparison of Systems," *IEEE Communications Magazine,* March 1997.

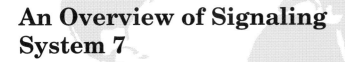

An Overview of Signaling System 7

17.1 INTRODUCTION

Signaling is the system that enables stored program control exchange, network databases, and other "intelligent" nodes of the network to exchange messages related to call setup, supervision, and teardown (call/connection control information). It also permits exchange of information required for distributed application processing (interprocess query/response, or user-to-user data) and NM information. SS7 is a general-purpose common channel signaling (CCS) system specified by the ITU-T.

In this chapter, we provide a brief overview of SS7 for the benefit of those of you who may not be familiar with SS7. Section 17.2 describes types of signaling. In section 17.3, we examine the type of signaling network structure. In section 17.4 we discuss MTP level 1, 2, and 3, along with SCCP. Section 17.5 covers signaling messages, and section 17.6 includes SS7 user parts: ISUP, TCAP, and OMAP. Section 17.7 discusses the application of SS7 in GSM. The performance objectives for SS7 are given in section 17.8.

17.2 TYPES OF SIGNALING

Signaling schemes that are used in practice include in-band signaling, out-of-band signaling, and signaling in ISDN.

17.2.1 In-band Signaling

In in-band signaling, signaling information is conveyed over the same channel that is used to carry speech. In-band signaling has these drawbacks:

☞ Long call setup times (10 to 20 seconds).

☞ Limited information can be transferred resulting in restrictive network routing capabilities.

☞ Call in progress cannot be "modified."

☞ The network is subject to fraud (black-box/blue-box tone generators).

While most telephone networks in the world have replaced in-band signaling with out-of-band signaling, in-band signaling still exists in some parts of the world.

17.2.2 Out-of-band Signaling

In out-of-band signaling, a common data channel is used to convey signaling information related to a number of trunks.

17.2.3 Signaling in ISDN

This type of signaling has two distinct components:

1. Signaling between user and network node to which the user is connected (access signaling)
2. Signaling between the network nodes (network signaling)

The protocol standard for ISDN access signaling is Digital Subscriber Signaling System No. 1 (DSSS1) and the protocol standard for ISDN network signaling is SS7.

17.3 TYPES OF SIGNALING NETWORK STRUCTURES

Signaling networks can use three different signaling modes where "mode" refers to the association between the path taken by the signaling message and its corresponding signaling relation.

☞ In the associated mode of signaling, the messages corresponding to a signaling relation between two points are conveyed over a link set directly interconnecting those two signaling points.

☞ In the nonassociated mode of signaling, the messages corresponding to a signaling relation between two points are conveyed over two or more link sets in tandem passing through one or more signaling points other than the origin and destination of the messages.

☞ The quasi-associated mode of signaling is a non-associated signaling mode where the path taken by the message through the signaling network is predetermined and fixed.

Figure 17.1 shows the quasistructure in which the Signaling Transfer Points (STPs) are mated on a pairwise basis. This type of structure is used in North America. The mesh network provides 100-percent redundancy; that is, for any single point of failure, the traffic can be diverted to alternate paths that do not increase the number of transfer points. The network is engineered so that each component can handle twice its peak load when there are no failures. Also, facility diversity requirements are placed on signaling links to meet availability requirements.

17.3.1 SS7 Performance

SS7 performance parameters are given in two ITU-T recommendations, Q.706 [1] and Q.709 [2]. The two main objectives to be satisfied in SS7 planning are:

1. To limit delay in signaling connections in the network
2. To achieve a high degree of availability of signaling connections

To specify signaling performance parameters and apportion them between national and international segments of an international signaling connection, a hypothetical signaling reference connection (HSRC) has been defined in ITU-T recommendation Q.709. Also, the performance requirements for the basic building block of an SS7 network—namely, Signaling Points (SPs), STPs, and signaling links—have been defined. Recommendation Q.706 deals with the performance of the MTP.

The HSRC defines a signaling connection for international working. The worldwide signaling network generally consists of two independent networks—the international signaling network and the national network. The HSRC includes two national segments and one international segment. The

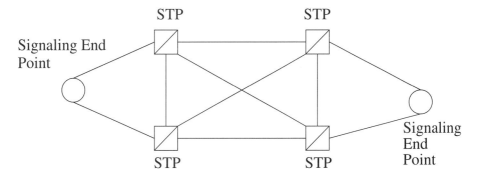

Fig. 17.1 Quasistructure of Signaling Network in North America

nodes of the HSRC are the application service parts (ASPS) and STPs, which are interconnected by signaling data links. The national segment contains the exchange (SPs) and the STPs in the national network involved in the connection but excludes the International Switching Center (ISC) in the country. The international segment includes international SPs and STPs. From the point of view of SS7, the ISC is considered as an international SP. In addition, the ISC may also possess STP capabilities. At least one signaling point in each country should provide the STP function for international SS7 traffic so that the provision of international signaling links can be optimized.

For national signaling, a distinction is made between an average-sized country and a large-sized country. An average-sized country is defined as the one that has a maximum distance of 1000 km between a subscriber and an ISC (in some exceptional cases this may go up to 1500 km) or where the number of subscribers is less than n x 10,000,000 (the value of n has not yet been specified). Countries with larger distances or number of subscribers are classified as large-sized countries. The maximum number of SPs and STPs allowed for international and national segments of the HSRC are given in Table 17.1.

Table 17.2 gives the breakup for the case of end-to-end signaling. End-to-end signaling involves a signaling relationship between the originating and terminating signaling points, often called Signaling End Points (SEPs). All other signaling nodes perform signal transfer functions. A distinction is made between signaling nodes involving message transfer through the MTP and a transfer function where both the SCCP and MTP are involved. The latter are signaling points with SCCP relay functions.

Table 17.1 Maximum Number of SPs and STPs in an HSRC—Link-by-Link Signaling

Size	Percentage of Signaling Connections	Maximum Number of		
For International Segment		**SPs**	**STPs**	**Nodes**
Large to Large	50	3	3	6
	95	3	4	7
Large to Average	50	2	2	4
	95	4	5	9
Average to Average	50	5	5	10
	95	5	7	12
For Each National Segment		**SPs**	**STPs**	**Nodes**
Average	50	2	2	4
	95	3	3	6
Large	50	3	3	6
	95	4	4	8

Table 17.2 Maximum Number of SEPs, STPs, and SPRs in an HSRC (End-to-End Signaling)

Size	Percentage of Signaling Connections	Maximum Number of			
For International Segment			**STPs**	**SPRs**	**Nodes**
Large to Large	50		4	2	6
	95		4	3	7
Large to Average	50		6	2	8
	95		6	3	9
Average to Average	50		8	2	10
	95		8	4	12
For Each National Segment		**SEPs**	**STPs**	**SPRs**	**Nodes**
Large	50	1	4	1	6
	95	1	5	2	8
Average	50	1	2	1	4
	95	1	4	1	6

17.4 NETWORK SERVICE PART (NSP)

SS7 is made of an NSP and user parts (UPs). Besides the use for signaling, SS7 is used for OAM&P of the GSM network and also for an alphanumeric messaging service known as SMS.

SS7 is the out-of-band signaling used throughout the world. It has evolved along the lines of OSI-7 layer network definition, where a layered structure (transparent from layer to layer) is used to provide network communications. But it is not fully OSI compliant.

Post-dial delay is one of the principal measures of performance of a signaling system. To minimize delay, the seven OSI layers were truncated at layer 4. CCITT (ITU-T) Rec. Q.709 specifies no more than 2.2 seconds of post-dial delay for 95 percent of calls. To achieve this goal, a limit is placed on the number of relay points, or STPs, that can be traversed by a signaling message and by the inherent design of SS7 as a four-layer system (see Figure 17.2).

The SS7 NSP consists of:

☞ MTP, i.e., MTP level 1, MTP level 2, and MTP level 3
☞ SCCP

17.4.1 MTP

MTP consists of levels 1, 2, and 3 of the SS7 protocol—signaling data link, signaling link, and signaling network functions. MTP provides a connectionless message transfer system to enable signaling information to be transferred

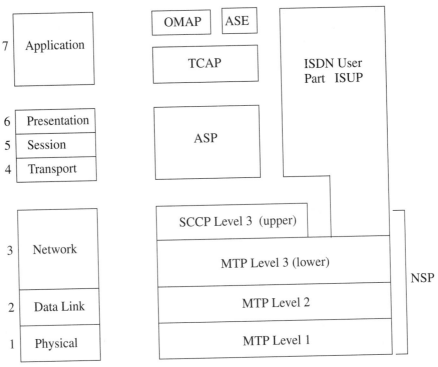

ASP: Application Service Part   TCAP: Transaction Capabilities Appli.
OMAP: Operation, Maint. & Admin. Part   SCCP: Signaling Conn. Control Part
ASE: Appl. Service Element   MTP: Message Transfer Part
                                                 NSP: Network Service Part

Fig. 17.2 Relationship between OSI and SS7 Protocol Model

across the network to its desired destination. MTP is tailored to the real-time needs of telephony applications. Connectionless (datagram) capability is used in MTP to provide a reliable transfer and delivery of signaling information across the signaling network and to take necessary actions in response to system and network failures to ensure that reliable transfer is maintained.

MTP level 1 defines the physical, electrical, and functional characteristics of the signaling data link. The signaling data link is the bidirectional transmission path for signaling. It consists of two data channels operating together in opposite directions at the same data rate. In the digital network, the recommended bit rate for the ANSI standard is 56 kbps and for the CCITT it is 64 kbps. Lower bit rates may be used, but message delay requirements of the UPs must be taken into consideration. The minimum bit rate allowed for telephone call control is 4.8 kbps.

MTP level 2 carries out the signaling link functions. It defines the functions and procedures for the transfer of signaling messages over one individual signaling data link. A signaling message is transferred over the signaling link in signal units of variable length. A signal unit consists of transfer control information in addition to the information contents of the signaling message. The signaling link functions are:

☞ Delamination of a signal unit by means of flags

☞ Flag imitation prevention by bit stuffing

☞ Error detection

☞ Error control by retransmission and signal unit sequence control by means of explicit sequence number of each signal unit and explicit continuous acknowledgment

☞ Signaling link failure detection by means of signal unit error monitoring and signaling link recovery by means of special procedures

MTP level 3 defines transport functions and procedures that are common to and independent of the individual signaling link. Two types of functions take place on level 3:

☞ Signaling message handling functions that direct the message to the proper signaling link or user part

☞ Signaling NM functions that control real-time routing, control, and network reconfiguration, if needed

17.4.2 SCCP

Each user part defines the functions and procedures peculiar to the particular user, whether telephone, data, ISDN, or mobile user part. SCCP is an MTP user part and is in the upper part of level 3 of SS7. SCCP provides additional functions to MTP for both connectionless and connection-oriented network services. SCCP satisfies the connectionless capability need. Dividing the OSI network layer functions in MTP level 3 and SCCP has advantages in the sense that the higher-overhead SCCP services can be used only when required. This division allows the more efficient MTP to serve the needs of those applications that can use a connectionless message transfer with limited addressing capability.

The addressing capability of MTP is limited to delivering a message to a node and using a 4-bit service indicator (a subfield of Service Information Octet [SIO]) to distribute a message within the node. SCCP supplements this capability by providing an addressing capability that uses Destination Point Codes (DPCs) and Subsystem Numbers (SSNs). The SSN is local addressing information used by SCCP to identify each of the SCCP users at a node. SCCP provides another addressing enhancement to MTP by giving it the ability to address messages with global titles, which are addresses, such as dialed digits, that do not explicitly contain information usable for routing by MTP. For global titles a translation capability is required in SCCP to translate the global title to a DPC plus SSN. This translation function can be performed at the originating point of the message or at another signaling point in the network (e.g., at an STP).

MTP and SCCP provide for services through an equivalent OSI layer model for ASP processes. OSI layers 4, 5, and 6 may be specified and developed in a common way for all such applications. These three layers of the protocol use the equivalent of the OSI transport, session, and presentation layer protocols. Thus, the ASP builds on the services of the MTP and SCCP by providing additional functions to meet the needs of the users of the signaling network. The ASE utilizes the capabilities of the MTP, SCCP, and ASP. One example of such application is the TCAP, which supports activities such as 800 or calling card services.

SCCP consists of four functional blocks.

1. SCCP connection-oriented control. This controls the establishment and release of signaling connections for data transfer on signaling connections.

2. SCCP connectionless control. This provides the connectionless transfer of data units.

3. SCCP management. This provides the capability, in addition to signal route management and lower control functions of the MTP, to handle the congestion or failure of either the SCCP user or signaling route to the SCCP user.

4. SCCP routing. On receipt of a message from the MTP or from the functions in 1, 2, or 3, SCCP routing either forwards the message to the MTP for transfer or passes the message to the function 1, 2, or 3.

17.4.3 Class of Network Services

The SCCP modifies the connectionless sequence transport service provided by MTP for those user parts requiring enriched connectionless or connection-oriented service to transfer signaling and other related information between nodes. SCCP therefore provides the combined service (MTP + SCCP) that is the NSP.

Five classes of network services are provided by SCCP (refer to Table 17.3).

Table 17.3 Class of Network Services

0	Basic sequence connectionless
1	Sequence connectionless (like MTP)
2	Basic connection oriented
3	Flow control, connection oriented
4	Error recovery & flow control, connection oriented

In class 0, the Network Service Data Unit (NSDU) is passed by higher layers to the SCCP in the node of origin. The NSDUs are transported independently and may be delivered out of sequence; thus, this class of service is purely connectionless.

In class 1 the features of class 0 are provided with an additional feature to allow a higher layer to indicate to SCCP that a particular stream of NSDUs should be delivered in sequence. SCCP does this by associating the stream members with a sequence control parameter and giving all messages in the stream the same Signaling Link Selection (SLS) code.

In class 2 a bidirectional transfer of NSDUs is performed by setting up a temporary or permanent signaling connection (virtual circuit through the signaling network). Messages belonging to the same signaling connection are given the same SLS code to ensure sequencing. This service class provides a segmentation and reassembly capability. With this capability, if NSDU is longer than 255 octets, it is split into multiple segments at the originating node. Then each NSDU segment is transported to the destination node in the data field of a data message; at destination node SCCP it reassembles the original NSDU.

In class 3 the capabilities of class 2 are provided with the addition of flow control. Also the detection of message loss and missequencing is provided. In the event of lost or missequenced messages, the signaling connection is reset and notification is given to the higher layers.

In class 4 the capabilities of class 3 are provided with the addition of error recovery.

17.5 SIGNALING MESSAGE

A signaling message is an assembly of information, defined at level 3 or 4, pertaining to a call, management transaction, etc. It is then transferred as an entity by an MTP. Each message contains service information, including a service indicator identifying the source user part and possibly whether a message relates to international or national application of the user part.

Signaling information of the message contains user information such as data or call control signals, management and maintenance information, and type and format of message. It also includes "label." The label enables the message to be routed by level 3 through the signaling network to its destination and directs the message to the desired user part or circuit.

A SS7 signaling message (refer to Figure 17.3) is bracketed by flag signals indicating the beginning and end of the message. Following the initial flag are the forward and backward sequence number (FSN, BSN) and forward and backward indicator bits (FIB, BIB). The FSN is the sequence number of the signal unit in which it is carried; the BSN is the sequence number of the signal unit being acknowledged. The FSN and BSN are numbers in binary code from a cyclic sequence ranging from 0 to 127.

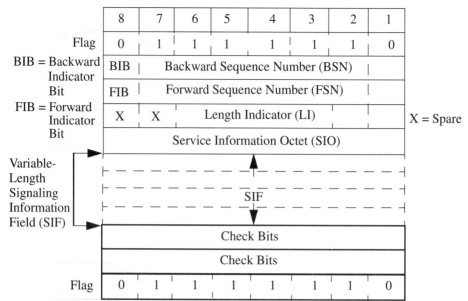

8	7	6	5	4	3	2	1

Flag 0 1 1 1 1 1 1 0

BIB | Backward Sequence Number (BSN) |

FIB | Forward Sequence Number (FSN) |

X X Length Indicator (LI)

Service Information Octet (SIO)

SIF

Check Bits

Check Bits

Flag 0 1 1 1 1 1 1 0

BIB = Backward Indicator Bit

FIB = Forward Indicator Bit

X = Spare

Variable-Length Signaling Information Field (SIF)

Fig. 17.3 SS7 Signaling Format

☞ The FIB and BIB, together with the FSN and BSN, are used to perform the signal unit sequence control and acknowledgment functions.

☞ The length indicator (LI) is used to indicate the number of octets following the length indicator octet and preceding the check bits.

☞ The SIO is used to associate signaling information with a particular UP and is present only in message signal units.

☞ The signaling information field (SIF) is variable length, and in national networks it may consist of as many as 272 octets (bytes). The format and codes of the signaling information field are defined for each UP, i.e., telephone, data, etc.

☞ The error detection function is performed by means of 16 check bits provided at the end of each signal unit followed by the closing flag to indicate the end of the message.

SS7 uses three types of signal units:

☞ Message signal unit (MSU) (see Figure 17.4)

☞ Link status signal unit (LSSU) (see Figure 17.5)

☞ Fill-in signal unit (FISU) (see Figure 17.6)

MSUs are retransmitted in case of error; LSSUs and FISUs are not. An MSU carries signaling information; an LSSU provides link status; an FISU is used during the link idle state.

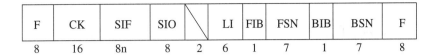

F	CK	SIF	SIO		LI	FIB	FSN	BIB	BSN	F
8	16	8n	8	2	6	1	7	1	7	8

Flag: 01111110

FIB and BIB together with FSN and BSN are used in basic error control method to perform signal unit sequence control and acknowledgment functions.

F: Flag
CK: Check bits
LI: Length Indicator
SF: Status Field
SIF: Signaling Info. Field
SIO: Service Info. Field
FSN: Forward Seq. Number
BSN: Backward Seq. Number
FIB: Forward indicator bit
BIB: Backward indicator bit

Fig. 17.4 Format of MSU

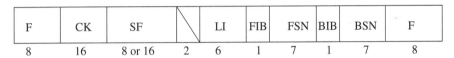

F	CK	SF		LI	FIB	FSN	BIB	BSN	F
8	16	8 or 16	2	6	1	7	1	7	8

LI = 0   Fill-in Signal Unit
LI = 1 or 2   Link Status Signal Unit
LI > 2   Message Signal Unit

Fig. 17.5 Format of LSSU

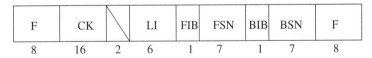

F	CK		LI	FIB	FSN	BIB	BSN	F
8	16	2	6	1	7	1	7	8

LI = 0   Fill-in Signal Unit
LI = 1 or 2   Link Status Signal Unit
LI > 2   Message Signal Unit

Fig. 17.6 Format of FISU

The SIO (Figure 17.4) is divided into the service indicator and subservice field. The service indicator is used to associate signaling information with a particular UP and is present only in MSUs. Each is four bits long. Example: a service indicator with a value 0100 relates to a telephone UP, 0101 to the ISDN UP. The subservice field portion of the SIO contains two network indicator bits and two spare bits. The network indicator bits discriminate between international and national signaling messages. They can also be used to discriminate between two national signaling networks.

The status field in LSSU (Figure 17.5) is made up of one or two octets. In case of a one-octet field, the first three bits (from right to left) are used and the remaining five bits are spare. The function of the first three bits is given in Table 17.4.

17.5.1 Routing Label

A standard routing label consists of 32 bits (4 octets) and appears at the beginning of signaling information field (refer to Figure 17.7).

An SLS field is used, where appropriate, in the performance of load sharing. Each signaling point will have routing information that allows to determine the signaling link over which a message has to be sent on the basis of DPC and SLS.

17.6 SS7 UPs

The UPs of SS7 utilize the transport services provided by MTP and SCCP: ISUP, TCAP, and OMAP.

Table 17.4 First 3 Bits Function

	Bits		Meaning
0	0	0	Out of alignment
0	0	1	Normal alignment
0	1	0	Emergency alignment
0	1	1	Out of service
1	0	0	Processor outage
1	0	1	Busy

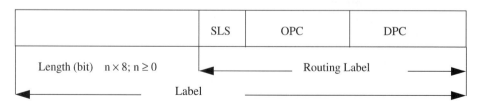

SLS: Signaling link selection
DPC: Destination point code
OPC: Origination point code

Fig. 17.7 Routing Label

17.6.1 ISUP

The ISUP of the SS7 protocol provides the signaling functions that are needed to support the basic bearer service and supplementary services for switched voice and nonvoice (e.g., data) applications in an ISDN environment. ISUP gives all the functions provided by the Telephone User Part (TUP) plus additional functions in support of nonvoice calls and advanced ISDN and Intelligent Network (IN) services. In North America, ISUP is used by all carriers, whereas most other telephone administrations in the world have implemented TUP as their call control signaling protocol. TUP and ISUP can coexist in the same exchange. Information exchange between ISUP and MTP (or SCCP) takes place through the use of parameters carried by interlayer service primitives. All ISUP messages include a routing level identifying the origin and destination of the message, a Circuit Identification Code (CIC), and a message-type code that uniquely defines the function and format of each ISUP message. In the mandatory fixed part of the message, the position, length, and order of parameters is uniquely determined by message type. The basic service offered by ISUP is the control of circuit-switched network connections between subscriber-line exchange terminations. Supplementary services supported by ISUP include user-to-user signaling, closed user group, calling line identification, and call forwarding.

17.6.2 TCAP

Transaction Capabilities (TC) refer to the set of protocols and functions used by a set of widely distributed applications in a network to communicate with one another. In SS7 terminology, TC refers to application layer protocols—i.e., TCAP—plus any transport, session, and presentation layer services and protocols that support it. TCAP directly uses the services of SCCP, which in turn uses the services of MTP, with transport, session, and presentation layers being null layers. TCAP provides a set of tools in a connectionless environment that can be used by an application at one node to invoke execution of a procedure at another node and exchange the results of such invocation. TCAP is closely related and aligned with OSI Remote Operation Service Element (ROSE) protocol (CCITT Recommendations X.219 and X.229). The primary use of TCAP is for invoking remote procedures in support of IN services like 800 service. A TC user ASE provides the specific information that a particular application needs (e.g., information for querying a remote database to convert an 800 number into a network-routeable telephone number). TCAP provides the tools needed by all applications that require remote operation. TCAP is divided into two sublayers: the component sublayer involved in exchange of components between TC users. These components contain either requests for action at the remote end or data indicating the response to the requested operation. The transaction sublayer deals with exchange of messages that contain such components. This involves establishment and management of a dialogue (transaction) between TC users.

17.6.3 OMAP

The OMAP of the SS7 provides the application protocols and procedures to monitor, coordinate, and control all the network resources that make communication based on SS7 possible. The collection of all monitoring, control, and coordination functions above the application layer is known as the Systems Management Application Process (SMAP). SMAP uses the services of TCAP. AP-ASE sits on the top of the TCAP component sublayer.

17.7 Use of SS7 by GSM

GSM does not introduce any new functions in MTP, but uses the protocols as defined by ITU-T. In GSM, signaling messages are carried over 64-kbps links.

When the length of a signaling message exceeds the maximum allowable length for frames, the message is segmented and transmitted over several frames. At the receiving end the message is reassembled. The receiver is provided with additional information to reconstruct the messages. This situation may occur on the A interface. The A interface transmits messages between the BSC and MSC as well as messages to and from the MS. The signaling over the A interface is done by BSSMAP and DTAP protocols that are transported by MTP. The maximum length of frames on the A interface is limited to 272 octets of information plus 6 octets for frame control without flags. The length is adequate to accommodate most of the signaling needs, and no segmentation procedure is required at this level.

MTP level 2 uses 16 redundancy bits in each frame according to the coding scheme selected for error detection. The combination of SCCP and MTP level 3 offers networking service on the A interface. MTP level 3 is used on the A interface to transfer messages between two contiguous entities (i.e., BSC and MSC). GSM uses only SCCP class 0 and class 2 services. The class 0 mode is used on the A interface for messages not directly related to a single MS. The class 2 mode allows setup of several independent connections. This feature is used on the A interface to distinguish a transaction with an individual MS (see chapter 9 for more details).

The HLR communicates with the VLR and MSC using the SS7 network as defined in the GSM recommendations. The transport layers of the signaling are handled according to TCAP protocol. The signaling between an MSC and PSTN or ISDN is based on SS7 TUP or ISUP.

The interface protocols used in GSM are the following (refer to Figure 17.8):

☞ MAP/B: MSC to VLR (if the VLR is located outside the MSC). The protocol is based on SS7.

☞ MAP/C: Interrogation of the HLR by the gateway MSC and used by the HLR to communicate with the SMS center.

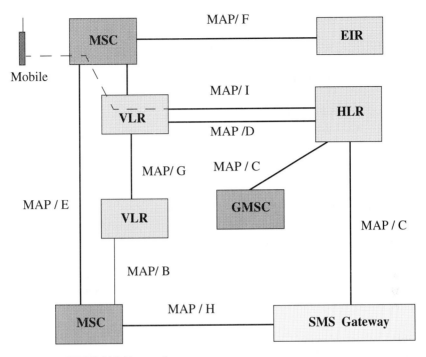

Fig. 17.8 GSM MAP Protocols

☞ MAP/D: MSC/VLR to HLR mainly for mobility management and also to provide call-related information for incoming calls.

☞ MAP/E: MSC to MSC for inter-MSC handovers.

☞ MAP/F: MSC/VLR to EIR.

☞ MAP/G: Used to access the mobile equipment identified by TMSI with reference to another MSC/VLR.

☞ MAP/H: Used to provide the routing between the mobile equipment and gateway MSC and also between MSC and SMS center.

☞ MAP/I: Carries encapsulated messages between the mobile equipment and HLR.

The STPs have a quad structure and are mated to provide full redundancy (see Figure 17.1). An SP is overlaid on every GSM node. Every SP of the GSM network is connected to an STP pair (see Figure 17.9).

TCAP is used at the OSI application layer to build a request/reply application. A TCAP message consists of a component portion and transaction portion. The component portion contains one or more components. A component is a request to perform an operation or a reply to such a request. It is characterized by a type. The component types are

☞ Invoke. Requests the performance of an operation.

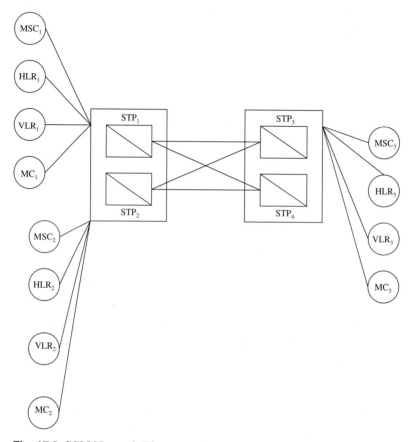

Fig. 17.9 GSM Network Elements Connected to STPs

☞ Invoke (last). Same as invoke. Last indicates that this component is last in the component portion.

☞ Return result. Returns the result of an invoked operation.

☞ Return result (last). Indicates that this component is last in the component portion.

☞ Return error. Reports the unsuccessful completion of an operation.

☞ Reject. Reports the receipt and rejection of a component that is syntactically incorrect.

The transaction portion specifies the package type. The package type represents the modality of the transaction. The following package types are defined:

☞ Query with permission. Initiates a transaction, indicating to the receiving node that it may end it.

☞ Query without permission. Initiates a transaction, indicating to the receiving node that it may not end it.

☞ Conversation with permission. Continues a transaction, indicating to the receiving node that it may end it.

☞ Conversation without permission. Continues a transaction, indicating to the receiving node that it may not end it.

☞ Response. Ends a transaction.

☞ Unidirectional. Indicates that no reply is expected.

☞ Abort. Aborts a transaction.

MAP specifies the SS7 messages and rules governing their exchanges. The messages are specified as TCAP messages. For each message, package types and component types are specified. For a message there is usually more than one component type and more than one package type. Query with permission, conversation with permission, conversation without permission, response, and unidirectional are the four package types used for SS7 messages. Invoke (last), return result (last), return error, and reject are the four component types. The following general rules govern the exchange.

☞ The component portion is limited to a maximum of one component. That is why neither invoke nor return result is used.

☞ Most of the transactions are limited to the exchange of two messages— query with permission and response. Unidirectional is seldom used.

☞ A timer is specified for execution of each operation. The course of action to take when the reply is received is specified for each possible answer— return result (last), return error, or reject. The course of action to take when a reply is not received before the expiration of the timer is also specified.

SS7 messages are specified to support OAM&P and SMS. SS7 is used for both circuit-related and non-circuit-related signaling. In the circuit-related case, the information conveyed by the SS7 messages is related to the specific traffic circuit and is identified by circuit number. In non-circuit-related signaling, there is no predetermined correlation between the information conveyed by SS7 messages and specific traffic circuits. However, information remains call related, although not directly. There is no correlation between the SS7 messages sent to a database to request translation and any specific circuit. However, the information is subsequently used to set up a call. GSM uses the following:

1. Circuit-related signaling is used for call setup/release between GSM and PSTN.

2. Circuit-related and non-circuit-related signaling are used for intersystem handover.

3. Non-circuit-related signaling is used for automatic roaming.

The nodes of the GSM network exchange OAM&P messages specified as SS7 messages and transported by SS7 NSP, which carries the signaling messages. The following messages are specified:

- ☞ Blocking is used by MSC to direct another MSC to remove a specific circuit from service.
- ☞ Unblocking is used by an MSC to direct another MSC to return to service a circuit that was previously blocked.
- ☞ Reset circuit is used by an MSC to direct another MSC to reset a specific circuit to the idle condition.
- ☞ Trunk test is used by an MSC to redirect another MSC to loop back a specific circuit.
- ☞ Trunk test disconnect is used by an MSC to direct another MSC to disconnect the loop back of a specific circuit.
- ☞ Unreliable data roamer directive is sent by an HLR to inform its associated serving systems that it has experienced a failure that has made its MS roamer data unreliable.

SMS allows users to receive and send alphanumeric messages using SS7 messages which are carried by SS7 NSP which carries the signaling messages and OAM&P messages. The submission and reception of the messages is done through the SME. The SME can compose and decompose SMS messages. It can be integrated with the MS to allow the user to use the same terminal for voice and SMS.

An SMS message is a one-shot message of limited size. No call setup or teardown is required. However, SMS users can have a dialogue by exchanging sequences of SMS messages. The delivery of the messages can be instantaneous or delayed. Delayed delivery is an example of a supplementary service. SMS delivery is based on recipient availability, and retries can be attempted. A message sent by an originator message center (MC) can cross many MCs and STPs on its way to the destination MC.

17.8 SS7 PERFORMANCE OBJECTIVES

In this section we'll explain these performance objectives—availability, dependability, ISUP dependability, delay, and message transfer time.

17.8.1 Availability

Availability is a measure of the reliability of the signaling network. For a signaling route set between an origination SP and a destination SP, the availability should be equal to or better than 99.9998 percent. This amounts to a maximum of 10 minutes of permissible downtime for a route set in one year.

The unavailability of a system is determined as the probability that the system is in failed state at a random point in time. This is expressed in terms of the system Mean-Time-to-Failure (MTTF) and Mean-Time-to-Repair (MTTR) as unavailability: MTTR/(MTTF + MTTR).

As an objective, the unavailability of an MTP signaling route set should not exceed 1.9×10^{-5}, which equates to an expected downtime of less than 10 minutes per year. As an objective, the unavailability of an SCCP relay point should not exceed 10^{-4}, which equates to an expected downtime of less than 53 minutes per year.

17.8.2 Dependability

Dependability objectives relate to the ability of the network to reliably transport messages and not cause malfunction. For MTP there are four objectives.

☞ **Undetected Error.** On each signaling link, not more than one in 10^{10} of all signal unit errors should be undetected by the MTP.

☞ **Lost Messages.** Not more than one in 10^7 messages should be lost due to failure of MTP.

☞ **Messages Out of Sequence.** Not more than one in 10^{10} messages should be delivered out of sequence to the user parts due to failures in the MTP. This includes duplicated messages.

☞ **Transmission Error Rate.** The signaling data link shall have a long-term bit error rate that does not exceed 10^{-6}.

17.8.3 ISUP Dependability Objectives

The ISUP dependability objectives are

☞ **Probability of False Operation.** Not more than one in 10^8 of all signal units transmitted should be accepted and, due to errors, cause false operation.

☞ **Probability of Signaling Malfunction.** Unsuccessful calls can be caused by undetected errors, loss of messages, or messages delivered out of sequence. Not more than two in 10^5 of all ISDN calls should be unsuccessful due to signaling malfunctions. Not more than one in 10^5 of all ISDN circuit connections should be unsuccessful due to signaling malfunction.

☞ **Delay.** Not more than one message out of 10^4 messages should be delayed by more than 300 ms due to error correction by retransmission.

☞ **Cross-office Signaling Transfer Time.** The cross-office signaling transfer time at a transit exchange should be as given in Table 17.5. (The cross-office time begins when the last bit of the message leaves the incoming signaling data link and ends when the last bit of the message enters the outgoing signaling data link.)

Table 17.5 Transit Exchange Cross-Office Signaling Delay

Message Type	Exchange Call Attempts	Cross-Office Signaling Transfer Time (ms)	
		Mean	95%
Simple messages, for example, an answer message (ANM)	Normal	110	220
	15% Overload	165	330
	30% Overload	275	550
Processing intensive messages, for example, an IAM	Normal	180	360
	15% Overload	270	540
	30% Overload	450	900

17.8.4 Delay Objectives

The maximum signaling delay is a function of several parameters, such as the number of SPs and STPs that may be involved in a connection, the delay at each of the SPs and STPs, and signaling propagation times on signaling links. Therefore, to keep the delays within the permissible limits, the signaling network should be designed to

☞ Limit the number of SPs and STPs involved in a signaling connection

☞ Limit the delays that may arise in each component of a signaling connection

Table 17.6 lists the maximum signaling delays in the international and national segments of an HSRC for link-by-link and end-to-end signaling.

17.8.5 Message Transfer Time

The message transfer time for a signaling relation is the interval between the time instants when the message leaves the user part at the origination SP and enters the user part at the destination SP. The message may pass through several intermediate STPs and interconnecting signaling data links.

The message transfer time is the sum of

☞ Handling time in the MTPs of the origination and destination SPs

☞ Queuing delays at the SPs

☞ STP message transfer time

☞ Propagation time on the signaling data links involved in signaling relation

ITU-T recommendations identify the various delay components that make up the MTP transfer time; the permissible values for these components are not specified, except for the value of STP transfer time.

Table 17.6 Maximum Signaling Delays (ms) in International and National Segments of an HSRC for Link-to-Link and End-to-End Signaling

		Message Type	
Network Component	**Percent of Connections**	**Simple**	**Complex**
For International Segment			
International (large to large)	50	390 (300)	600 (440)[*]
	95	410 (410)	620 (620)
International (large to average)	50	520 (340)	800 (480)
	95	540 (450)	820 (660)
International (average to average)	50	650 (380)	1000 (520)
	95	690 (600)	1040 (880)
For Each National Segment			
National (large)	50	390 (300)	600 (440)
	95	520 (430)	800 (640)
National (average)	50	260 (260)	400 (400)
	95	390 (300)	600 (400)

*Note: A number outside parentheses is an end-to-end delay and a number inside parentheses is a link-to-link delay.

The STP message transfer time is the interval that begins when the last bit of an MSU leaves the incoming signaling data link; it ends when the last bit of the MSU enters the outgoing signaling data link for the first time. Thus, retransmission of the MSU is not included. ITU-T and American National Standards Institute (ANSI) standards specify the message transfer time at an STP based on a mean signaling load of 0.2 Erlang. Tables 17.7 and 17.8 list the message transfer time at an STP for ITU-T and ANSI recommendations.

Table 17.7 Message Transfer Time at an STP (ITU-T Standards)

	Message Transfer Time (ms)	
Signaling Traffic Load	**Mean**	**95%**
Normal	20	40
+15%	40	80
+30%	100	200

Table 17.8 Message Transfer Time at an STP (ANSI Standards)

Signaling Traffic Load	Message Transfer Time (ms)	
	Mean	95%
Normal	45	80
$2 \times$ Normal	55	90

These tables are based on TUP messages with an average length of 15 octets. For user parts other than TUP, the STP message transfer time is calculated by adding the STP processing time (T_{ph}) and the outgoing signaling link delay (T_{od}). STP processing begins when the last of the MSU leaves the incoming signaling data link and ends when the last bit of the MSU enters the level 2 transmission buffer associated with the outgoing signaling data link. The outgoing signaling link delay begins when the last bit of the MSU enters the outgoing level 2 buffer and ends when the last bit of the MSU leaves the outgoing signaling data link. The STP processor handling time specified in ITU-T Recommendation Q.706 is given in Table 17.9. The values of signaling link delay have been calculated in Q.706 for various combinations of the following parameters:

☞ Link loads of 0.2 and 0.4 Erlang
☞ Basic and preventive cyclic retransmission (PCR) methods of error correction
☞ Presence and absence of disturbances (errors)
☞ Loop delays for terrestrial links of 30 ms and 600 ms (for satellite links)
☞ Mean message lengths of 15, 23, 50, 140, and 279 bytes

Tables 17.10 and 17.11 list the T_{od} values in the absence of disturbances for basic and PCR methods of error correction, respectively. Table 17.12 lists the T_{od} values with basic and PCR methods in the presence of errors.

Table 17.9 STP Processor Handling Time (T_{ph})

Processor Load	Value	STP Processor Handling Time (T_{ph}) for Various Length (bytes)			
		23	50	140	279
Normal	Mean	19	22	33	55
	95%	35	40	50	75
+30%	Mean	60	70	100	160
	95%	120	140	200	320

Table 17.10 Outgoing Link Delay T_{od} with Basic Error Correction Method in Absence of Disturbances

Traffic (Erlang)	Value	Outgoing Link Delay (ms) for Various MSU Lengths (bytes)				
		15	23	50	140	279
0.2	Mean	2.7	4.0	8.3	21.5	39.6
	95%	9.3	14.0	30.1	66.0	61.5
0.4	Mean	3.5	5.2	10.8	27.6	46.9
	95%	12.2	18.6	40.0	88.7	87.1

Table 17.11 Outgoing Link Delay T_{od} with PCR Method in Absence of Disturbances

Traffic (Erlang)	Loop Delay Terrestrial Link (ms)	Value	Outgoing Link Delay (ms) for Various MSU Lengths (bytes)				
			15	23	50	140	279
0.2	30	Mean	4.2	6.4	11.3	23.1	38.7
		95%	12.5	18.9	38.1	71.1	58.6
	600	Mean	4.2	6.5	14.1	35.7	56.0
		95%	12.6	19.4	42.2	93.1	86.2
0.4	30	Mean	5.0	7.6	15.3	29.9	43.8
		95%	15.0	22.9	48.3	93.8	79.8
	600	Mean	5.0	7.7	16.7	41.8	63.9
		95%	15.0	23.0	50.0	111.4	108.9

E X A M P L E 1 7 – 1

Problem Statement

Consider an SS7 network with an average link loading of 0.2 Erlang, a BER of 10^{-7}, and a mean MSU length of 50 bytes. Calculate the mean STP message transfer time using ITU-T recommendations.

Solution

STP message transfer time = STP processor handling time (T_{ph}) + outgoing link delay (T_{od})

- T_{ph} from Table 17.9 = 22.0 ms
- T_{od} from Table 17.12 = 8.2 ms

STP Message Transfer Time = 22.0 + 8.2 = 30.2 ms

Table 17.12 Outgoing Link Delay T_{od} with Basic and PCR Methods in Presence of Errors

Traffic (Erlang)	Loop Delay Terrestrial Link (ms)	Value	Error	Outgoing Link Delay (ms) for Various MSU Lengths (bytes)				
				15	23	50	140	279
0.4	30	Mean	10^{-5}	3.8 (5.6)	5.4 (8.7)	11.2 (18.9)	28.3 (47.5)	48.1 (81.1)
			10^{-7}	3.5 (5.0)	5.2 (7.6)	10.8 (15.3)	27.6 (30.1)	46.9 (44.2)
		95%	10^{-5}	14.3 (26.3)	20.4 (42.5)	41.7 (106.2)	91.4 (310.9)	91.3 (448.2)
			10^{-7}	12.3 (15.2)	18.6 (23.3)	40.0 (49.7)	88.7 (101.1)	87.1 (96.1)
	600	Mean	10^{-5}	86.3 (47.2)	88.5 (55.7)	96.9 (84.8)	121.9 (183.5)	152.0 (282.5)
			10^{-7}	4.3 (5.3)	6.0 (8.0)	11.6 (17.1)	28.5 (42.8)	47.9 (65.3)
		95%	10^{-5}	490.1 (379.7)	496.4 (422.7)	521.2 (586.2)	586.6 (1103.0)	626.9 (1470.0)
			10^{-7}	44.0 (30.2)	48.4 (38.1)	64.5 (67.1)	107.4 (141.5)	106.9 (153.9)
0.2	30	Mean	10^{-5}	2.8 (4.2)	4.1 (6.2)	8.4 (11.4)	21.9 (24.0)	40.4 (41.3)
			10^{-7}	2.7 (4.2)	4.0 (6.2)	8.2 (11.3)	21.5 (23.1)	39.6 (38.7)
		95%	10^{-5}	10.8 (12.8)	15.4 (19.3)	31.4 (39.5)	68.0 (80.4)	64.8 (84.5)
			10^{-7}	9.3 (12.8)	14.1 (18.9)	30.1 (38.0)	66.0 (71.3)	61.5 (59.0)
	600	Mean	10^{-5}	29.6 (5.0)	31.2 (7.4)	36.9 (15.5)	55.0 (39.3)	80.3 (63.0)
			10^{-7}	3.0 (4.2)	4.3 (6.5)	8.5 (14.1)	21.9 (35.8)	40.1 (56.1)
		95%	10^{-5}	248.4 (27.8)	254.2 (34.9)	275.0 (60.3)	329.4 (127.2)	363.8 (140.9)
			10^{-7}	28.9 (13.0)	32.7 (19.7)	46.0 (42.5)	79.2 (93.7)	76.9 (87.4)

EXAMPLE   17 – 2

Problem Statement

Using the information in Example 1 and additional data given in Table 17.13, calculate the number of circuits for en bloc signaling and overlap signaling.

- Speech circuit occupancy during busy hour = 0.8 Erlang
- Ratio of successful calls to total calls (S) = 0.70
- Ratio of unsuccessful calls to total calls (U) = 1 – S = 0.3
- Average call holding time for successful calls (H_s) = 150 s
- Average call holding time for unsuccessful calls (H_u) = 20 s

Solution

Case I: En Bloc Signaling

Mean duration of a call (D) = $S \times H_s + U \times H_u$ = 0.7 × 150 + 0.3 × 20 = 121 s

Number of Busy Hour Call Attempts (BHCA) per circuit = $\dfrac{3600 \times 0.8}{121} \approx 24$ calls per circuit

Mean number of digits dialed per call (N) = $N_s \times S + N_u \times U$ = 11 × 0.7 + 4 × 0.3 = 8.9 ≈ 9

where:
N_s = Number of digits dialed for successful call = 11
N_u = Number of digits dialed for unsuccessful calls = 4

Table 17.13 Data for Signaling Messages[*]

Message Name	Length (octets)	Probability of Occurrence
IAM ----------->	29	100%
ACM <-----------	17	90%
ANM <-----------	15	
or CON <-----------	17	5%
ACM ----------->	17	5%
or CPG <-----------	16	
ANM <-----------	15	
REL ----------->	19	100%
RLC <-------------	15	100%

Note: ACM = Address Complete Message; ANM = Answer Message; CON = Connect Message; CPG = Call Progress Message; IAM = Initial Address Message; REL = Release Message; RLC = Release Complete Message.

[*]A successful call consists of an IAM message plus (ACM + ANM), CON, and an (ACM + CPG + ANM) message along with the REL and RLC messages. An unsuccessful call includes an IAM message plus the REL and RLC messages.

Total signaling information per successful call = $1 \times 29 + 0.9 \times 32 + 0.05 \times 17 + 0.05 \times 48 + 1 \times 19 + 1 \times 15 = 95.05 \approx 95$ octets.

Total signaling information per unsuccessful call = IAM + REL + RLC = 29 + 19 + 15 = 63 octets.

Mean octets for successful call in each direction = 95.0/2 = 47.5 octets.

Mean octets for unsuccessful call in each direction = 63.0/2 = 31.5 octets.

Mean number of octets in each direction for a call = $47.5 \times 0.7 + 31.5 \times 0.3 = 42.70 \approx 43$ octets.

Information transfer in each direction on a signaling link per circuit = $43 \times 24 = 1032$ octets.

At a 64-kbps signaling rate (or 8000 octets per second), the number of circuits that can be serviced by a signaling link with 0.2 Erlang will be:

$$\frac{8000 \times 3600 \times 0.2}{1032} = 5582$$

This figure depends on the mean signaling traffic generated by each speech circuit. Higher signaling traffic will reduce the number of speech circuits that can be serviced by a signaling link. The signaling load generated by a speech circuit increases with the increase in call attempts, percentage of successful calls, mean number of digits dialed per call, and amount of signaling information transferred per call.

Case II: Overlap Signaling

It is assumed that the IAM message is followed by a single subsequent address message (SAM). The sequence of the message is the same as in the previous case, except the IAM is reduced (27 octets assumed) and the SAM (assumed value of 21 octets) follows the IAM.

Total signaling information for successful call = $1 \times 27 + 1 \times 21 + 0.9 \times 32 + 0.05 \times 17 + 0.05 \times 48 + 1 \times 19 + 1 \times 15 = 114.05 \approx 114$ octets.

Mean number of octets in each direction for a call =

$$\frac{114}{2} \times 0.7 + 31.5 \times 0.3 = 49.35 \approx 49 \text{ octets.}$$

Information transfer in each direction on a signaling link per circuit = $49 \times 24 = 1176$ octets.

Number of circuits that can be served by a signaling link with 0.2 Erlang =

$$\frac{8000 \times 3600 \times 0.2}{1176} = 4889.$$

It is obvious that en bloc signaling is more efficient than overlap signaling, because more circuits can be served by en bloc signaling at 0.2 Erlang traffic.

17.9 SUMMARY

This chapter presented a brief description of the SS7 network. SS7 plays a vital role in signaling between various elements of the GSM network. GSM did not introduce any new functions in MTP, but uses the same protocols as defined by CCITT. SCCP class 0 and class 2 services are used in GSM. GSM uses BSSMAP and DTAP protocols over the A interface. For signaling between an MSC and PSTN, either TUP or ISUP protocol is used. We also presented two numerical examples to demonstrate the calculations for signaling load.

17.10 REFERENCES

1. ITU-T Recommendation Q.706, rev. 1, "Specifications of Signaling System No. 7: Message Transfer Part Signaling Performance," 1993.

2. ITU-T Recommendation Q.709, rev. 1, "Specifications of Signaling System No. 7: Message Transfer Part. Hypothetical Signaling Reference Connection," 1993.

3. Modarressi, R. A., and Skoog, R. A., "Signaling System No. 7: A Tutorial," *IEEE Communication Magazine* 28 (7), July 1990, pp. 19–35.

4. CCITT Study Group XI, "Specifications of Signaling System No. 7," CCITT Blue Book, vol. 6, Fascicle VI.9, Geneva, Switzerland, 1989.

5. William, G., and Kuhn, P. J., "Performance Modeling of Signaling System No. 7," *IEEE Communication Magazine* 28 (7), July 1990, pp. 44–56.

6. Bhatnagar, P. K., *Engineering Network for Synchronization, CCS7, and ISDN*, IEEE Press, New York, NY, 1997.

CHAPTER 18

Telecommunication Traffic Engineering

18.1 Introduction

The capacity of a switch system is usually expressed as the maximum number of originating plus incoming (O+I) calls that can be processed in a high-traffic hour while meeting the dial-tone delay requirements. The call volume offered to a switch depends on geographical area, class of service mix, and time of day. A switch is required to process calls while serving a representative complement of features. An accurate call capacity estimation of a switch is needed to properly plan, engineer, and administer its use. The call capacity of a switch varies with the subscriber call mix, feature mix, and equipment configuration. Typically, processor capacity limits the switch call capacity. Under specific conditions, though, the switching network, peripheral equipment, trunk terminations, or directory numbers can limit switch call capacity.

In this chapter we provide definitions of the terms often used in teletraffic engineering and provide several numerical examples to illustrate applications for calculating MSC traffic. In the latter part of the chapter we concentrate on Erlang's and Poisson's blocking formulas that are used to calculate the GoS. Typical call mixes for systems in different environments are also given.

18.2 SERVICE LEVEL

Service level for telecommunication traffic can be divided into two main areas: the delay in receiving a dial tone[*] and the probability of service denial. Dial-tone delay is the maximum amount of time a user must wait to hear a dial tone after removing the handset from the hook. Dial-tone delay has the following characteristics:

☞ A large number of users compete for a small number of servers (dial-tone connections, dial-tone generators).

☞ An assumption that the user will wait until a server is available.

Service denial, or the probability that the service trunk will not be available, is similar to dial-tone delay, with several additional characteristics:

☞ A large number of users competing for a small number of trunks.

☞ An assumption that no delay will be encountered. The user is either given access to a trunk or is advised by a busy signal or recording that none are available.

☞ The user may frequently reinitiate the call attempt after receiving a busy signal.

In either case, the basic measure of performance is the probability that service delay will exceed some specified value or the probability that the call will be blocked (the *blocking probability*). In a system that discards calls when serving trunks are not available (a *loss system*), the blocking probability is the performance measure.

18.3 TRAFFIC USAGE

The usage of a traffic path is defined by two parameters:

☞ **Calling rate**, or the number of times a route or traffic path is used per unit time; more properly defined, the call intensity per traffic path during busy hour

☞ **Holding time**, or the average duration of occupancy of a traffic path by a call.

A traffic path is a channel, time slot, frequency band, line, trunk, switch, or circuit over which individual communications pass in sequence. The *carried* traffic is the volume of traffic actually carried *by* a switch, and *offered* traffic is the volume of traffic offered *to* a switch.

$$\text{offered load} = \text{carried load} + \text{overflow} \tag{18.1}$$

---

[*]For a cellular system, radio signaling delays correspond to dial-tone delay.

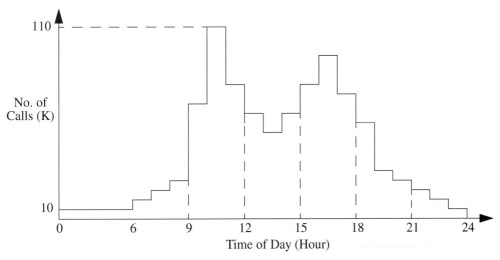

Fig. 18.1 Example of Voice Traffic Variation Hour by Hour

Figure 18.1 shows a typical hour-by-hour voice traffic variation for a serving switch in the United States. We observe that the busiest period—the busy hour (BH)—is between 10 A.M. and 11 A.M. We define the BH as the time-consistent hour span of time (not necessarily a clock hour) that has the highest average traffic load for the business day throughout the busy season. The peak hour is defined as the clock hour with highest traffic load for a single day.

Since the traffic also varies from month to month, we define the average busy season (ABS) as the three months (not necessarily consecutive) with the highest average BH traffic load per access line. Telephone systems are not engineered for maximum peak loads, but for some typical BH load. The blocking probability is defined as the average ratio of blocked calls to total calls and is referred to as the **grade of service**.

18.4 TRAFFIC MEASUREMENTS UNITS

Traffic is measured in either Erlangs, percentage of occupancy, 100 call seconds (CCS), or peg count [3].

☞ **Erlangs.** *Traffic intensity* is the average number of calls simultaneously in progress during a particular period of time. It is measured either in units of Erlangs or CCS. An average of one call in progress during an hour represents a traffic intensity of 1 Erlang or 1 Erlang = 1×3600 call seconds = 36 CCS. Traffic intensity can be obtained as:

Traffic Intensity = (the sum of circuit holding time) / (the duration of the monitoring period)

$$I = \frac{\sum\limits_{i=1}^{N_c} h_i}{T} = \frac{N_c \bar{h}}{T} = n_c \bar{h} \qquad (18.2)$$

where:
I = traffic intensity,
T = duration of monitoring period,
h_i = the holding time of the i-th individual call,
N_c = the total number of calls in monitoring period,
$\bar{h}$ = average call holding time, and
n_c = number of calls per unit time.

☞ **Percentage of occupancy.** Percentage of time a server is busy.

☞ **Peg count.** The number of attempts to use a piece of equipment.

The usage (U), peg count per time period (PC), overflow per period (O), and average holding time $\bar{h}$ are related as:

$$U = (PC - O) \cdot \bar{h} \qquad (18.3)$$

Usage and average holding time are in the same units of time. The percentage of occupancy is defined as (measured usage)/(maximum usage).

E X A M P L E   1 8 – 1

Problem Statement
In a switching office an equipment component with an average holding time of 5 seconds has a peg count of 450 for a one-hour period. Assuming that there was no overflow (i.e., the system handled all calls), how much usage in seconds, CCS, and Erlangs has accumulated on the piece of the equipment?

Solution

$$U = (450 - 0) \times \frac{5}{3600} = 0.625 \text{ Erlangs}$$

0.625 Erlangs × 36 CCS/Erlangs = 22.5 CCS = 2250 seconds

E X A M P L E   1 8 – 2

Problem Statement
If the carried load for a component is 2900 CCS at 5% blocking, what is the offered load?

Solution

$$\text{Offered Load} = \frac{2900}{(1 - 0.05)} = 3050 \text{ CCS}$$

EXAMPLE 18 – 3

Problem Statement

In a voice network each subscriber generates two calls/hour on average and a typical call holding time is 120 seconds. What is the traffic intensity?

Solution

$$I = \frac{2 \times 120}{3600} = 0.0667 \text{ Erlangs} = 2.4 \text{ CCS}$$

EXAMPLE 18 – 4

Problem Statement

To determine voice traffic on a line, the following data was collected during a period of 90 minutes (refer to Table 18.1). Calculate the traffic intensity in Erlangs and CCS.

Solution

$$\text{Call Arrival Rate} = \frac{10 \text{ calls}}{1.5 \text{ hours}} = 6.667 \text{ Calls/hour}$$

Average Holding Time:

$$\bar{h} = \frac{60 + 74 + 80 + 90 + 92 + 70 + 96 + 48 + 64 + 126}{10} = 80 \text{ seconds}$$

$$I = \frac{6.667 \times 80}{3600} = 0.148 \text{ Erlangs} = 5.33 \text{ CCS}$$

Table 18.1 Traffic Data Used to Estimate Traffic Intensity

Call No.	Duration of Call (s)
1	60
2	74
3	80
4	90
5	92
6	70
7	96
8	48
9	64
10	126

Table 18.2 Traffic on Customer Line between 9:00 A.M. and 4:00 P.M.

Call No.	Call Started	Call Terminated
1	9:15	9:18
2	9:31	9:41
3	10:17	10:24
4	10:24	10:34
5	10:37	10:42
6	10:55	11:00
7	12:01	12:02
8	2:09	2:14
9	3:15	3:30
10	4:01	4:35
11	4:38	4:43

E X A M P L E   1 8 – 5

Problem Statement

The data in Table 18.2 was recorded by observing the activity of a single customer line during the eight-hour period from 9.00 A.M. to 5.00 P.M. Determine the traffic intensity during the eight-hour period, during the BH (which is assumed to be between 4:00 P.M. and 5:00 P.M.).

Solution

$$\text{Call Arrival Rate} : \lambda = \frac{11 \text{ calls}}{8 \text{ hours}} = 1.375 \text{ calls/hour}$$

Total call minutes = 3 + 10 + 7 + 10 + 5 + 5 + 1 + 5 + 15 + 34 + 5 = 100 minutes

If statistical stationary is assumed, then the calling rate and average holding time could be converted to a one-hour time period, with the same result holding for the offered traffic load. That is:

$$\bar{h} = \frac{100 \text{ min}}{11 \text{ calls}} \times \frac{1 \text{ hour}}{60 \text{ minutes}} = 0.1515 \text{ hour/call}$$

The traffic intensity is:

$$I = 1.375 \times 0.1515 = 0.208 \text{ Erlangs} = 7.5 \text{ CCS}$$

The busy hour is 4:00 P.M. to 5:00 P.M. Since there are only two calls between this period, we can write:

$$\lambda = 2 \text{ calls/hour}$$

The average call holding time during busy hour:

$$\bar{h} = \frac{(34 + 5)\ \text{min}}{2\ \text{calls}} = 19.5\ \text{minutes/call} = 0.325\ \text{hour/call}$$

The traffic load in the BH is:

$$I = 2 \times 0.325 = 0.65\ \text{Erlang} = 23.4\ \text{CCS}$$

18.5 DEFINITION OF CALL CAPACITY

Call capacity is defined with respect to a view of the switch. In general there are two basic approaches to view the system.

☞ *Global View*. The entire switch is considered as a single unit. Each request to the switch for service is counted as an attempt. This approach is applicable to central processors involved in call processing. In the global view, we represent the call volume of interest as the sum of originating and incoming (O+I) calls.

1. Originating Calls (O)
   - Partial dial calls—calls with partial time-outs and abandons
   - Intraoffice calls—all calls handled entirely by the switch from line originating to line termination
   - Outgoing calls—all calls that originate from a line on the switch but terminate on a different switch

2. Incoming Calls (I)
   - Incoming-terminating calls—all calls that terminate on a line but originate from a different switch
   - Tandem calls—trunk-to-trunk calls through the switch
   - Direct inward dialing (DID)—calls to PBX system

☞ *Component View*. The component of interest is considered as a subsystem. Each request to the component for service is counted as an attempt. This view applies to peripheral processors involved in call processing. In the component view, the call volume of interest is expressed as the sum of O + terminating (T) half-calls.

1. Originating Half-Calls. One originating half-call is for each originating call, because two peripheral equipment connections are required for a completed call. If a component serves both lines and trunks, incoming and outgoing half-calls are added to the total half-call volume.

2. Terminating Half-Calls. One terminating half-call is for each incoming-terminating call and each intraoffice call.

It should be noted that false starts and permanent signals (ineffective attempts) are not counted as calls in either case. However, partial dial attempts, attempts receiving busy treatment, and attempts not answered are all considered calls.

The primary determinant of a Stored Program Control (SPC) system's call handling capacity is the processor real time available for call processing. Processor call capacity is the maximum number of calls per hour that the processor can handle while still meeting the service criteria. Central processors have high day BH (HDBH) O+I call capacities, whereas peripheral processors have HDBH O+T call capacities.

We define the following call types (refer to Figure 18.2):

☞ *Originating call (O)*. The call placed by subscriber of the office

☞ *Terminating call (T)*. The call received by subscriber of the office

☞ *Outgoing call*. The call going out of the office

☞ *Incoming call (I)*. The call coming from outside of the office

☞ *Intraoffice call*. The originating plus terminating calls that are within the same office

☞ *Tandem calls*. The calls that come in on a trunk from another switch and go out over a trunk to a different switch

☞ *O+T*. The measure of traffic load on the line side of the switch both from within the office and from other offices

☞ *O+I*. The measure of incoming trunk-circuit traffic load and traffic load on the switch

Bell Communication Research (Bellcore) [1] suggests the use of the standard call mixes given in Tables 18.3 and 18.4 for calculating call capacity of central processors (in terms of O+I calls) and peripheral processors (in terms of O+T half-calls), respectively. These call mixes do not represent best and

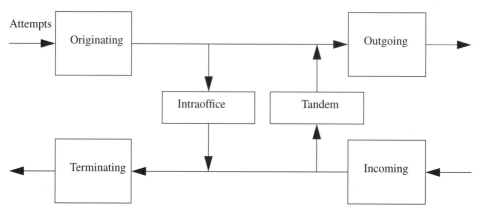

Fig. 18.2 Call Types in a Telephone Switch

worst cases for call capacity, but they do represent average or typical cases for each environment. These mixes reflect the differences found in the following traffic environments:

☞ Metropolitan (metro). Class 5 office in a major metropolitan area with a high proportion of business traffic

☞ Single System City (SSC). Class 4/5 office serving a medium-sized town with a number of outlying Community Dial Offices (CDOs) homing on it

☞ Suburban. Class 5 office residing in the suburbs of a major city

Table 18.5 provides the typical traffic intensity for metro, suburban, and rural environments in the United States.

18.6 TRAFFIC DISTRIBUTION

A typical traffic distribution in a U.S. metro environment for mobile application is given in Table 18.6.

Table 18.3 Standard Call Mixes for Central Processors (percentage of total traffic)

Type	Traffic Environments		
	Metro	SSC	Suburban
Ineffective			
False Start	10	12	18
Permanent Signal	1	2	2
Partial Dial			
Abandon	3	4	5
Time-Out	1	1	1
Intraoffice			
Answered	12	24	17
No Answer	2	2	3
Busy	2	4	4
Outgoing			
Answered	34	17	28
No Answer	5	3	4
Busy	5	3	5
Incoming-Terminating			
Answered	27	18	24
No Answer	5	2	5
Busy	4	3	4
Tandem	0	19	0

Table 18.4 Standard Half-Call Mixes for Peripheral Processors Controlling Subscriber Lines (percentage of total traffic)

	Traffic Environments		
Type	Metro	SSC	Suburban
Ineffective			
False Start	8	11	14
Permanent Signal	1	2	2
Partial Dial			
Abandon	2	3	4
Time-Out	1	1	1
Originating			
Answered	40	37	36
No Answer	6	5	6
Busy	8	6	7
Terminating			
Answered	34	38	33
No Answer	6	4	6
Busy	5	6	7

Table 18.5 Traffic Intensity for O+I

Environment	O+I Call/Line
Metro	3.5–4.0
Suburban	2.0–2.5
Rural	1.2–1.5

Table 18.6 Typical Traffic Distribution in U.S. Metro Environment

Application	Traffic Type	Distribution
Mobile	Mobile to land	65%
	Mobile to mobile	5%
	Land to mobile	30%

18.7 DEFINITIONS OF TERMS

These are terms used for mobile systems.

☞ **Number of calls attempted.** The total number of calls attempted (also called the number of bids) is the best measure of unconstrained customer demand.

☞ **Number of calls completed.** The number of calls completed in a network sense (i.e., reaching ringing tone or answered), when compared with number of calls attempted, provides a measure of the state of network congestion.

☞ **GoS.** {(No. of BH call attempts) – (No. of BH calls completed) / {(No. of BH call attempts)}

The number of answered calls is lower than the number of completed calls by the network, since some calls are bound to encounter either a "subscriber-busy" state or a "ring tone/no reply" condition. The proportion of answered calls to attempted calls is called the **answer bid ratio** (ABR) and the proportion of answered calls to *seizures* is called the **answer seizure ratio** (ASR).

$$ABR = (\text{No. of calls answered})/(\text{No. of calls attempted})$$

$$ASR = (\text{No. of calls answered})/(\text{No. of seizures})$$

ABR and ASR are measured over relatively short periods of time (5–15 minutes). Both are good indicators of instantaneous network congestion. Lower values of ABR or ASR indicate higher network congestion. However higher ABR or ASR values do not mean lower network congestion since calls may remain unanswered for a range of other reasons.

18.8 DATA COLLECTION CATEGORIES

The following data are collected on a switch [5]:

☞ **Peg count.** One peg count for each of the categories—call attempted, trunk-group seizure attempt, test made for dial tone speed, call queued

☞ **Overflow.** Overflow for each attempt collected for universal tone decoders, trunk groups, etc.

☞ **Traffic Usage.** Measured for trunks, decoders, etc.

☞ **Maintenance Usage.** Customer usage = total usage – maintenance usage

18.9 OFFICE ENGINEERING CONSIDERATIONS

The following steps are used in typical office engineering of a wireline or wireless office.

1. Switches are engineered and administered based on the traffic load during the average BH of the busy season.
2. The office BH is used for the overall administration, engineering, and maintenance of an office.

3. The component BH is used to establish trends, make projections, set capacities, and derive future requirements.

4. Dial-tone speed delay is recorded whenever a test call does not receive a dial tone within 3 seconds.

5. Terminating blockage is recorded whenever a terminating call is unable to complete because of a lack of an available path to the called line.

6. Trunk-group BH is the time-consistent hour during which maximum trunk-group load occurs. Trunk-group BH data is used to provide an adequate trunk base to meet service requirements.

7. Traffic data is collected for one or two weeks by the half-hour during all parts of the day that may produce high loads (e.g., 8 A.M. to 11 P.M.).

8. Five days of the week with the heaviest load are determined; this is the *business week* of the office.

9. The hour (on the clock hour or on the half-hour) with the highest total load for the business week is determined; this is the *office busy hour*.

10. Traffic data is collected for the BH for the months likely to be parts of the year that may produce high loads.

11. The three months, not necessarily consecutive, having the highest BH load are determined; this is the busy season.

12. The average load for the BH for the busy season's business day is (ABS/BH).

13. Use the following approximate relation to estimate design traffic:

- O+T Call: $(HD)/(ABS) \approx 1.4\text{–}1.5$
- O+I Call : $(HD)/(ABS) \approx 1.6\text{–}1.7$
- High Day (HD) origination attempt per call ≈ 1.45

E X A M P L E   1 8 – 6

Problem Statement

Calculate ABS/BH switch calling rate and CCS for a switch located in a large metropolitan area. The switch carries 80,000 lines. The distribution of lines on the switch is as follows:

- Residential lines: 12,000
- Business lines: 64,000
- PBX, WATS, and Foreign Exchange (FX) lines: 4000

The ABS/BH call rates for residential, business, and high-usage customers are 2, 3, and 10 calls per line, respectively. The average holding times for these customers are 140, 160, and 200 seconds, respectively. Assuming HD/ABS for the switch = 1.5, calculate the design call capacity for the switch and the design Erlangs.

Solution

Percentage of residential lines: 12,000/80,000 = 0.15 = 15%

Percentage of business lines: 64,000/80,000 = 0.80 = 80%

Percentage of high-usage lines: 4000/80000 = 0.05 = 5%

CCS per residential line: $2 \times 140/100 = 2.8$

CCS per business line: $3 \times 160/100 = 4.8$

CCS per high-usage line: $10 \times 200/100 = 20$

Switching calling rate: $2 \times 0.15 + 3 \times 0.8 + 10 \times 0.05 = 3.2$ calls per line

Switch CCS rate: $2.8 \times 0.15 + 4.8 \times 0.8 + 20 \times 0.05 = 5.26$ CCS per line

Average holding time per line for the switch: $\frac{5.26}{3.2} \times 100 = 164$ seconds

ABS/BH calls: $3.2 \times 80,000 = 256,000$

ABS/BH usage: $\frac{5.26 \times 80,000}{36} = 11,689$ Erlangs

Design call capacity based on HD: $1.5 \times ABS/BH = 1.5 \times 256,000 = 384,000$

Design Erlangs based on HD: $1.5 \times 11,689 = 17,584$

18.10 BLOCKING FORMULAS

18.10.1 Erlang B Formula

The Erlang B formula [2] provides the probability of blockage at the switch due to congestion or to "all trunks busy." This is expressed as GoS or the probability of finding N channels busy.

Assumptions are:

- ☞ Traffic originates from an infinite number of sources.
- ☞ Lost calls are cleared assuming a zero holding time.
- ☞ Number of trunks or serving channels is limited.
- ☞ Full availability exists.
- ☞ Interarrival times of call requests are independent of each other.
- ☞ The probability of a user occupying a channel (called the *service time*) is based on exponential distribution.
- ☞ Traffic requests are represented by a Poisson distribution implying exponentially distributed call interarrival times.

$$B(N, A) = \frac{\dfrac{A^N}{A!}}{\displaystyle\sum_{i=0}^{N} \dfrac{A^i}{i!}} \tag{18.4}$$

where:
N = number of serving channels,
A = offered load, and
$B(N, A)$ = blocking probability.

18.10.2 Poisson's Formula

The Poisson formula is used to design trunks for a given GoS.

$$p_b = e^{-A} \sum_{i=N}^{\infty} \frac{A^i}{i!} \tag{18.5}$$

where:
p_b = probability of blocking
A = offered load, and
N = number of trunks

18.10.3 Comparison of Erlang B and Poisson's Formulas

A comparison between the Erlang B and Poisson's blocking formulas [2] shows that Poisson's formula results in higher blocking than that obtained by the Erlang B formula for a given traffic load.

The Erlang B and Poisson's formulas are commonly used to calculate the blocking probabilities (or GoS) of the wireline or wireless system. For Erlang loss system, the carried traffic A' will be:

$$A' = A[1 - B(N, A)] \tag{18.6}$$

where:
A' = carried traffic load

The carried traffic equals that portion of the offered traffic A that is not lost, and $A \cdot B[N, A]$ is the lost traffic.

18.10.4 Erlang C Formula

The Erlang C [2] formula assumes that a queue is formed to hold all requested calls that cannot be served immediately. Customers who find all N servers busy join a queue and wait as long as necessary to receive service. This means that blocked customers are delayed. No server remains idle if a customer is waiting.

The assumptions in Erlang C formula are:

☞ Infinite sources

☞ Poisson input

☞ Lost calls delayed

☞ Exponential holding time

☞ Calls served in order of arrival

$$C[N, A] = \frac{\dfrac{A^N}{N!(1-A/N)}}{\displaystyle\sum_{i=0}^{N-1} \dfrac{A^i}{i!} + \dfrac{A^N}{N!(1-A/N)}} \qquad (18.7)$$

where:
N = number of serving channels,
A = offered load, and
C $[N, A]$ = blocking probability.

18.10.5 Binomial Formula

The assumptions for the binomial formula are:

☞ Finite source

☞ Equal traffic density per source

☞ Lost calls held

$$p_b = \left(\frac{s-D}{s}\right)^{s-1} \sum_{i=N}^{s-1} \binom{s-1}{N} \left(\frac{D}{s-D}\right)^i \qquad (18.8)$$

where:
D = expected traffic density,
p_b = blocking probability,
N = number of channels in the group of channels, and
s = number of sources in group of sources.

E X A M P L E   1 8 – 7

Problem Statement
The maximum calls per hour in one cell equal 4000 and the average call holding time is 150 seconds. If the blocking probability is 2%, find the offered load A. How many channels are required to handle the load?

Solution
$$A = \frac{4000 \times 150}{3600} = 166.67 \ \text{Erlangs}$$

Using the Erlang B table in appendix A, N = 182 channels giving 168.3 Erlangs at 2% blocking.

E X A M P L E   1 8 – 8

Problem Statement
How many users can be supported with 50 channels at 2% GoS? Assume average call holding time equals 120 seconds and average busy hour call per user is 1.2 calls per hour.

Solution

From the Erlang B table in appendix A, for 50 channels at 2% blocking, the capacity = 40.26 Erlangs

$$\text{Average traffic per user} = \frac{1.2 \times 120}{3600} = 0.04 \ \text{Erlangs}$$

$$\text{No. of users} = \frac{40.26}{0.04} \approx 1{,}006$$

18.11 SUMMARY

In this chapter we provided some of the basic principles of teletraffic engineering that are used to engineer and administer a wireline or wireless switch. We also presented a systematic procedure to determine the design call capacity of a switching system. We discussed the Erlang B and Poisson's blocking formulas for calculating blocking probabilities or GoS of a system. Finally, we concluded the chapter by presenting the Erlang C and binomial formulas that apply to a queued system. Several numerical examples were given to illustrate the applications of various blocking formulas.

18.12 PROBLEMS

1. Estimate the number of subscribers that can be supported by a cell with 400 radio channels. Assume each subscriber generates on an average 2.5 calls per hour with average call holding time of 120 seconds.

2. If there are 60 radio channels in a cell to handle all the calls and average call holding time is 120 seconds, how many calls are handled in this cell with a GoS of 2%?

3. What is the average holding time in seconds of a group of circuits that has accumulated 0.4445 Erlangs of usage in an hour, based on 330 call attempts with 10 calls overflowing?

4. A trunk accumulated 0.75 Erlangs of usage while 9 calls were carried in a hour with no overflow. What is the average holding time per call in seconds, CCS, and Erlangs?

5. A group of servers was engineered for 1% blocking criterion. Measured data was as follows: Usage = 21,643 CCS; Peg count = 11,574 CCS; Overflow = 208 CCS. Was the criterion met? What are the offered and carried loads in Erlangs?

18.13 REFERENCES

1. Bellcore, "LATA Switching Systems Generic Requirements—Traffic Capacity and Environment," Technical Reference, TR-TSY-000517, Issue 3, March 1989.

2. Copper, R. B., *Introduction to Queuing Theory,* North-Holland, New York, 1981.

3. Rappaport, T. S., *Wireless Communications,* Prentice Hall, Upper Saddle River, NJ, 1996.

4. Rapp, Y., "Planning of Junction Network in a Multiexchange Area," *Ericsson Technics* 20 (1), 1964, pp. 77–130.

5. Sharma, R. L., et al., *Network Systems,* Van Nostrand Reinhold Co., New York, 1982.

Comparison of TDMA Systems for Cellular/PCS

19.1 INTRODUCTION

In 1989 the Digital Advanced Mobile Phone System (DAMPS), IS-54 [1], was standardized following many precedents established by the European GSM communications [2]. DAMPS accommodates three higher-quality digital channels in a conventional 30-kHz analog AMPS channel. The unique feature of DAMPS is that it uses a 2-bit per symbol nonbinary modem, which can tolerate a more severe propagation environment than that of the GSM system. Currently, there is much activity worldwide to define the PCN in Europe and PCS in North America and Japan. In the European community's RACE program, there were two projects dedicated to establish the most suitable multiple access scheme between TDMA and CDMA for the PCN. In North America, cellular standards have made a transition into standards for PCS. Six new air interfaces have been approved for PCS by the Joint Technical Committee (JTC) of the TIA TR-46 and the ATIS T1P1 committees: DAMPS-1900 (a derivative of IS-136), IS-95 (CDMA), Personal Access Communication System (PACS), Composite TDMA/CDMA, PCS-2000 (wideband CDMA), and PCS-1900.

DAMPS-1900 is based on the TIA IS-136 [3] air interface standard which is an extension of the TIA IS-54 standard. PCS-1900 is a derivative of GSM. DAMPS-1900 has been specified in North America as a PCS technology for both the cellular (800-MHz) and PCS (1900-MHz) bands. The availability of dual-band/dual-mode 800/1900-MHz handsets ensures a quick and low-cost start-up for PCS services. DAMPS-1900 has been designed to allow seamless interworking and infrastructure sharing with DAMPS networks at 800 MHz as well as with the analog AMPS networks. This enables new PCS service providers to offer full, wide-area coverage from day one through infrastructure sharing or roaming agreements with 800-MHz service providers in the same geographical area. Also, since approximately 50 percent of the world's cellular subscribers are connected to an AMPS/DAMPS system, the potential of international roaming exists. A recent approval of DAMPS (IS-136) to incorporate DCCH enables service providers to easily introduce fully digital features and services in their existing infrastructure.

Japan PDC [4], which is similar to North American DAMPS, has been adopted as an access technology to provide cellular/PCS services.The PDC system does not provide backward compatibility with the existing analog cellular system because

☞ Spectrum for a digital cellular system lies in the 800-MHz and 1.5-GHz bands.

☞ Backward compatibility is not required in the Japanese cellular market because terminals for analog cellular systems are supplied only on a lease basis.

All digital cellular operators in Japan employing the PDC standard (RCR STD-27) are required to support the following services:

☞ Voice (full-rate and half-rate codec)

☞ Supplementary services (e.g., call waiting, voice mail, three-party call, and call forwarding)

☞ Nonvoice data (up to 9.6 kbps)

☞ Packet-switched wireless data

In this chapter we focus mainly on the three TDMA-based PCS/cellular systems (DAMPS-1900 [IS-136], PCS-1900, and PDC) that have been deployed or are being deployed in North America and Japan. We compare these systems with respect to the logical channel structure, framing, speech coding, modulation schemes, and system capacity. In section 19.2, a brief description of the access method used in the three systems is given, whereas in section 19.3, we provide a summary of the basic parameters used in the three systems and present the technical details about speech coding, modulation, logical channel structure, among other topics.

19.2 DAMPS-1900, PCS-1900, AND PDC

DAMPS-1900, PCS-1900, and PDC are TDMA/FDD systems. In a TDMA system, data from each subscriber is conveyed in time intervals referred to as slots. Several slots make up a frame. Most TDMA systems use TDM which incorporates several fixed-rate bit streams time-division multiplexed onto a TDMA bit stream. The data is transmitted via radio carrier from a BS to several active MSs in the downlink. In the reverse direction (uplink), transmission from MSs to BSs is time sequenced and synchronized on a common frequency for TDMA.

In a TDMA/FDD system, an identical frame structure is used for both uplink (MS to BS) and downlink (BS to MS) transmission, but the carrier frequencies are different for the uplink and downlink. Uplink and downlink frequencies are separated enough to prevent crosstalk between the two links. In general, in a TDMA/FDD system a delay of several time slots is intentionally induced between the uplink and downlink time slots of the particular user to avoid the use of duplexers in the MS. Table 19.1 gives a summary of the basic parameters used to compare the three systems.

19.2.1 Logical Channels and Framing

Bursts, which carry different types of information, form logical channels that are organized into frames and higher structures. The useful end-to-end information exchanged through the network is voice or data. This information is transmitted over a logical channel called the **traffic channel**. A set of control channels are used to manage and control the link between the mobile terminal and the infrastructure. Each TCH is permanently associated with SACCH for nonurgent messages. For urgent messages during a call setup or release process, a TCH is temporarily used for control and becomes an FACCH.

In PCS-1900, eight other channels are used besides the TCH. These are synchronization, broadcast control, paging, stand-alone dedicated control, access grant, cell broadcast, associated control, and random access channel. All these logical channels can be mapped on the physical channels in many different ways. Comparatively, the logical channel structure used in DAMPS-1900 and PDC is similar than PCS-1900 (see Figures 19.1, 19.2, and Figure 7.1). Each channel in the DAMPS-1900 and PCS-1900 is allocated a convolutional code. The time slot structure for PDC, DAMPS-1900, and PCS-1900 is shown in Figures 19.3, 19.4, and 7.6. Note that the DAMPS-1900 and PDC time slots (overhead $52/324 = 16.1$ percent and $52/280 = 18.6$ percent, respectively) are more efficient than the PCS-1900 time slot (overhead $42.25/156.25 = 27.1$ percent).

19.2.2 Modulation

A modulation scheme in the mobile environment must use the transmitted power and RF channel bandwidth as efficiently as possible because the mobile radio channel is both power and band limited. To conserve power, an efficient encoding scheme is often used, but it is at the expense of bandwidth. On the other hand to save spectrum in band-limited systems, a spectrally efficient modulation scheme is required, the objective of which is to maximize band-

Table 19.1 Basic Parameters of DAMPS-1900, PCS-1900, and PDC Air Interface

Parameters	DAMPS-1900	PDC	PCS-1900
Frequencies: • Uplink (MS to BS)	A- Band: 1850–1865 MHz	940–956 MHz 1477–1513 MHz	A-Band 1850–1865 MHz
• Downlink (BS to MS)	1930–1945 MHz	810–826 MHz 1429–1465 MHz	1930–1945 MHz
Channel Spacing	30 kHz	25 kHz	200 kHz
Multiple Access Technique	TDMA/FDMA/FDD	TDMA/FDMA/FDD	TDMA/FDMA/FDD
Modulation	$\pi/4$-DQPSK	$\pi/4$-DQPSK	GMSK, BT = 0.3
Data Transmission Rate	48.6 kbps	42 kbps	270.833 kbps
Spectral Efficiency	1.62 bps/Hz	1.68 bps/Hz	1.35 bps/Hz
Number of Channels per Carrier	3 (full rate) 6 (half rate)	3 (full rate) 6 (half rate)	8 (full rate) 16 (half rate)
Total Number of Voice Channels	500	640 1440	75
TDMA Frame Duration	20 ms	20 ms	4.62 ms
Portable Terminal Power (Maximum)	600 mW	Class III: 800 mW Class IV: 300 mW	1000 mW
Frequency Band Separation	45 MHz	130 MHz 48 MHz	45 MHz
Equalizer	Adaptive	Adaptive	Adaptive
Time-Slot Duration	6.667 ms	6.667 ms	0.58 ms
Channel Coding	7 bits CRC with half rate, $k = 6$, Convolutional	Different CRC bits are used for different slots (see section 19.2.4)	3 bits CRC with half rate, $k = 5$, Convolutional
Speech Coding	VSELP (13.0 kbps) Source (7.95 kbps); FEC (5.05 kbps) ACELP (13 kbps) Source (7.45 kbps); FEC (5.55 kbps)	VSELP (11.2 kbps) Source (6.7 kbps); FEC (4.5 kbps)	RELP (22.8 kbps) Source (13 kbps); FEC (9.8 kbps)
Handover	Mobile-Assisted Handover	Mobile-Assisted Handover	Mobile-Assisted Handover

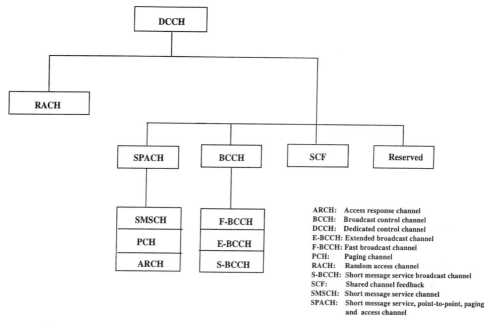

ARCH: Access response channel
BCCH: Broadcast control channel
DCCH: Dedicated control channel
E-BCCH: Extended broadcast channel
F-BCCH: Fast broadcast channel
PCH: Paging channel
RACH: Random access channel
S-BCCH: Short message service broadcast channel
SCF: Shared channel feedback
SMSCH: Short message service channel
SPACH: Short message service, point-to-point, paging
         and access channel

Fig. 19.1 Logical Channel Structure for DAMPS-1900

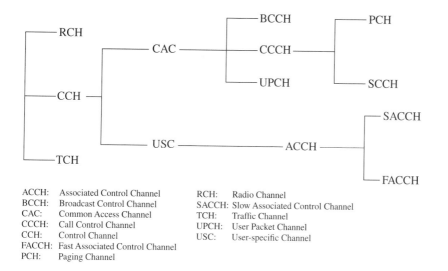

ACCH: Associated Control Channel   RCH: Radio Channel
BCCH: Broadcast Control Channel   SACCH: Slow Associated Control Channel
CAC: Common Access Channel   TCH: Traffic Channel
CCCH: Call Control Channel   UPCH: User Packet Channel
CCH: Control Channel   USC: User-specific Channel
FACCH: Fast Associated Control Channel
PCH: Paging Channel

Fig. 19.2 Logical Channel Structure for PDC

Uplink

G 54	R 4	P 48	SW 32	#1 21	#2 21	#3 21	Q 1	G 78

Downlink

R 4	P 102	SW 32	#1 21	#2 21	#3 21	Q 1	Post 78

Synchronization Slot Format (PDC)

Uplink

R 4	P 2	TCH(FACCH) 112	SW 20	CC 8	SF 1	SACCH (RCH) 15	TCH (FACCH) 112	G 6

Downlink

R 4	P 2	TCH(FACCH) 112	SW 20	CC 8	SF 1	SACCH (RCH) 21	TCH (FACCH) 112

TCH, FACCH, SACCH Slot Format (PDC)

Uplink (first burst)

R 4	P 48	CAC 66	SW 20	CC 8	CAC 116	G 14

Uplink (second & later burst)

R 4	P 2	CAC 112	SW 20	CC 8	CAC 116	G 18

Downlink

R 4	P 2	CAC 112	SW 20	CC 8	CAC 112	E 22

CAC (BCCH, SCCH and PCH) Slot Format (PDC)

Fig. 19.3 PDC Slot Formats

6	6	16	28	122	12	12	122
G	R	Data	Sync	Data	SACCH	CDVCC	Data

Format from mobile to cell site

28	12	130	12	130	12
Sync	SACCH	Data	CDVCC	Data	RSVD

Format from cell site to mobile

CDVCC:	Coded digital verification color code
G:	Guard bit
R:	Ramp bit
RSVD:	Reserved field
SACCH:	Slow associated control channel
Sync:	Synchronization

Fig. 19.4 Time Slot Structure of DAMPS-1900

width efficiency. It is desirable to achieve bandwidth efficiency at a given BER with minimum transmitted power. The large variation in signal strength experienced in the mobile environment due to Raleigh fading renders amplitude modulation schemes almost inoperative. There are essentially two basic classes of modulation schemes that have been emphasized for use in the mobile environment—PSK- and MSK-derived modulation schemes.

☞ **Linear Modulation Schemes.** The family of linear modulation schemes requires a high degree of linearity in modulating the carrier by baseband signal and RF power amplification before transmission. Linear modulation is more spectrally efficient, but requires a linear amplifier to avoids changing the signal amplitude variations. The important linear modulation schemes are QPSK, Offset-QPSK (OQPSK), and higher-level PSK.

☞ **Continuous Phase Modulation (CPM) Schemes.** CPM schemes avoid linearity requirements, thereby reducing the cost of amplification. Modulation schemes derived from the CPM family have quite narrow power spectra. On the other hand, the spectral efficiency is somewhat lower than that seen in linear modulation schemes. Among the CPM or constant envelope schemes currently used in the mobile environment are MSK and GMSK. A comparison of modulation schemes is given in Table 19.2. The required SNRs for a BER equal to 1×10^{-6} are also shown.

Spectral efficiency affects the spectrum occupancy in a mobile radio system. Theoretically, an increase in the number of modulation levels results in higher spectral efficiency. However, the precision required at the demodulator to detect the phase and frequency changes also increases significantly. This results in higher SNR requirements to achieve the same BER performance.

In Table 19.3, we have summarized the performance of various modulation schemes that have been used in second-generation cellular and CT systems. The linear modulation schemes offer better spectral efficiency. GMSK modulation in PCS-1900 with a product of channel width and symbol duration equal to 0.3 is a compromise modulation scheme. The choice between the CPM and linear modulation schemes is influenced by the best spectral efficiency

Table 19.2 Comparison of Different Modulation Schemes

Modulation Scheme	Spectral Efficiency (bps/Hz)	Required SNR (dB)
Binary PSK (BPSK)	1	11.1
QPSK	2	14.0
PSK (16-level)	4	26.0
MSK (2-level)	1	10.6
MSK (4-level)	2	13.8

Table 19.3 Comparison of Modulation Schemes Used in Second-Generation Systems

System	Modulation	Channel Width (kHz)	Data Rate (kbps)	Spectral Efficiency (bps/Hz)
DAMPS-1900	π/4-DQPSK	30	48.6	1.62
PCS-1900	GMSK	200	270.833	1.35
PDC	π/4-DQPSK	25	42.0	1.68
CT-2	GMSK	100	72	0.72
DECT	GMSK	1728	1572.0	0.67

that must be achieved by trading off cost, size, and adjacent channel selectivity constraints.

One of the effects produced by a mobile environment is the distortion of the signal due to delay spread. This results in ISI. It is known that, if the effects of delay spread are ignored, spectral efficiency improves with a higher level of modulation. This is realized at the expense of a high SNR. Simulation studies indicate that no significant performance improvement is achieved as the modulation level exceeds four. BER performance depends strongly on the rms value of delay spread. Thus, a four-level modulation scheme appears to be the best choice when delay spread is significant.

In PCS-1900, the channel bandwidth has been selected to achieve a constant envelope modulation, whereas in DAMPS-1900, this approach was not feasible because of the requirement for a dual-mode system. PCS-1900 uses GMSK modulation in which the data rate is comparable to radio channel bandwidth; severe ISI may result. PCS-1900 requires an equalizer to overcome the problem of ISI. ISI occurs when the delay spread is a major part of the symbol duration. In DAMPS-1900, the receiver has been designed to handle delay spread of the same length as the symbol duration (41 μs). The corresponding values in PCS-1900 and PDC are 16 μs and 10 μs, respectively. Thus, DAMPS-1900 system is less vulnerable to ISI than PCS-1900 and PDC systems.

The π/4-DQPSK modulation used in DAMPS-1900 is a compromise between the QPSK and OQPSK in terms of the allowed maximum phase shift. In π/4-DQPSK, the maximum phase shift change is restricted to ±135°, as compared to 180° for the QPSK and 90° for the OQPSK. Thus, the π/4-DQPSK signal preserves the constant envelope property better than the band-limited QPSK, but it tends to show more variations than OQPSK. The π/4-DQPSK modulation offers a simplification in the receiver design. Its performance in multipath spread and fading is superior to OQPSK. GMSK modulation is easier to implement than π/4 -DQPSK. The π/4-DQPSK carries 2 bits in a symbol, making it twice as efficient as GMSK. However, the spectrally efficient solution in DAMPS-1900 and PDC leads to a low power efficiency and has an impact on battery lifetime in portable devices.

19.2.3 Speech Coding

A low-bit coding scheme is used for transmission of speech signals in digital mobile communications to efficiently use the limited frequency resource. However, there is a tendency for a lower bit rate to increase sensitivity to channel errors. Therefore, a convolutional code is used as an error correction code for speech data in digital mobile communications.

DAMPS-1900 and PDC use the VSELP codec (see Figure 19.4). The VSELP algorithm uses a codebook with a predefined structure to reduce the number of computations. The output of the VSELP codecs for DAMPS-1900 and PDC are 7.95 kbps and 6.7 kbps, respectively. They produce a speech frame every 20 ms containing 159 bits and 134 bits, respectively.

Recently the ACELP [5] codec has been selected to replace the VSELP codec in DAMPS-1900. This codec has an output bit rate of 7.40 kbps.

The PCS-1900 full-rate speech codec is based on the Residually Excited Linear Predictive (RELP) scheme [6], which is enhanced by including an LTP (refer to Figure 8.3). The codec provides 260 bits for a 20-ms block of speech, yielding a bit rate of 13 kbps. Table 19.4 compares the transmission rate among the systems.

The subjective quality of speech codecs is compared in Table 19.5 based on the MOS using a scale of 1 to 5, where 5 indicates excellent speech quality. The performance of the ACELP codec is superior to the VSELP and RELP. The ACELP codec operates at a lower speech rate than the codec used in the PCS-1900 system and has a lower delay.

19.2.4 Channel Coding

Mobile communication link performance is improved by channel coding which adds redundant bits into the transmitted message. A channel codec maps a digital message sequence into another specific code sequence that contains a

Table 19.4 Comparison of Speech Codec in DAMPS-1900, PDC, and PCS-1900 Systems

System/Speech Codec	Transmission Rate (kbps)		
	Speech Codec (kbps)	Error Correction	Total
DAMPS-1900			
- VSELP	7.95	5.05	13.0
- ACELP	7.40	5.60	13.0
PDC			
- VSELP	6.7	4.5	11.2
PCS-1900			
- RELP	13.0	9.8	22.8

Table 19.5 Comparison of Speech Quality in Codecs

Codec	Speech Codec (kbps)	Voice Quality (MOS)	Delay (ms)
VSELP (DAMP-1900)	7.95	3.70	20
ACELP (DAMP-1900)	7.40	3.85	~ 10
VSELP (PDC)	6.7	3.45	20
RELP (PCS-1900)	13.0	3.80	20

greater number of bits than the original number of bits in the message. The coded message is then modulated for transmission over the wireless channel.

The receiver uses channel coding to detect or correct some (or all) of the errors that are introduced in the channel in a particular sequence of message bits. The additional coding bits reduce the raw data transmission rate through the channel. There are two general types of channel codes—block codes and convolutional codes. Block codes are FEC codes that enable a limited number of errors to be detected and corrected without retransmission. In block codes, parity bits are added to a block of message bits to make code blocks. In a block encoder, k information bits are encoded into n code bits. A total of $n - k$ redundant bits are added to the k information bits to detect and correct errors. The block code is referred to as an (n, k) code and its rate is defined as $R = k/n$.

Convolutional codes are different from block codes in that information sequences are not divided into distinct blocks. A continuous sequence of information bits is mapped into a continuous sequence of encoder output bits. This mapping is highly structured to enable a decoding method considerably different from the block codes. A convolutional code can achieve a larger coding gain than the one achieved using a block code with the same complexity. A convolutional code is generated by passing the information sequence through a finite-state shift register. The shift register contains k bits stages and m linear algebraic function generators based on the generator polynomial. The input is shifted into and along the shift register k bits at a time. The number of output bits for each k input is n bits. The code rate R is equal to k/n. The parameter k is called the constraint length and indicates the number of input data bits on which the current output depends. The k determines how powerful and complex the code is.

In DAMPS-1900, 159 bits within a speech codec frame (see Figure 19.5) are divided into 77 class I and 82 class II bits. The class I bits are the most significant bits and are error protected using a half-rate convolutional code of constraint length $k = 6$. In addition to convolutional coding, the 12 most significant bits among the class I bits are block coded using a 7 bits CRC error detection code to ensure that the most important speech codec bits are detected with a high degree of probability at the receiver. The class II bits, being less significant, do not have error protection. After channel coding, 159

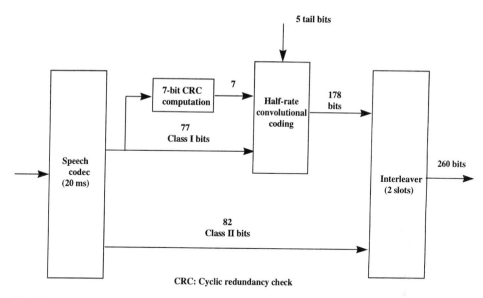

Fig. 19.5 DAMPS Channel Coding—Full Rate

bits in each speech code frame are represented by 260 channel-coded bits. The gross bit rate of the speech codec with the added bits is 13 kbps.

Before transmission, the channel-coded data is interleaved over two time slots using a 26 x 10 interleaving array. In other words, each time slot contains exactly half of the data from each of the two sequenced speech codec frames.

FACCH data is different from that used for speech-coded data. A FACCH data block contains 49 bits per 20-ms frame. A 16-bit CRC code word is added to each FACCH data block to provide a coded FACCH word of 65 bits. The 65-bit word is then passed through a quarter-rate convolutional codec of length k = 6 to yield 260 bits of FACCH data for each 20-ms frame. A FACCH data block occupies the same amount of bandwidth as a single frame of coded speech. Therefore, speech data on the Digital Traffic Channel (DTC) can be replaced with coded FACCH data. Interleaving of FACCH data is handled in the same manner as DTC speech data.

An SACCH data word consists of 6 bits during each 20-ms speech frame. Each raw SACCH data word is passed through a half-rate convolutional codec of length k = 5 to give 12 coded bits during a 20-ms interval, or 24 bits during each frame.

In the ACELP channel codec, 148 bits in a speech frame are divided into 48 class Ia bits, 48 class Ib bits, and 52 class II bits. The class Ia and class Ib bits are error protected using a half-rate convolutional code of constraint length k = 5. In addition to convolutional code, class Ia bits are also block coded using a 7-bit CRC error detection code. The class II bits are transmitted without any protection. A light puncturing (8 out of 216 coded bits) is used in the output of the convolutional encoding in order to get all except the 52 least-sensitive bits under protection. Thus, the total number of bits per frame is 260 bits every 20 ms (i.e., 13 kbps).

In the PCS-1900 system, the output bits of the speech codec (see Figure 19.6) are put into groups for error protection based upon their significance in contributing to speech quality. Out of 260 bits in a frame, the most important 50 bits, called class Ia bits, have 3 parity check (CRC) bits added to them. The next 132 bits along with the first 53 (50 class Ia + 3 CRC) are reordered and appended by 4 tailing zero bits to provide a data block of 189 bits. This data block is then encoded for error protection using a half-rate convolutional encoder with length $k = 5$ to provide 378 bits. The least important 78 bits do not have any error protection and are concatenated to the existing sequence to form a block of 456 bits in a 20-ms frame. The error protection coding scheme increases the gross data rate of the speech signal to 22.8 kbps.

The coding provided for 9.6 kbps data rate is based on handling 60 bits of user data at 5-ms intervals in accordance with the modified CCITT V.110 modem standard. Two hundred forty bits of user data (see Figure 19.7) are applied with 4 tail bits to the half-rate punctured convolutional codec with length $k = 5$. The resulting 488 coded bits are reduced to 456 encoded data bits through puncturing (32 bits are not transmitted), and data is separated into 114-bit data bursts that are applied in an interleaved manner to consecutive time slots.

In PCS-1900, control messages are 184 bits long. They are encoded using a shortened binary cyclic fire code, followed by a half-rate convolutional codec (Figure 19.8). This produces 184 message bits followed by 40 parity bits. Four tail bits are added to clear the convolutional code that follows, yielding a 228-

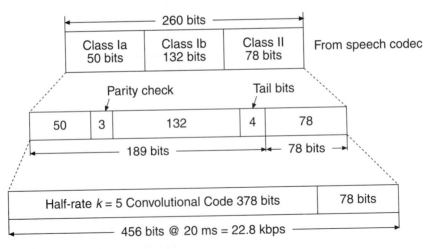

Fig. 19.6 PCS-1900 Channel Coding Full Rate

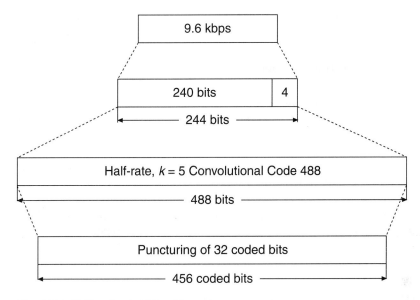

Fig. 19.7 Coding for 9.6 kbps Data Rate

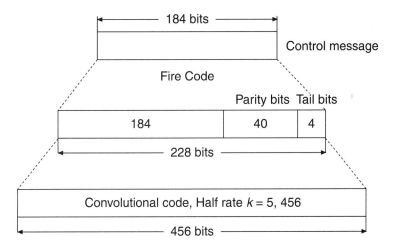

Fig. 19.8 Control Channels Coding

bit of data block. The block is applied to half-rate, $k = 5$ convolutional code. The resulting 456 encoded bits are interleaved in the same manner as DTC speech data.

The total 456 encoded bits in each 20-ms speech frame or control message frame are subdivided into eight 57-bit sub-blocks. These 8 sub-blocks that make up a single speech frame are spread over eight consecutive TCH time slots. If a burst is lost due to interference or fading, channel coding ensures that enough bits will still be received correctly to allow the error correction to work.

Table 19.6 shows a comparison of the key characteristics of channel codecs used in DAMPS-1900, PDC, and PCS-1900. Several observations are worth mentioning. Overall, the DAMPS-1900 channel codec is much more robust than the PCS-1900 channel codec for several reasons. First, it provides better error protection (1 bit for 11 bits in VSELP as compared to 1 bit for 16.7 bits in RELP). The encoded data rate of the RELP codec is higher than either the VSELP or ACELP codec. Also, the overhead in RELP is higher than in VSELP (75.4 percent vs. 63.5 percent). In DAMPS-1900, the 260-bit data block is interleaved diagonally and thus self-synchronized over the two time slots. Channel coding and interleaving in DAMPS-1900 are continuous processes and do not require block synchronization. PCS-1900 has a better interleaving scheme than DAMPS-1900. PCS-1900 provides interleaving over eight time slots, whereas in DAMPS-1900 it is limited to only two time slots.

Table 19.6 Comparison of Channel Codecs

System/Codec Type	DAMPS-1900		PDC	PCS-1900
	VSELP	ACELP	VSELP	RELP
TCH Raw Data Rate	7.95 kbps	7.40 kbps	6.7 kbps	13.0 kbps
Input Bits Distribution	Class I: 77, Class II: 82	Class Ia: 48, Class Ib: 48, Class II: 52		Class Ia: 50, Class Ib: 132, Class II: 78
Type of Channel Code	Half-rate convolutional, $k = 6$	Half-rate convolutional, $k = 5$	BCH	Half-rate convolutional, $k = 5$
CRC	7 bits on 77 bits or 1 bit per 11 bits	7 bits on 48 bits or 1 bit per 6.9 bits	Different CRC used for different slots	3 bits on 50 bits or 1 bit per 16.7 bits
Encoded Data Rate	13 kbps	13 kbps	11.2 kbps	22.8 kbps
Overhead Due to Channel Coding	5.05/7.95 = 63.5%	5.55/7.45 = 75.5%	4.5/6.7 = 67.2%	9.8/13.0 = 75.4%
Interleaving	over 2 time slots	over 2 time slots	vary for slots	over 8 time slots
Control Channel Type of Channel Code	Quarter-rate convolutional, $k = 6$	Quarter-rate convolutional, $k = 6$	BCH	Half-rate convolutional, $k = 5$
CRC	12 bits on 49 bits	12 bits on 49 bits	vary for slots	40 bits on 184 bits

In PDC, BCCH, PCH, and SCCH are time-division multiplexed on a sub-program. In the superframe, full-rate transmission is applied in each frame. One superframe consists of 36 frames and corresponds to 720 ms. BCCH slots are located in the beginning of each superframe. PCH slots in a superframe are divided into several clusters, and each terminal is permitted to have access only to one of the groups determined by their terminal numbers.

TCH and ACCH or RCH are time-division multiplexed in each TDMA slot, and SACCH and Radio Channel (RCH) are time-division multiplexed on a superframe. The TCH is transmitted every 20 ms in case of full rate and every 40 ms in case of half rate.

The SACCH and RCH are multiplexed on a superframe with a frame length of 720 ms. Thus, one superframe consists of 36 slots for full-rate and 18 slots for half-rate transmissions. RCH is transmitted using two consecutive slots every 360 ms in the case of full-rate transmission and using two consecutive slots every 720 ms in case of half-rate transmission. SACCH is transmitted using 16 consecutive slots every 360 ms for the full-rate transmission and using 16 consecutive slots every 720 ms in the case of half-rate transmission.

The first Common Access Channel (CAC) slot in the uplink direction contains 8 bits of message configuration information word (W), 104 bits of information, and 2 dummy bits, i.e., filler. W includes information on whether the slot is the first, last, or other slot, as well as how many slots are left to be transmitted. The multiplexed bits are encoded using 16-bit CRC. The CRC-encoded sequence is then divided into 13 blocks; each block contains 10 bits of CRC-encoded data and is encoded by 1-bit shortened (14, 10) BCH code. The BCH-encoded data sequence of 182 bits is interleaved using (13, 14) low-column conversion-type interleaver. The sequence of 182 bits is divided into 66- and 116-bit groups and mapped onto the physical channel slot. The slot format for the second or later CAC slots in the uplink consists of 128 bits of information and 8 bits of W. The encoding process is almost the same as that of the first slot except that FEC is 3-bit shortened (12, 8) BCH and the interleaver size is (9, 12). The downlink slot of the CAC consists of 136 bits of information and 8 bits of W. It is similar to the first uplink slot except the interleaver size is (16, 14) instead of (13, 14).

BCCH, SCCH, and PCH data mapping onto the physical channel with a superframe structure in the downlink and uplink is assumed to be 1000 bits. In the downlink direction, 136 bits of signaling information are transmitted in the eighth (last) superframe. In the uplink direction, the first superframe transmits only 104 bits, and the latter superframe transmits 128 bits of signaling information.

For RCH, in the uplink direction 14 bits of source bits are encoded using 8-bit CRC. The CRC-encoded 22-bit data sequence is divided into 2 blocks; each of them is encoded by (15, 11) BCH code, interleaved by the (2, 15) interleaver, and mapped onto the RCH bit location in the assigned User Specific Channel (USC) slot. In the downlink direction, 22 bits of the source data are CRC encoded using 8-bit CRC with the same generator polynomial as that for the uplink to produce 30 bits of encoded sequence. The sequence is divided into 3 blocks, each of which is encoded by 1-bit shortened (14, 10) BCH code, interleaved by the (2, 14) interleaver, and mapped onto the RCH bit location in two consecutive frames.

For SACCH, in the uplink direction, 64 source bits associated with 8 bits of W are encoded by 16-bit CRC. The CRC-encoded 88-bit data sequence is divided into 8 blocks, each of which is encoded by (15, 11) BCH code, interleaved by (8, 15) interleaver, and mapped onto the SACCH bit location in the assigned slot. In the downlink direction, 96 bits of source data with 8 bits of W are CRC encoded using 16-bit CRC to produce 120 bits of encoded sequence. The sequence is divided into 12 blocks, each of which is encoded by 1-bit shortened (14, 10) BCH code, interleaved by (12, 14) interleaver, and mapped onto the SACCH bit location in the assigned slots.

For FACCH, the same encoding scheme is used for both the uplink and downlink directions. A message unit consisting of 8 bits W and 88 bits FACCH message is CRC encoded using 16-bit CRC. The encoded data sequence is divided into 28 blocks of 4-bit data. The blocked 4-bit data is encoded by the 1-bit shortened (8, 4) BCH code, and interleaved by the (28, 8) interleaver; 112 bits are mapped onto the former part of the FACCH section in the present frame, and the latter 112 bits are mapped onto the later part of the next frame.

19.2.5 Link Budget, Coverage, and Capacity

Table 19.7 shows a comparison of link budget and coverage for the DAMPS-1900 and PCS-1900 systems. Evidently, the DAMPS-1900 system provides better coverage in all environments as compared to the PCS-1900 system. In Table 19.8, we provide a comparison between the capacities of the DAMPS-1900 and PCS-1900 systems with respect to the AMPS system. For the sake of comparison we have assumed a total spectrum of 25 MHz and calculated the traffic in Erlangs per square km for the two systems. Notice that DAMPS-1900 has a better capacity than PCS-1900.

19.3 SUMMARY

DAMPS-1900 (IS-136) is the upbanded version of the DAMPS (IS-54) system that operates at 800 MHz and has been in operation in North America since 1992. DAMPS-1900 is dual-band/dual-mode PCS technology for the 800-MHz and 1900-MHz bands. This allows for dual-band terminals to support 800 MHz and 1900 MHz with full national/international roaming feature transparency. DAMPS-1900 has been designed to allow existing IS-54-based DAMPS networks to be upgraded with a simple software addition. This solution provides an optimum evolution path on which AMPS operators can migrate into digital cellular and then into the PCS arena while protecting

Table 19.7 Comparison of Link Budget and Coverage

System Parameter	DAMPS-1900	PCS-1900
Transmitted Power (dBm)	27.8	30
Receiver Sensitivity (dBm); $kT = -174$ dBm/Hz; F = Noise Figure; B_w = Channel Bandwidth; S/I = Required S/I Ratio	−105.8	−101.0
Antenna Gain (dBi)	20	20
Log-Normal Fade Margin (dB)	10	10
Max. Allowable Path Loss (dB)	143.6 $27.8 - (-105.8) + 20 - 10 = 143.6$ dB	141.0 $30 - (-101) + 20 - 10 = 141$ dB
Urban Coverage (40-m Antenna)	1.20 km	1.0 km
Suburban Coverage (50-m Antenna)	3.0 km	2.5 km

Table 19.8 Capacity Comparison of DAMPS-1900 and PCS-1900 with AMPS

	AMP	DAMP-1900	PCS-1900
Bandwidth (MHz)	25	25	25
Channel Width (kHz)	30	30	200
Users/Channel	1	3	8
Total No. of Channels	833	2500	1000
Frequency Reuse Factor	7	7 or 4	4 or 3
Channel per Site	119	357 or 625	250 or 333
Traffic (Erlangs per sq. km)	11.9	41.0 or 74.8	27.7 or 40.0
Capacity Gain w.r.t AMPS	1.0	3.5 or 6.3	2.3 or 3.4

their capital investment. Certainly, PCS-1900 cannot support this strategy. The DAMPS-1900 supports TDMA PCS to analog, cellular TDMA to TDMA PCS, and TDMA PCS to cellular TDMA handovers. In addition, the DAMPS-1900 system appears to have a better modulation, better speech coding, a low handset power requirement, superior spectral efficiency, an efficient time slot structure, and a simpler logical channel structure as compared to the PCS-1900 system. The PDC system is very similar to DAMPS-1900 except in its speech and channel coding. Because of the low rate of speech codec in the

PDC, the voice quality of PDC is somewhat inferior to the DAMPS-1900 and PCS-1900.

19.4 REFERENCES

1. EIA/TIA IS-54, "Dual-Mode Mobile Station-Base Station Compatibility Standard," December 1989.

2. ETSI/GSM, Series 01-12, "GSM Specifications."

3. EIA/TIA J-STD-010, "IS-136 Based Air Interface Compatibility Standard."

4. RCR, RCR STD-27, "Personal Digital Cellular Telecommunication System," April 1991.

5. EIA/TIA IS-641, "TDMA Cellular/PCS-Radio Interface-Enhanced Full-Rate Speech Codec," May 1996.

6. Elwood, K. P., et al. "Speech Codec for European Mobile Radio System," IEEE Vehicular Technology Conference, 1989, pp. 1065–69.

7. Sampei, S., *Applications of Digital Wireless Technologies to Global Wireless Communications,* Prentice Hall, Upper Saddle River, NJ, 1997.

8. Garg, V. K., and Sneed, E. L., "North American TDMA Systems for PCS Networks," *Bell Labs Journal*, October 1997.

Future Wireless Services

20.1 INTRODUCTION

The explosion in wireless phone usage throughout the late 1990s has resulted in worldwide activities to improve their operation and the services that support them. Work has progressed in two main standards bodies. In Europe, ETSI has done extensive planning for next-generation GSM. In the United States, the TIA has similar work activities for next-generation systems, services, and phones. The ITU is planning IMT-2000, the next generation of mobile services.

Existing cellular service has several problems that are currently being addressed by the various standards groups. At present there is no one worldwide frequency allocation for cellular service. In Europe, the primary cellular bands are at 900 and 1800 MHz for digital GSM service and at 450 and 800 MHz for the various older incompatible analog systems. In the United States and some other areas of the world, cellular and PCS service is offered in the 800- and 1900-MHz bands. In Asia, 800-, 900-, and 1700-MHz bands are used. While there is some overlap between the European 1800-MHz band and the U.S. 1900-MHz band, there is no single band that can be used worldwide. Thus, worldwide roaming requires a phone that can support multiple frequency bands and costs more to manufacture than a single-band phone. The proliferation of different technologies (CDMA, GSM, and TDMA) also complicates roaming issues.

A second problem with current cellular phones is the lack of high-speed data services. The explosion of the Internet has created a demand for higher and higher data speeds on the wireline network. As users have become accustomed to the higher speeds, they demand the same speeds on their wireless networks. A business user receiving an e-mail message with an attached 1-megabyte document can easily retrieve the document in the office over an LAN. Unfortunately, the time and cost of retrieving the same document from a wireless phone is exceedingly high. New wireless services must offer data rates and costs comparable to the wireline network.

A third problem is access to a roaming mobile subscriber and the provision of services to that subscriber. Users have a rich set of features on their home and office wireline phones. They may have speed dialing lists, three-way calling and conferencing, voice mail, call forwarding, etc. Currently, home, office, and wireless voice mailboxes in most service areas are not integrated. Thus, a user attempting to retrieve voice mail may have to call into three separate voice mailboxes, each with a different reach number and user interface. While each service may be provided by a different service provider, it is desirable to have the services appear seamless to the user. The activities on intelligent networks and wireless intelligent networks hold the promise of fully integrated wireless and wireline networks.

In this chapter we will examine the future of wireless services. Our primary focus will be on the extensions of GSM proposed by ETSI, but the activities around the world are converging on a common solution rather than national solutions. First we will examine the service needs of the providers. Then we will examine the network and radio issues that must be solved to meet these service needs. The work is in flux, and you will want to consult the most recent documents of ETSI and TIA for the latest developments.

20.2 SERVICE NEEDS OF FUTURE WIRELESS SERVICES

While cellular phones were originally considered a premium service for businesses and wealthy individuals, the current worldwide market for wireless phones demonstrates that wireless has more of a mass appeal than was originally envisioned. With the explosion of wireless phone usage, service providers, regulatory agencies, and users consider wireless phones as alternatives to wireline phones. With this new concept of the use of wireless comes a demand for more service capabilities. Some of the more important of these needs are [1]

- ☞ **Anywhere-anytime service.** As users roam the world, they will no longer tolerate the poor service provided when they are not in their home system. The full range of services and features must be available anywhere in the world. The same number must be used to reach the phone, independent of its location.

- ☞ **Services equivalent to wireline.** While the user may always pay a premium for use of a wireless phone, the services should be identical to wireline phones. A user accessing the Internet or a business intranet when at home or in the office should be able to access the same services at the same speed and reliability over the wireless phone.

☞ **A unified and seamless wireless network.** Many service providers today are operating multiple networks. There are paging networks, two-way dispatch networks, cellular networks, and PCS networks. Each network has a different technology, different frequency band, user device, and so on. The cost of network equipment, user equipment, and employee training is too high to tolerate multiple networks any longer. One network must support all types of wireless services that a customer may desire. The user will no longer carry multiple phones, pagers, and two-way radios.

☞ **Wireless-wireline integration.** While the wireless and wireline companies may be two divisions of the same company or direct competitors, the user doesn't care. Users want to get their home or office calls seamlessly forwarded to the wireless phone without dialing 50 digits every hour to inform the network of their location. The technology exists on paper to perform the integration; the service providers will need to deploy networks that support it.

☞ **Rapid introduction of new technology where it is needed.** While some countries have more than one phone for each person in the country, other countries have one phone for every 100 or more persons. The use of WLL systems that offer fixed and mobile wireless services offers the possibility of rapid upgrading of the telecommunications structure of many countries. Even in fully developed countries, WLL offers the possibility of providing competition in a area previously thought to be served by a monopoly.

☞ **High-speed backbone network.** Current wireless systems are interconnected via SS7 networks that operate at low speeds. With the explosion of voice and data services, the networks must be interconnected with Asynchronous Transfer Mode (ATM) or other higher-speed transport mechanisms.

☞ **Intelligent terminals.** Today's handsets have their programs in ROM or in some cases flash ROM. When new features are added, or when bugs are found in the software, the handsets must be returned to the service provider for costly reprogramming. Future handsets may have their software upgraded over the wireless network. As the user roams from one service area to another, the software may be modified to meet many of the service needs or to adapt the terminal to different network infrastructures. The possibility of malicious reprogramming must be considered and security issues must be solved for this concept to fully work.

Once these service features are provided in the network, service providers can give the user a rich set of services, features, and user interfaces that will be identical in the home or office and on the road. Thus the concept of a "virtual home environment" evolves. This concept could, of course, be considered a negative since businesses will no longer need to provide a separate and unique office or cubical for each employee. Employees could work at home, in the office, or in a visiting office at any desk and see the same environment. As an employee arrives at work, he or she could be assigned a cubical for the day (or hour) and would then set up the virtual office in a featureless allocation of space. The amount of office space could then be assigned on traffic engineering principles instead of a fixed amount for each employee. Lost time because of traffic jams or adverse weather no longer occurs because the worker can work anywhere, anytime. The savings to a company could be considerable. The company intranet might need a chat room and virtual water cooler, but hard disk space is smaller and cheaper than office space.

None of these new services can be developed in a vacuum. The worldwide market for wireless phones is huge. If all users are migrated to the network of the future, the current infrastructure must be retired. The cost of the migration will be more than the operators and users are willing to tolerate. Therefore, operators may have a mixed network for years. Any proposed solutions must allow for a migration plan.

Closely related to the service provider's needs is the regulatory environment [2]. While the radio frequencies are allocated on a regional basis under the banner of the World Radio Conference (WRC), each nation has its own set of rules that govern the ownership and operation of wireless devices. Table 20.1 shows worldwide band usage and Table 20.2 shows the use of CDMA and GSM throughout the world.[*] For roaming to be truly worldwide, a frequency

Table 20.1 Cellular/PCS Bands in Each Region of the World

Frequency Band	Africa	Asia	Australia	Europe	North America	South America
824–849 869–894	x	x	x	x[*]	x	x
880–915 925–960	x	x	x	x	—	x
1710–1785 1805–1880	—	x	—	x	—	—
1850–1910 1930–1990	—	x	—	x[†]	x	x

[*]In Europe, only Russia and Poland have adopted the U.S. cellular frequency band.
[†]In Europe, only Russia has adopted the U.S. PCS frequency band.

---

[*]There are exceptions to all general cases. Consult the frequency plans for each nation to determine which frequencies are available.

Table 20.2 CDMA and GSM Usage in Each Region of the World

Frequency Band	Africa	Asia	Australia	Europe	North America	South America
CDMA-800	x	x	—	x[*]	x	x
CDMA-1700	—	x	—	—	—	—
CDMA-1900	—	x	—	x[†]	x	x
GSM-900	x	x	x	x	—	—
DCS-1800	—	x	—	x	—	—
DCS-1900	—	—	—	—	x	—

[*]Russia and Poland only.
[†]Russia only.

allocation is needed that is identical in every region of the world and every country of the world. Also rules on handset ownership and operation need to be identical worldwide. Since all frequencies from 3 kHz to at least 60 GHz have been assigned, worldwide frequency allocations will require movement of some current users of some bands. Clearly, from a political, regulatory, and technical viewpoint, this process will take years. Eventually, a common frequency allocation for wireless services will emerge and true worldwide roaming will be possible.

20.3 FUTURE NETWORK ARCHITECTURES

For the wireline network, Bellcore has defined a set of protocols called Intelligent Network [7] that enable a rich set of new telephony capabilities to be generated without additional software development in the central switch. As IN has grown in popularity for the wireline network, the wireless network is also embracing the concept. IN improves the ability to locate and efficiently direct calls to roaming MSs and provides other advanced features. In this section we will examine the work of the international standards bodies to add IN to wireless systems.

An examination of the needs of IN and the mobility aspects of wireless communications shows several similarities. The wireline user should have one phone number anywhere in the world. Anyone dialing that number would have their calls routed to the owner's destination independent of the number's owner's location. The services that a user has available in the home or office should also be available in a visiting home or office and at public phones.

The functions of the HLR/VLR in a wireless network are similar to the functions of a Switching Control Point (SCP) in a wireline network.

The ITU has adapted IN for use in wireless networks. Question 8 of Study Group 11 has defined a communications control plane (Figure 20.1) and a radio resource control plane (Figure 20.2) using IN concepts.

On the network side of the communications control plane, the following functional elements are defined:

☞ The **Bearer Control Function (BCF)** provides those bearer functions needed to process handovers. A conference bridge to support soft handovers is a common example.

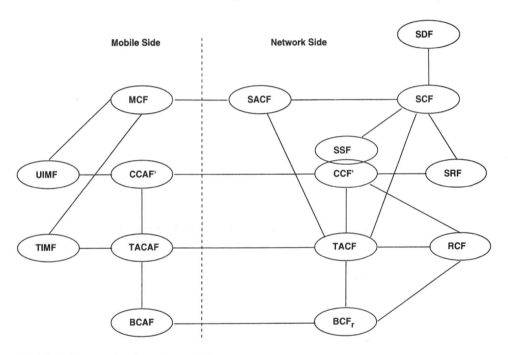

Fig. 20.1 Communications Control Plane

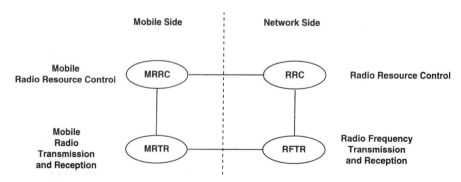

Fig. 20.2 Radio Resource Control Plane

☞ The **Bearer Control Function for the Radio Bearer (BCF$_r$)** provides the functions necessary to select bearer functions and radio resources. It also detects and responds to pages from the network and performs handover processing. Some example bearer functions are PCM voice, ADPCM voice, packet data, and circuit-switched data.

☞ The **Call Control Function Enhanced (CCF')** provides the call and connection control in the network. CCF' establishes, maintains, and releases call instances requested by the CCAF'; provides IN triggers to the SSF, and controls bearer connection elements in the network.

☞ The **Service Access Control Function (SACF)** provides the network side of mobility management functions. Some examples are registration and location of the MS.

☞ The **Service Control Function (SCF)** contains the service and mobility control logic and call processing to support the functions of a mobile terminal.

☞ The **Service Data Function (SDF)** provides data storage and data access in support of mobility management and security data for the network.

☞ The **Specialized Resource Function (SRF)** provides the specialized functions needed to support execution of IN services. Some examples are dialed digit receivers, conference bridges, and announcement generators.

☞ The **Service Switching Function (SSF)** provides the functions required for interaction between the CCF' and the SCF. It supports extensions of the CCF' logic to recognize IN triggers and interact with the SCF. It manages the signaling between the CCF' and the SCF and modifies functions in the CCF' to process IN services under control of the SCF.

☞ The **Terminal Access Control Function (TACF)** provides control of the connection between the MS and the network. It provides MS paging, page response handling, handover decision, and completion. It also provides trigger access to IN functionality.

On the mobile side, the following functional elements are defined:

☞ The **Bearer Control Agent Function (BCAF)** establishes, maintains, modifies, and releases bearer connections between the MS and the network.

☞ The **Call Control Agent Function-Enhanced (CCAF')** supports the call processing functions of the MS.

☞ The **Mobile Control Function (MCF)** supports the mobility management functions of the MS.

☞ The **Terminal Access Control Agent Function (TACAF)** provides the functions necessary to select bearer functions and radio resources. It also detects and responds to pages from the network and performs handover processing.

☞ The **Terminal Identification Management Function (TIMF)** stores terminal-related security information. It provides terminal identification to other functional elements and provides terminal authentication and cryptographic calculations.

☞ The **User Identification Management Function (UIMF)** provides user-related security information similar to the TIMF.

Both the TIMF and the UIMF can be stored in either the MS or a separate security module often implemented in a smart card.

The radio resource control plane (Figure 20.2) is in charge of assigning and supervising radio resources. Four function entities (two on the mobile side and two on the network side) perform the functions of the radio access subsystem.

☞ The **Radio Resource Control (RRC)** provides functionality in the network to select radio resources (channels, spreading codes, etc.), make handover decisions, control the RF power of the MS, and provide system information broadcasting.

☞ The **Radio Frequency Transmission and Reception (RFTR)** provides the network side of the radio channel. It provides the radio channel encryption and decryption (if used) and channel quality estimation (data error rates for digital channels). It sets the RF power of the MS and detects the MS accessing the system.

☞ The **Mobile Radio Resource Control (MRRC)** processes the mobile side of the radio resource selection. It provides BS selection during start-up, mobile-assisted handover control, and system access control.

☞ The **Mobile Radio Transmission and Reception (MRTR)** provides the mobile side of the radio channel and performs functions similar to the RFTR.

The two control planes interact to provide services to the MS and the network.

The TR-46 PCS network reference model working group has generated a simplified version of the ITU model and has defined explicit operations functions for the model. The IN function reference model for PCS (Figure 20.3) has many of the same functional element names as the ITU model. Those that are different include:

☞ The **Radio Terminal Function (RTF)** contains all functionality of the mobile side of the reference model.

☞ The **Radio Access Control Function (RACF)** is similar to the SACF and provides mobility management functions.

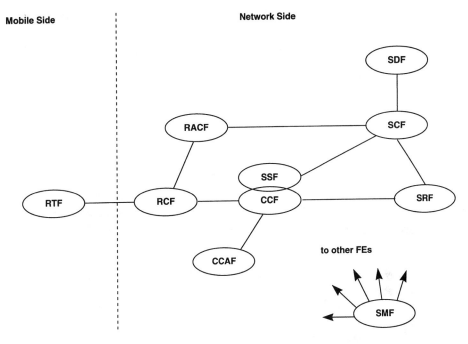

Fig. 20.3 IN Functional Reference Model for PCS

☞ The **Radio Control Functions (RCF)** provides the capabilities of the TACF, the BCF$_r$, and the BCF in the ITU model. It provides the radio ports and the radio port controller capabilities in the PCS network.

☞ The **Call Control Agent Function (CCAF)** provides access to the wireless network by wireline users.

☞ The **Service Management Function (SMF)** provides network management functions for each functional element.

With the ITU reference model, the functionality to support all of the features and capabilities for a wireless network can be partitioned into different functional elements and still meet the variety of national and worldwide standards. Therefore no exact partitioning of the functions of the network reference models described in chapter 5 can be made. We encourage you to examine the standards and implementations of the various manufacturers for details.

Two common MAPs are used universally. GSM systems deployed worldwide use the GSM MAP (chapter 9). In the systems that were originally deployed in the United States, the IS-41 MAP is used. The use of two different MAPs makes roaming between the two systems difficult, as we have discussed in one of our previous books [8]. The use of the IN application part with enhancements will offer the opportunity for a common MAP for all mobile and wireline services.

20.4 FUTURE RADIO STANDARDS

Many of the services envisioned for next-generation wireless systems require radio technologies that can support data rates of up to at least 1 Mbps. Internet access and one- and two-way video services require the highest speeds. When wireless systems replace wireline systems, the delays on a vocoded wireless-to-wireless call are unacceptable. Thus, the radio system needs the service quality that can be delivered by PCM or ADPCM. If a large number of users, each requiring large data bands, are to be served on a BS, the bandwidths of the radio link must be increased. Work is under way in various standards committees examining the next generation of wireless systems. In the United States under the TIA, those operators that use IS-136 TDMA are examining a wideband version that uses a wider bandwidth TDMA system. The operators that currently use IS-95 CDMA are examining various forms of wideband CDMA. In Europe under ETSI, GSM operators and manufacturers are examining various forms of TDMA with wideband CDMA overlays. In this section we will focus on the work in ETSI to replace the GSM TDMA radio link with an upgraded link that uses some form of TDMA and CDMA. The next section will examine how the network infrastructure can support multiple radio link protocols.

There also have been proposals to extend IS-136 by adding a 1- to 2-Mbps overlay packet system [25,26]. The radio physical layer would consist of 120 subchannels. Each subchannel would transmit data at a rate of 5 kbaud with a channel spacing of 6.35 kHz. Thus, the total data rate of the system would be 600 kbaud and with a QPSK modulator the signaling rate would be 1.2 Mbps. The total bandwidth occupied by the channel would be about 1 MHz in each direction for a total of 2 MHz.

The OFDM system encodes the transmitted data user an Inverse Fast Fourier Transform (IFFT). The IFFT data is sent on multiple (120) parallel channels. In the receiver, the data is decoded by performing an FFT. Since each channel has a low data rate, equalization is not needed. The goal is for the system to support a data rate of 384 kbps. Adding control and signaling information raises the channel data rate to approximately 600 kbps.

The packet system would not share the spectrum with the existing IS-136 users. Therefore, less spectrum would be available for voice and low-speed data usage. The range of the system is designed so that the packet users could share the same cell tower locations as the voice users. No new towers would be needed. Of course, the radio link is incompatible with current IS-136, so new BS hardware and software would be needed.

Proposals for extending IS-95 to higher speeds have also been made. The current IS-95 system uses Walsh codes on the downlink and pseudonoise codes on the uplink. A pilot signal is used on the downlink for synchronization but is not used on the uplink. In many ways, IS-95 is an asymmetric system since the uplink and downlink have many differences [27]. The current bandwidth of IS-95 is 1.25 MHz, supporting data rates of up to 14.4 kbps using one code and 115.2 kbps using multiple codes in one handset. The use of multiple codes to achieve higher data rates reduces system capacity drastically—in some cases, only one high-speed data user per cell can be supported.

In the proposals to extend IS-95, the bandwidth would be increased to 5 or 10 MHz and the link would be made symmetric in both directions. The channel would still use QPSK, but a pilot signal would be sent in both directions. The specifics of channel data rates, pilot signals, spreading codes, etc., are being worked on in TIA standards committees established at the end of 1997.

In 1997 ETSI established five committees to examine five proposals for a future GSM radio link protocol. The committees were code-named

☞ Alpha—a system based on wideband CDMA [28–30]

☞ Beta—a system based on OFDM [31–33]

☞ Gamma—a system based on wideband TDMA [34–36]

☞ Delta—a system based on TDMA with spreading [37–39]

☞ Epsilon—a system based on packet access and called Opportunity Division Multiple Access (ODMA)

Each of these systems has advantages and disadvantages, and each committee examined the proposed system in detail. The epsilon proposal concept was merged into the alpha and beta concepts, resulting in four concepts for evaluation:

☞ **Concept alpha** is a wideband CDMA system with carrier spacing of about 5 MHz. Table 20.3 summarizes the key technical characteristics of the WCDMA radio interface.

Table 20.3 WCDMA Radio Interface

Parameter	Characteristic
Multiple Access Scheme	DS-CDMA
Duplex Scheme	FDD/TDD
Chip Rate	4.096 Mcps (expandable to 8.192 Mcps and 16.384 Mcps)
Carrier Spacing (4.096 Mcps)	Flexible in the range 4.4–5.2 MHz (200 kHz carrier raster)
Frame Length	10 ms
Inter-BS Synchronization	FDD mode: No accurate synchronization needed TDD mode: Synchronization needed
Multirate/Variable-Rate Scheme	Variable spreading factor + Multicode
Channel Coding Scheme	Convolutional coding (rate 1/2–1/3) Optional outer RS coding (rate 4/5)
Packet Access	Dual mode (common and dedicated channel)

☞ **Concept beta** is a TDMA system that uses OFDM. The minimum-bandwidth system uses 100 kHz, and extensions to 1.6 MHz are possible. Table 20.4 summarizes some of the characteristics of the TDMA/OFDM system.

☞ **Concept gamma** uses frequency hopping and a time hopping protocol to achieve multiple access. There are 16 or 64 slots in each TDMA frame. Table 20.5 summarizes the characteristics of concept gamma.

Table 20.4 TDMA/OFDM Characteristics

Parameter	Characteristic
Multiple Access Scheme	SFH-TDMA and OFDM
Duplex Method	FDD (and TDD)
OFDM Carrier Spacing	100 kHz/24 = 4.17 kHz
OFDM Symbol Duration	240 µs
Modulation Time/Guard Time	278 µs/38 µs
Time Slot Length	288 µs
Band Slot Width (Minimum Bandwidth)	100 kHz (24 subcarriers)
Data Frame Length	4.615 ms (16 slot/frame)
Bandwidth	100 kHz, 200 kHz, 400 kHz, 800 kHz, 1.6 MHz (flexible)
Frequency Hopping	1 (hop/burst) = 867 hop/s (no hopping option)
Channel Coding	Convolutional coding, rate 1/3–2/3, Optional outer RS coding (rate: 4/5)
Interleave	Typical 18.46 ms for speech
Subcarrier Modulation Scheme	Frequency domain DQPSK, frequency domain D8PSK, coherent modulation schemes supported
Bit Rates	Typical 11.6 kbps per time slot/band slot (coding = 1/3)
Frequency Reuse	1 (fractional load = 30%), 3 (load = 100%)
Maximum Use Bit Rate	No limitation (depends on system BW)
Power Control	Open loop & closed loop
Power Control Step, Period	1 dB, 1.153 ms/control
Frequency Deployment Step	100 kHz
Handover	Hard handover; soft handover not required

Table 20.5 Frequency/Time Hopping

Parameter	Characteristic
Multiple Access Method	TDMA
Duplexing Method	FDD and TDD
Channel Spacing	1.6 MHz
Carrier Bit Rate	2.6 Mbps/5.2 Mbps
Time Slot Structure	16 or 64 slots/TDMA frame
Frame Length	4.615 ms
FEC Codes	Rate-compatible punctured convolutional codes, turbo codes, Reed Solomon codes
ARQ Scheme	Type II hybrid ARQ
Interleaving	Interslot and intraslot interleaving
Modulation	B-OQAM / Q-OQAM
Pulse Shaping	Root raised cosine, roll-off = 0.35
Detection	Coherent, based on midamble

☞ **Concept delta** uses 8 TDMA time slots with 8 CDMA spreading codes. The bandwidth is 1.2 MHz (equivalent to 6 GSM carriers). Each MS transmits on a time slot and spreading code. The system can support data rates from 8 kbps to 2 Mbps by the use of multiple time slots and spreading codes. The system uses 16-point constellation of Quadrature Amplitude Modulation (16-QAM). Since the system uses CDMA, multiuser detection can be used to reduce intracell interference [40]. Table 20.6 summarizes some of the characteristics of the TDMA/CDMA system.

20.5 SOFTWARE RADIOS

The worldwide standards bodies have developed multiple standards for high- and low-tier cellular and PCS. While each carrier has plans for offering service based on a specific standard, the multiplicity of standards results in the reduced ability for customers to roam into different areas. As new standards for IMT-2000 are developed, still additional radio link interfaces are needed. Clearly, a service provider needs a solution to support multiple standards and the growth to a common worldwide standard. Even when (or if) there is one standard, it will not be static but will have revisions. Solutions to a multiplicity of radio link standards are needed if users are to roam from one system to another.

Table 20.6 WB-TDMA/CDMA Key Technical Characteristics

Parameter	Characteristic
Multiple Access Method	TDMA and CDMA
Duplex Method	FDD and TDD
Channel Spacing	1.2 MHz
Carrier Chip/Bit Rate	2.167 Mcps
Time Slot Structure	8 slots/TDMA frame
Spreading	Orthogonal, 16 chips/symbol
Frame Length	4.615 ms
FEC Codes	$R = 1/8 \ldots 1$ (convolutional, punctured)
Modulation	QPSK/16QAM
Burst Types	2 different burst types: -burst 1: for long-delay-spread environments -burst 2: for short-delay-spread environments
Detection	Coherent, based on midamble
Power Control	Slow
Handover	Mobile-assisted hard handover
Intracell Interference	Suppressed by multiuser detection
Intercell Interference	Like in other clustered systems

There are two possible approaches for solving the roaming problem. Each customer could have a multimode phone that supports several different air interfaces. The second solution is for each BS to transmit pilot signals that would allow any handset to access the network. While some industry attention has been focused on multimode handsets, most of the attention has been on AMPS/one digital technology handsets. While it is possible to make a handset that supports AMPS/CDMA/TDMA, AMPS/CDMA/GSM, GSM/IMT-2000, or even AMPS/TDMA/CDMA/GSM/IMT-2000, the cost of those handsets will always be higher than the cost of a single-mode handset. Thus, not everyone will have a multimode handset.

By making the BS multimode, any roamer that enters the system can be offered service. The BS must transmit a pilot signal for each type of air interface that is supported on the BS. TCHs would then be assigned depending on the number of users that request calls on each air interface. This type of BS could be designed in hardware. Unfortunately sufficient hardware (channel elements) would be needed to permit all channels to be assigned to a particular air interface. By designing and implementing the BS in software, each channel element could be dynamically reconfigured to the appropriate air interface as needed.

As service providers acquire holdings around the world, there is intense pressure to integrate these operations. The level of integration could be as simple as routing all operations functions to a common central point. It also could be as complicated as using the same set of hardware in multiple systems. A software BS that can use a common hardware platform to support multiple technologies has advantages of common spare parts as well as common operations support. Figure 20.4 shows a representative example of how such a system might be configured. In city B, the BS has software loaded for UMT-2000; in city A, the same BS hardware has software loaded for GSM. A common set of spare parts and a common operations support system interface would be available, even though two different systems are supported. In this implementation, we are not necessarily proposing that either system support the air interface of the other system.

Figure 20.5 shows the architecture of a traditional hardware BS. The receive path has a common filter and low noise amplifier (LNA). All further fil-

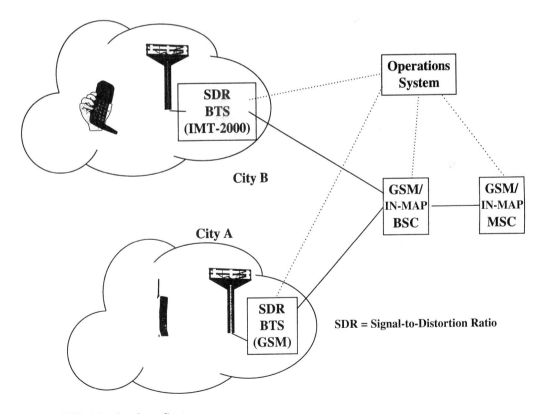

Fig. 20.4 Multitechnology System

Antenna

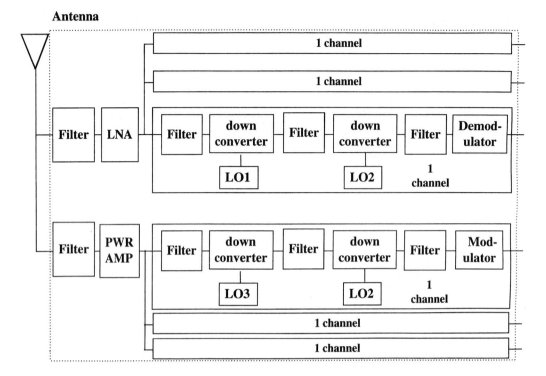

Fig. 20.5 Traditional Hardware BS Architecture

tering, frequency conversion, and demodulation are done on a channel basis. If more than one channel is needed, the hardware items inside the box labeled "1 channel" are duplicated. The transmitter is also duplicated on a channel-by-channel basis. The output of each transmitter is amplified by a power amplifier, filtered, and sent to the antenna. While the architecture shows a common antenna for the transmitter and receiver, in many cases separate antennas are used.

The traditional hardware approach has a dedicated channel element for each transmit and receive channel if the type of modulation is changed. For example, as several GSM channels are converted to wideband CDMA, the channel elements must be removed and replaced with a channel element that supports the new air interface. The channel element is then used elsewhere in the system, put into inventory as a spare, or sold as a surplus unit for its salvage value.

The design of a software-defined BS involves many trade-offs. In the long term (as speeds of the computing section of the BS improve), the BS will consist of (Figure 20.6):

☞ For the receiver—an input filter, an LNA, an analog-to-digital (A/D) converter, and a Digital Signal Processor (DSP)

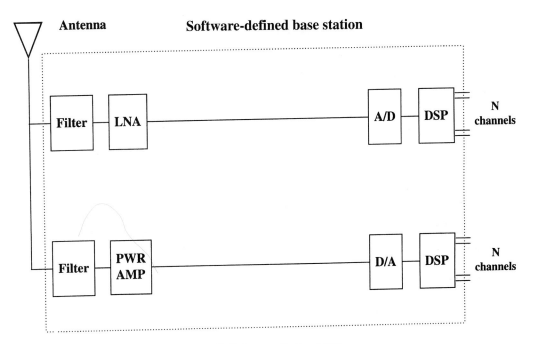

Fig. 20.6 Architecture of Future (Ideal) Software-Defined BS

☞ For the transmitter—a DSP, a digital-to-analog (D/A) converter, a power amplifier and filter

Unfortunately current A/D, D/A, and DSP speeds do not permit this design for PCS frequencies. It is almost possible, though, for shortwave frequencies and will ultimately also be possible for PCS frequencies.

A more realistic design uses a superheterodyne receiver and several stages of down conversion so that the resultant signal is within the range of available A/D, D/A, and DSP products on the market. Thus, a current design for a BS would have one or two stages of down conversion and filtering for the receiver and up conversion and filtering for the transmitter (Figure 20.7). In addition, the blocks labeled A/D, D/A, and DSP may actually be implemented in multiple stages depending on the speed of the parts.

Designers of the software-defined BS must examine the many issues. For the transmitter these include

☞ Transmit antenna filters

☞ Power amplifier gain and linearity

☞ Choice of frequencies for the first and second Intermediate Frequencies (IFs) and thus choice of the frequencies for the three local oscillators (LO1, LO2, and LO3)

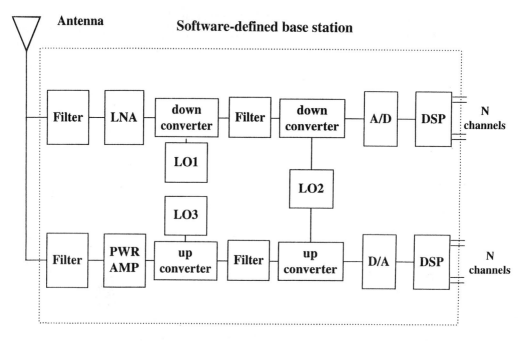

Fig. 20.7 Current Architecture of Software-Defined BS

☞ Number of D/As needed; the sample rate, number of bits, and linearity for each D/A

☞ Number of DSPs and the DSP speed needed to modulate a PCM signal into the signal needed for each channel

☞ Overall gain of the analog section of the transmitter

For the receiver, design issues include:

☞ Receive antenna filters

☞ LNA noise figure and gain

☞ Choice of frequencies for the first and second IFs and thus choice of the frequencies for the three local oscillators (LO1, LO2, and LO3)

☞ Number of A/Ds needed; the sample rate, number of bits, and linearity for each A/D

☞ Number of DSPs and the DSP speed needed to demodulate each channel into a PCM signal

☞ Overall gain of the analog section of the receiver

The choice of values for one element in the design will affect the other elements in the design. Several iterations may be needed to establish a reasonable set of conditions for all elements.

The software radio offers a solution to one of the IMT-2000 problems—the need to evolve the network while still offering service to users of older technology. For additional information on software radio see references [10–24].

20.6 SUMMARY

In this chapter we discussed future concepts for wireless communications. The material presented here was prepared in early 1998 and updated just before publication of this book. As with any high technology area, changes occur rapidly. We caution you to read the latest standards contributions and publications before making any decisions about the state of technology.

In this chapter we first discussed the future service needs of operators and users. The fundamental need is for the wireless network to provide the same types of services as the wireline network to a user on the move anywhere in the world. Thus, in the ideal, a user would not know that the services were delivered via radio and would think that he or she is at home or in the office. Next we examined the network architecture needs for future services and discussed the concepts of wireless intelligent networking. Current systems have difficulty with roaming because of the support of two MAPs in the world (GSM and IS-41) and the lack of a common worldwide frequency allocation. With wireless intelligent networks and extensions to the IN MAP, true worldwide roaming may become possible once a common frequency band is available.

The three dominant radio technologies (CDMA, GSM, and TDMA) each have proposals for extensions to support high-data-rate services. We briefly discussed extensions to CDMA and TDMA and focused on efforts to extend GSM. ETSI has agreed to use concept alpha (wideband CDMA) for outdoor systems and concept delta (CDMA/TDMA) for indoor systems.

A critical component in the growth to third-generation systems is the ability to simultaneously support several second- and third-generation technologies on the same BS. A software radio or software BS may solve the problem. Therefore, we examined some of the techniques used to remove the radio hardware from a BS and replace it with software. While a future BS may consist of an antenna and a computer (or digital signal processor), until processor speeds exceed the carrier frequency of PCS, a hybrid approach will be needed.

20.7 REFERENCES

1. Buchnan, K., et al., "IMT-2000: Service Provider's Perspective," *IEEE Personal Communications*, August 1997.

2. Leite, F., et al., "Regulatory Considerations Relating to IMT-2000," *IEEE Personal Communications*, August 1997.

3. Pandya, R., et al., "IMT-2000 Standards: Network Aspects," *IEEE Personal Communications*, August 1997.

4. Carsello, R. D., et al., "IMT-2000 Standards: Radio Aspects," *IEEE Personal Communications*, August 1997.

5. Markoulidakis, J. G., et al., "Mobility Modeling in Third Generation Mobile Telecommunications Systems," *IEEE Personal Communications*, August 1997.

6. Bellcore, "Advanced Intelligent Network (AIN) Release 1 Switching System Generic Requirements," TA-NWT-001123, Issue 1, May 1991.

7. ITU Study Group 11, "Version 1.1.0 of Draft New Recommendation Q.FNA, Network Functional Model for FPLMTS," Document Q8/TYO-50, September 15, 1995.

8. Garg, V. K., and Wilkes, J. E., *Wireless and Personal Communications Systems*, Prentice Hall, Upper Saddle River, NJ, 1996.

9. Garg, V. K., and Wilkes, J. E., "Interworking and Interoperability in North American PCS," *IEEE Communications Magazine*, January 1996.

10. Turletti, T., and Tennenhouse, D. L., "Estimating the Computational Requirements of a Software GSM Base Station," Proc. ICC'97, Montreal, June 8–12, 1997.

11. Tennenhouse, D. L., Turletti, T., and Bose, V. G., "The SpectrumWare Testbed for ATM-based Software Radios," Proc. ICUCP'96, Boston, MA, September 29–October 2, 1996.

12. Frerking, M. E., *Digital Signal Processing in Communications Systems*, Chapman & Hall, London, 1994.

13. Proakis, J. G., and Manolakis, D. G., *Introduction to Digital Signal Processing*, Macmillan, New York, 1988.

14. Tsui, J., *Digital Techniques for Wideband Receivers*, Artech House, Boston, 1995.

15. Mitola, J., "The Software Radio Architecture," *IEEE Communications Magazine* 33 (5), May 1995.

16. Wepman, J. A., "Analog-to-Digital Converters and Their Applications in Radio Receivers," *IEEE Communications Magazine* 33 (5), May 1995.

17. Baines, R., "The DSP Bottleneck," *IEEE Communications Magazine* 33 (5), May 1995.

18. Lackey, R. J., and Upmal, D. W., "Speakeasy: The Military Software Radio," *IEEE Communications Magazine* 33 (5), May 1995.

19. Kennedy, J., and Sullivan, M. C., "Direction Finding and 'Smart Antennas' Using Software Radio Architectures," *IEEE Communications Magazine* 33 (5), May 1995.

20. Lober, R. M., "Wideband, Software-Definable Base Station Technology: Approaches, Benefits, and Applications," 7th Annual Virginia Tech/MPRG Symposium on Wireless Communications, June 1997.

21. Scholnik, D. P., and Coleman, J. O., "Integrated I-Q Demodulation, Matched Filtering, and Symbol Rate Sampling Using Minimum Rate IF Sampling," 7th Annual Virginia Tech/MPRG Symposium on Wireless Communications, June 1997.

22. Lansford, J., "Universal Cordless Telephone Transceivers Using DSP," 7th Annual Virginia Tech/MPRG Symposium on Wireless Communications, June 1997.

23. The European Commission, DGXIII-B Workshop on Software Radio Technology, June 1997.

    - "Software Radio Technology Challenges and Opportunities," J. Mitola III, MITRE.
    - "Modular Multifunctional Information Transfer System Forum," J. Hoffmeyer, Chair MMITS Forum, U.S. Department of Commerce.
    - "Advances in Semiconductor Technologies—Enabling Software Radio," B. Kraemer, Harris Corporation.
    - "Operators Perspective and Market Trends," M. Swinburne, Orange.
    - "Software Radio Challenges," K. Rissanen, Nokia Research Centre.
    - "Technical Challenges in Introducing Software Radio for Mobile Telephony Base Stations," B. Hedberg, Ericsson Radio Systems.
    - "Flexible Integrated Radio Systems Technology," C. Taylor, ERA.
    - "Software Radio: The Standards Perspective," B. Robinson, Motorola.
    - "The Impact of Software Radio," W. Tuttlebee, Roke Manor Research.
    - "Protocol and Architectural Issues for Software Radio," G. Colombo, CSELT.
    - "Implementing Terminal Configurability in the Network," S. Pike, Lucent Technologies.

24. Wepman, J. A., and Hoffman, J. R., "RF and IF Digitization in Radio Receivers: Theory, Concepts, and Examples," NTIA, March 1996.

25. Sollenberger, N. R., Sehadri, N., and Cox, R., "The Evolution of IS-136 TDMA for Third Generation Services," to be published.

26. Cimini, L. J., Jr., Chuang, J. C., and Sollenberger, N. R., "Advanced Cellular Internet Services (ACIS)," *IEEE Communications Magazine*, September 1993.

27. Garg, V. K., Smolik, K., and Wilkes, J. E., *Applications of CDMA to Wireless/Personal Communications*, Prentice Hall, Upper Saddle River, NJ, 1997.

28. ETSI Working Document, SMG#24 Tdoc 903/97, "Concept Group Alpha—Wideband Direct-Sequence CDMA: System Description Summary," December 1997.

29. ETSI Working Document, SMG#24 Tdoc 904/97, "Concept Group Alpha—Wideband Direct-Sequence CDMA: Evaluation Summary," December 1997.

30. ETSI Working Document, SMG#24 Tdoc 905/97, "Concept Group Alpha—Wideband Direct-Sequence CDMA (WCDMA) Evaluation Document (3.0), Parts 1–4," December 1997.

31. ETSI Working Document, SMG#24 Tdoc 894/97, "Summary of the Concept Description of the Beta Concept," December 1997.

32. ETSI Working Document, SMG#24 Tdoc 895/97, "Summary of the Concept Evaluation of the Beta Concept," December 1997.

33. "ETSI Working Document, SMG#24" Tdoc 896/97, December 1997.

34. ETSI Working Document, SMG#24 Tdoc 900/97, "Concept Group Gamma—WB-TDMA: System Description Summary," December 1997.

35. ETSI Working Document, SMG#24 Tdoc 901/97, "Concept Group Gamma—WB-TDMA: Evaluation Summary," December 1997.

36. "ETSI Working Document, SMG#24" Tdoc 902/97, December 1997.

37. ETSI Working Document, SMG#24 Tdoc 897/97, "Concept Group Delta—WB-TDMA/CDMA: System Description Summary," December 1997.

38. ETSI Working Document, SMG#24 Tdoc 898/97, "Concept Group Delta—WB-TDMA/CDMA: Evaluation Summary," December 1997.

39. "ETSI Working Document, SMG#24" Tdoc 899/97, December 1997.

40. Duel-Hallen, et al., "Multiuser Detection for CDMA Systems," *IEEE Personal Communications*, April 1993.

41. ETSI Working Document, SMG#24 Tdoc 1019/97, "Technical Analysis and Comparison of ULTRA Concepts," December 1997.

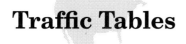

Traffic Tables

This appendix provides traffic tables for a variety of blocking probabilities and channels.[*] The blocked-calls-cleared (Erlang B) call model is used. In Erlang B, when traffic arrives in the system, we assume that it either is served, with probability from the table, or is lost to the system. A customer attempting to place a call will therefore either see a call completion or will be blocked and will abandon the call. This assumption is acceptable for low blocking probabilities. In some cases, the call will be placed again after a short period of time. If too many calls reappear in the system after a short delay, the Erlang B model will no longer hold.

In addition to the data provided in Table A.1 for 1–100 servers and blocking probabilities from 0.5 to 10 percent, we have also provided a program to calculate the offered load for servers from 1–130 for any blocking probability from 0 to 100 percent.

---

*The data in the tables was supplied by P. J. Wilkes.

Table A.1 Offered Load for Given Blocking Probability

Number of Servers	Blocking Probability							
	0.005	0.01	0.015	0.02	0.03	0.05	0.07	0.1
1	—	0.01011	0.01524	0.02041	0.03093	0.05264	0.07527	0.1111
2	—	0.1527	0.1904	0.2235	0.2816	0.3814	0.4705	0.5955
3	—	0.4556	0.5352	0.6022	0.7152	0.8994	1.057	1.271
4	—	0.8693	0.9919	1.092	1.259	1.525	1.748	2.045
5	—	1.361	1.524	1.657	1.875	2.219	2.504	2.881
6	—	1.909	2.112	2.276	2.543	2.961	3.305	3.759
7	—	2.501	2.741	2.936	3.25	3.738	4.139	4.666
8	—	3.127	3.405	3.627	3.987	4.543	4.999	5.597
9	—	3.783	4.095	4.345	4.748	5.371	5.88	6.547
10	—	4.462	4.808	5.084	5.53	6.216	6.777	7.511
11	—	5.159	5.54	5.842	6.328	7.076	7.688	8.487
12	—	5.877	6.288	6.615	7.142	7.95	8.61	9.474
13	—	6.607	7.05	7.401	7.967	8.835	9.543	10.47
14	—	7.352	7.825	8.201	8.804	9.73	10.49	11.47
15	—	8.109	8.61	9.01	9.651	10.63	11.43	12.48
16	—	8.876	9.406	9.829	10.51	11.54	12.39	13.5
17	—	9.653	10.21	10.66	11.37	12.46	13.35	14.52
18	—	10.44	11.03	11.49	12.24	13.39	14.32	15.55
19	—	11.23	11.85	12.33	13.12	14.31	15.29	16.58
20	—	12.03	12.67	13.18	14	15.25	16.27	17.61
21	11.86	12.84	13.51	14.04	14.89	16.19	17.25	18.65
22	12.63	13.65	14.35	14.9	15.78	17.13	18.24	19.69
23	13.42	14.47	15.19	15.76	16.68	18.08	19.23	20.74
24	14.2	15.3	16.04	16.63	17.58	19.03	20.22	21.78
25	15	16.13	16.9	17.5	18.48	19.99	21.22	22.83
26	15.8	16.96	17.75	18.38	19.39	20.94	22.21	23.89
27	16.6	17.8	18.62	19.27	20.31	21.9	23.21	24.94
28	17.41	18.64	19.48	20.15	21.22	22.87	24.22	26
29	18.22	19.49	20.35	21.04	22.14	23.83	25.22	27.05
30	19.04	20.34	21.23	21.93	23.06	24.8	26.23	28.11

Table A.1 Offered Load for Given Blocking Probability

Number of Servers	Blocking Probability							
	0.005	0.01	0.015	0.02	0.03	0.05	0.07	0.1
31	19.85	21.19	22.1	22.83	23.99	25.77	27.24	29.17
32	20.68	22.05	22.98	23.73	24.92	26.75	28.25	30.24
33	21.51	22.91	23.87	24.63	25.85	27.72	29.26	31.3
34	22.34	23.77	24.75	25.53	26.78	28.7	30.28	32.37
35	23.17	24.64	25.64	26.44	27.71	29.68	31.29	33.43
36	24.01	25.51	26.53	27.34	28.65	30.66	32.31	34.5
37	24.85	26.38	27.42	28.26	29.59	31.64	33.33	35.57
38	25.69	27.25	28.32	29.17	30.53	32.62	34.35	36.64
39	26.54	28.13	29.22	30.08	31.47	33.61	35.37	37.72
40	27.38	29.01	30.12	31	32.41	34.6	36.4	38.79
41	28.23	29.89	31.02	31.92	33.36	35.59	37.42	39.86
42	29.09	30.77	31.92	32.84	34.31	36.57	38.45	40.94
43	29.94	31.66	32.83	33.76	35.25	37.57	39.47	42.01
44	30.8	32.55	33.74	34.68	36.2	38.56	40.5	43.09
45	31.66	33.43	34.65	35.61	37.16	39.55	41.53	44.17
46	32.52	34.32	35.56	36.54	38.11	40.55	42.56	45.24
47	33.38	35.22	36.47	37.46	39.06	41.54	43.59	46.32
48	34.25	36.11	37.38	38.39	40.02	42.54	44.62	47.4
49	35.12	37.01	38.3	39.32	40.98	43.54	45.65	48.48
50	35.99	37.9	39.21	40.25	41.93	44.53	46.69	49.56
51	36.86	38.8	40.13	41.19	42.89	45.53	47.72	50.64
52	37.73	39.7	41.05	42.12	43.85	46.53	48.76	51.73
53	38.6	40.61	41.97	43.06	44.81	47.54	49.79	52.81
54	39.47	41.51	42.9	44	45.78	48.54	50.83	53.89
55	40.36	42.41	43.82	44.94	46.74	49.54	51.86	54.98
56	41.23	43.32	44.74	45.88	47.71	50.54	52.9	56.06
57	42.11	44.23	45.67	46.82	48.67	51.55	53.94	57.15
58	43	45.13	46.6	47.76	49.64	52.55	54.98	58.23
59	43.87	46.04	47.52	48.7	50.6	53.56	56.02	59.32
60	44.75	46.95	48.45	49.65	51.57	54.57	57.06	60.4

Table A.1 Offered Load for Given Blocking Probability

Number of Servers	Blocking Probability							
	0.005	0.01	0.015	0.02	0.03	0.05	0.07	0.1
61	45.65	47.86	49.38	50.59	52.54	55.57	58.1	61.49
62	46.53	48.78	50.31	51.54	53.51	56.58	59.14	62.58
63	47.42	49.69	51.24	52.48	54.48	57.59	60.18	63.66
64	48.3	50.6	52.18	53.43	55.45	58.6	61.22	64.75
65	49.2	51.52	53.11	54.38	56.42	59.61	62.27	65.84
66	50.09	52.43	54.05	55.33	57.4	60.62	63.31	66.93
67	50.98	53.36	54.98	56.28	58.37	61.63	64.35	68.02
68	51.87	54.27	55.92	57.23	59.34	62.64	65.4	69.11
69	52.76	55.2	56.85	58.18	60.32	63.66	66.44	70.2
70	53.66	56.12	57.79	59.13	61.29	64.67	67.49	71.29
71	54.55	57.04	58.73	60.08	62.27	65.68	68.53	72.38
72	55.46	57.96	59.67	61.04	63.25	66.69	69.58	73.47
73	56.35	58.88	60.61	61.99	64.22	67.71	70.63	74.56
74	57.26	59.81	61.55	62.95	65.2	68.72	71.67	75.65
75	58.15	60.73	62.5	63.9	66.18	69.74	72.72	76.74
76	59.06	61.66	63.44	64.86	67.16	70.75	73.77	77.83
77	59.96	62.58	64.38	65.82	68.14	71.77	74.81	78.93
78	60.87	63.51	65.32	66.77	69.12	72.79	75.86	80.02
79	61.77	64.44	66.27	67.73	70.1	73.8	76.91	81.11
80	62.67	65.36	67.21	68.69	71.08	74.82	77.96	82.2
81	63.58	66.3	68.16	69.65	72.06	75.84	79.01	83.3
82	64.47	67.23	69.11	70.61	73.04	76.86	80.06	84.39
83	65.38	68.16	70.05	71.57	74.03	77.88	81.11	85.49
84	66.3	69.09	71	72.53	75.01	78.89	82.16	86.58
85	67.2	70.02	71.95	73.49	75.99	79.91	83.21	87.67
86	68.11	70.95	72.9	74.46	76.98	80.93	84.26	88.77
87	69.03	71.89	73.85	75.42	77.96	81.95	85.31	89.86
88	69.93	72.81	74.8	76.38	78.95	82.97	86.36	90.96
89	70.85	73.75	75.74	77.34	79.93	83.99	87.41	92.05
90	71.76	74.69	76.7	78.31	80.92	85.02	88.46	93.15

Table A.1 Offered Load for Given Blocking Probability

Number of Servers	Blocking Probability							
	0.005	0.01	0.015	0.02	0.03	0.05	0.07	0.1
91	72.67	75.62	77.65	79.27	81.9	86.04	89.52	94.24
92	73.58	76.56	78.6	80.24	82.89	87.06	90.57	95.34
93	74.5	77.5	79.56	81.2	83.88	88.08	91.62	96.43
94	75.41	78.43	80.51	82.17	84.86	89.1	92.67	97.53
95	76.33	79.37	81.46	83.14	85.85	90.13	93.73	98.63
96	77.25	80.3	82.42	84.1	86.84	91.15	94.78	99.72
97	78.16	81.25	83.37	85.07	87.83	92.17	95.83	100.8
98	79.08	82.19	84.33	86.03	88.82	93.19	96.89	101.9
99	80	83.12	85.28	87.01	89.81	94.22	97.94	103
100	80.92	84.07	86.24	87.98	90.8	95.24	99	104.1

Abbreviations

3PTY	Third Party
16-QAM	16 (level) Quadrature Amplitude Modulation
A/D	Analog to Digital
A3	Authentication Algorithm A3
A38	A single algorithm performing the functions of A3 and A8
A5/1	Encryption Algorithm A5/1
A5/2	Encryption Algorithm A5/2
A5/X	Encryption Algorithm A5/0–7
A8	Ciphering Key Generating Algorithm A8
ABR	Answer Bid Ratio
ABS	Analysis-by-Synthesis
ABS	Average (load) Busy Seasons (business day)
ACA	Adaptive Channel Allocation
ACCH	Access Channel
ACCH	Associated Control Channel
ACELP	Algebraic Code Book Excited Linear Prediction
ACK	Acknowledgment

ACM	Accumulated Call Meter
ACM	Address Complete Message
ACSE	Association Control Service Element
ADPCM	Adaptive Differential Pulse Code Modulation
AGCH	Access Grant Channel
AMPS	Advanced Mobile Phone Service
AMR	Available Multirate (codec)
ANM	Answer Message
ANSI	American National Standards Institute
AOC	Advice of Charge
APBBS	AntiPodal BaseBand Signaling
ARFCN	Absolute Radio Frequency Channel Number
ARQ	Automatic Repeat Request
ASE	Application Service Entity
ASK	Amplitude Shift Keying
ASN.1	Abstract Syntax Notation - 1
ASP	Application Service Part
ASR	Answer Seizure Ratio
AT	Prefix for dialing using a modem
ATC	Adaptive Transform Coding
ATIS	Alliance for Telecommunications Industry Solutions
ATM	Asynchronous Transfer Mode
AUC	Authentication Center
AWGN	Additive White Gaussian Noise
B	Byte
B-CDMA	Broadband CDMA
B-ISDN	Broadband Integrated-Services Digital Network
BAIC	Barring of All Incoming Calls Supplementary Service
BAOC	Barring of All Outgoing Calls Supplementary Service
BCC	Base Transceiver Color Code
BCCH	Broadcast Control Channel
BCFr	Barring Call Forwarding (while) Roaming
BER	Bit Error Rate
BH	Busy Hour
BHCA	Busy Hour Call Attempts
BIB	Backward Indicator Bits
BIC-Roam	Barring Incoming Call (when) Roaming
BML	Business Management Layer
BMS	Business Management System
BOIC	Barring of Outgoing International Calls Supplementary Service
BOIC-exHC	BOIC except Home Country

bps	bits per second
BPSK	Binary Phase-Shift Keying
BRAT	Basic Rate Adaptation Service
BS	Base Station
BSC	Base-Station Controller
BSIC	Base Station Identity Code
BSN	Backward Sequence Number
BSS	Base Station System
BSSAP	Base Station System Application Part
BSSGP	Base Station Subsystem GPRS Protocol
BSSMAP	Base Station System Management Application Part
BSSOMAP	Base Station System Operations and Maintenance Application Part
BT	Bandwidth (times) Time (product)
BTS	Base Transceiver Station
BW	Bandwidth
CAC	Common Access Channel
CBCH	Cell Broadcast Channel
CBWL	Channel Borrowing Without Locking
CC	Call Control
CC	Country Code
CCAF	Call Control Agent Function
CCAF'	Call Control Agent Function-enhanced
CCBS	Completion of Calls to Busy Subscriber Supplementary Service
CCCH	Common Control Channel
CCF	Conditional Call Forwarding
CCH	Control Channel
CCIR	Consultative Committee for International Radio Communications
CCITT	Comité Consultatif International Télégraphique et Téléphonique (The International Telegraph and Telephone Consultative Committee)
CCS	Common Channel Signaling
CCS	Hundred Call Seconds
CD	Compact Disk
CDMA	Code-Division Multiple Access
CDO	Community Dial Offices
CDVCC	Coded Digital Verification Color Code
CELP	Code-Excited Linear Prediction
CEPT	Conférence des Administrations Européennes des Postes et Telecommunications
CFB	Call Forwarding on Mobile Subscriber Busy Supplementary Service

CFNR	Call Forwarding Mobile Subscriber Not Reachable
CFNRc	Call Forwarding–Mobile Subscriber Not Reachable
CFNRy	Call Forwarding–Mobile Subscriber No Reply
CFU	Call Forwarding Unconditional Supplementary Service
CH	Channel
CHV	Card Holder Verification Information
CI	Cell Identity
CI	Common Interface
CIC	Circuit Identification Code
CISS	Call Independent Supplementary Services
CLIP	Calling Line Identification Presentation Supplementary Service
CLIR	Calling Line Identification Restriction Supplementary Service
CLMS	Connectionless Message Service
CM	Communication Management
CMIP	Common Management Information Protocol
CMIS	Common Management Information Service
CMISE	Common Management Service Element
CNIP	Calling Number Identification Presentation
CNIR	Calling Number Identification Restriction
CNOP	Connected Number Identification Presentation
CNOR	Connected Number Identification Restriction
CO	Central Office
COLP	Connected Line Identification Presentation Supplementary Service
COLR	Connected Line Identification Restriction
COMS	Connection-Oriented Message Service
CON	Connect (Message)
CONF	Conference
COST	Committee on Standards and Technology
CPG	Call Progress Message
CPM	Continuous Phase Modulation
CRC	Cyclic Redundancy Code
CRSS	Call Related Supplementary Services
CS-1	Capability Set One (of IN)
CSMA	Carrier-Sense Multiple Access
CSMA/CD	CSMA with Collision Detection
CT	Cordless Telephone
CT-2 (also CT [ver. 2])	Cordless Telephone (Version 2)
CTA	Cordless Terminal Adapter
CTIA	Cellular Telecommunication Industry Association
CUG	Closed User Group Supplementary Service
CW	Call Waiting Supplementary Service

D/A	Digital-to-Analog
DAMPS	Digital Advanced Mobile Phone System
DAMPS-1900	Digital Advanced Mobile Phone System (at 1900 MHz)
dB	Decibels
dBm	Decibels with respect to 1 milliwatt
dBw	Decibels with respect to 1 watt
DC	Direct Current
DCA	Dynamic Channel Assignment
DCCH	Dedicated Control Channel
DCN	Data Communication Network
DCS	Digital Cellular System
DCS-1800	Digital Cellular System at 1800 MHz
DDI	Direct Dialing In
DECT	Digital Enhanced Cordless Telecommunications
DF	Dedicated File
DID	Direct Inward Dialing
DLCI	Data Link Connection Identifier
DPC	Destination Point Code
DPSK	Differential Phase Shift Keying
DQPSK	Differential Quadrature Phase Shift Keying
DS	Direct Sequence
DSP	Digital Signal Processing
DSS1	Digital Subscriber Signaling System No. 1
DSSS	Direct Sequence, Spread Spectrum
DTAP	Direct Transfer Application Part
DTC	Digital Traffic Channel
DTMF	Dual-Tone Multifrequency
DTX	Discontinuous Transmission (Mechanism)
E	Electric (Field)
E-3	Industry Standard for ATM (34.736 Mbps)
ECT	Explicit Call Transfer
EF	Elementary File
EFRC	Enhanced Full Rate Codec
EIR	Equipment Identity Register
EIRP	Effective Isotropic Radiated Power
EM	Element Manager
EML	Element Manager Layer
EMS	Element Management System
ERP	Effective Radiated Power
ESC	Escape
ETSI	European Telecommunications Standards Institute

FAC	Final Assembly Code
FACCH	Fast Associated Control Channel
FACCH/F	Fast Associated Control Channel/Full Rate
FACCH/H	Fast Associated Control Channel/Half Rate
FAF	Floor Attenuation Factor
FCA	Fixed Channel Assignment
FCC	Federal Communication Commission
FCCH	Frequency Correction Channel
FDD	Frequency Division Duplex
FDMA	Frequency-Division Multiplex Access
FE	Functional Entity
FEC	Forward Error Correction
FFT	Fast Fourier Transform
FH	Frequency Hopping
FHMA	Frequency Hopping Multiple Access
FHSS	Frequency Hopping, Spread Spectrum
FIB	Forward Indicator Bits
FISU	Fill-in Signal Unit
FM	Frequency Modulation
FN	Frame Number
FP	Fixed Part
FPH	Free Phone
FPLMT	Future Public Land Mobile Telephone
FREL	Frame Relay Service
FSK	Frequency Shift Keying
FSN	Forward Sequence Number
FSWI	Frame Switching Service
GDMO	Guideline for Definition of Managed Objects
GF	Galois field
GHz	Gigahertz
GMSC	Gateway Mobile Switching Center
GMSK	Gaussian Minimum Shift Keying
GoS (also GOS)	Grade of Service
GPA	GSM PLMN Area
GPRS	General Packet Radio Service
GSM	Global System for Mobile Communications
GSM-NA	GSM for North America
GT	Global Title
H	Magnetic (Field)
HCA	Hybrid Channel Assignment
HDBH	High Day Busy Hour
HDLC	High-Level Data Link Control

HLR	Home Location Register
HO	Handover
HOLD	Call Hold Supplementary Service
HPLMN	Home Public Land Mobile Network
HSN	Hopping Sequence Number
HSRC	Hypothetical Signaling Reference Connection
Hz	Hertz
I	Interoffice
I/W	Interworking
IAM	Initial Address Message
ID	Identification
IF	Intermediate Frequency
IFFT	Inverse Fast Fourier Transform
IMEI	International Mobile Station Equipment Identity
IMSI	International Mobile Subscriber Identity
IMT	Improved Mobile Telephone
IMT-2000	Improved Mobile Telephone (for the year 2000)
IMTS	Improved Mobile Telephone Services
IN	Intelligent Network
INAP	Intelligent Network Application Part
IP	Internet Protocol
IrDA	Infrared Data Association
IS-54	Interim Standard 54 (Dual-Mode TDMA/AMPS)
IS-95	Interim Standard 95 (Dual-Mode CDMA/AMPS)
ISC	International Switching Center
ISDN	Integrated Services Digital Network
ISI	Intersymbol Interference
ISO	International Organization for Standardization
ISUP	ISDN User Part (of Signaling System 7)
ITU	International Telecommunications Union
J	Joules
JTC	Joint Technical Committee
k	Constraint Length of the Convolutional Code
K	Kelvin
kbaud	kilobaud
kbps	kilobits per seconds
K_c	Ciphering Key
kcps	kilochips per second
kHz	kiloHertz
K_i	Encryption Key
km	kilometers

km/h	kilometers per hour
L2RBOP	Layer 2 Radio Bit Oriented Protocol
LA	Location Area
LAC	Location Area Code
LAI	Location Area Identity
LAN	Local Area Network
LAPB	Link Access Protocol Balanced
LAPDm	Link Access Protocol on the Dm Channel
LI	Length Indicator
LLA	Logical Layer Architecture
LLC	Logical Link Control
LLC	Low Layer Compatibility
LMSI	Local Mobile Station Identity
LNA	Low-Noise Amplifier
LOS	Line of Sight
LP	Linear Predictor
LPAS	Linear-Prediction-based Analysis-by-Synthesis
LPC-RPE	Linear Predictive Coding–Regular Pulse Excitation
LPF	Low Pass Filter
LSSU	Link Status Signal Unit
LTP	Long-Term Predictor
LTPD	Long-Term Predictor Delay
LU	Location Updates
m	meter
M.3010,3020	M Series of Recommendations
MA	Multiple Access
MAC	Media Access Control
MAH	Mobile Access Hunting Supplementary Service
MAI	Mobile Allocation Index
MAIDT	Mean Accumulated Intrinsic Down Time
MAIO	Mobile Allocation Index Offset
MAP	Mobile Application Part
Mbps (also MBS)	megabits per second
MC	Message Center
MCC	Mobile Country Code
MCF	Mobile Control Function
MCI	Malicious Call Identification Supplementary Service
MCID	Malicious Call Identification
Mcps	Mega chips per second
MD	Mediation Devices
MF	Master File

MHz	MegaHertz
MIB	Management Information Base
MIPS	Million Instructions per Second
MIT	Management Information Tree
MM	Mobility Management
mm	millimeter
MMI	Man Machine Interface
MNC	Mobile Network Code
MO	Managed Object
MOC	Managed Object Classes
Modem	Modulator/Demodulator
MOPS	Million Operations per Second
MOS	Mean Opinion Score
MOU	Memorandum of Understanding
mph	miles per hour
MPT	Ministry of Posts and Telecommunications
MRRC	Mobile Radio Resource Control
MRTR	Mobile Radio Transmission and Reception
MS	Mobile Station
ms	millisecond
MSC	Mobile Switching Center (or Centre)
MSIN	Mobile Subscriber Identification
MSISDN	Mobile Station International ISDN Number
MSK	Minimum Shift Keying
MSN	Multiple Subscriber Number
MSRN	Mobile Station Roaming Number
MSS	Mobile Switching Subsystem
MSU	Message Signal Unit
MT	Mobile Termination
MT2	Mobile Termination (Type) 2
MTP	Message Transfer Part
MTTF	Mean Time to Failure
MTTR	Mean Time to Repair
Mw	milliwatt
NCC	Network Color Code
NE	Network Element
NEL	Network Element Layer
NLOS	No Line of Sight
NM	Network Management
NMC	Network Management Center
NMF	Network Management Forum
NML	Network Management Layer
NMS	Network Management System
NMSI	National Mobile Subscriber Identity

NMT	Nordic Mobile Telephone
NRZ	Nonreturn to Zero
NSDU	Network Service Data Unit
nsec	nanosecond
NSP	Network Service Part
NSS	Network and Switching Subsystem
NTT	Nippon Telephone and Telegraph
O	Originating
O+I	Originating + Interoffice
O+T	Originating + Terminating
OA&M	Operations, Administration, and Maintenance
OAM&P	Operations, Administration, Maintenance, and Planning
ODMA	Opportunity Division Multiple Access
ODP	Open Distributed Processing
OFDM	Orthogonal Frequency Division Modulation
OMAP	Operations and Maintenance Application Part
OMC	Operations and Maintenance Center
OMSS	Operations and Maintenance Subsystem
OPC	Origination Point Code
OQPSK	Offset Quadrature Phase Shift Keying
OS	Operations System
OSI	Open System Interconnection
OSS	Operations Support System
PABX	Private Automatic Branch Exchange
PACCH	Packet Access Control Channel
PAD	Packet Assembly/Disassembly
PAGCH	Packet Access Grant Channel
PBCCH	Packet Broadcast Control Channel
PBX	Private Branch Exchange
PCCCH	Packet Common Control Channel
PCH	Paging Channel
PCM	Pulse Code Modulation
PCN	Personal Communications Network
PCR	Preventive Cyclic Retransmission
PCS	Personal Communications Services
PCS	Personal Communications System
PCS-1900	Personal Communications System (at 1900 MHz)
PDC	Personal Digital Cordless
PDF	Probability Density Function
PDS	Packet Data Service
PDSS1	Packet Data Service–Service (type) 1

PDSS2	Packet Data Service–Service (type) 2
PDSS2-SN	Packet Data Service–Service (type) 2- Service Node
PDTCH	Packet Data Traffic Channel
PHS	Personal Handyphone System
PIN	Personal Identification Number
PLL	Phase-Locked Loop
PLMN	Public Land Mobile Network
PMR	Private Mobile Radio
PNCH	Packet Notification Channel
PP	Portable Part
PPCH	Packet Paging Channel
PRACH	Packet Random Access Channel
PSD	Power Spectral Density
PSI	Pitch Synchronous Innovation
PSK	Phase Shift Keying
PSPDN	Packet-Switched Public Data Network
PSTN	Public Switched Telephone Network
PTCH	Packet Traffic Channel
PTN	Personal Telecommunication Number
PTT	Push to Talk
PWT	Personal Wireless Telecommunications
PWT-E	Personal Wireless Telecommunications–Enhanced
QA	Q (Interface)-Adapter
QAM	Quadrature Amplitude Modulation
QoS	Quality of Service
QPSK	Quadrature Phase Shift Keying
RACF	Radio Access Control Function
RACH	Random Access Channel
RAM	Random Access Memory
RAND	Random Number (Used for Authentication)
RCELP	Radio Control Functions
RCH	Radio Channel
RE	Radio Exchange
REL	Release
RELP	Residual Excited Linear Predictor
REVC	Reverse Charging
RF	Radio Frequency
RFN	Reduced TDMA Frame Number
RFP	Radio Fixed Part
RFTR	Radio Frequency Transmission and Reception
RLC	Release Complete Message
RLP	Radio Link Protocol

rms	Root Mean Square
ROM	Read Only Memory
ROSE	Remote Operation Service Element
ROT	Roll Out Time
RP	Radio Ports
RPE	Residual Pulse Excitation
RR	Radio Resource
RS	Radio System
RSC	Radio Resource Control
RSVD	ReSerVeD
RTF	Radio Terminal Function
RTMS	Radio Telephone Mobile System
RXLEV	Received Signal Level
RXQUAL	Received Signal Quality
S-BCCH	Slow Broadcast Control Channel
S/I	Signal to Interference
SABM	Set Asynchronous Balanced Mode
SACCH	Slow Associated Control Channel
SACCH/C4	Slow Associated Control Channel/SDCCH/4
SACCH/C8	Slow Associated Control Channel/SDCCH/8
SACCH/TF	Slow Associated Control Channel/Traffic Channel Full Rate
SACCH/TH	Slow Associated Control Channel/Traffic Channel Half Rate
SACF	Service Access Control Function
SAM	Subsequent Address Message
SBC	Sub-band coding
SC	Service Center (Used for SMS)
SCCP	Signaling Connection Control Point
SCF	Service Control Function
SCH	Synchronization Channel
SCP	Switching Control Point
SCT	Single-Step Call Transfer
SDCCH	Standalone Dedicated Control Channel
SDCCH/C4	Standalone Dedicated Control Channel (4 control channels shared with traffic channels)
SDCCH/C8	Standalone Dedicated Control Channel (8 control channels shared with traffic channels)
SDF	Service Data Function
SDR	Signal-to-Distortion Ratio
SDU	Service Data Unit
SEP	Signaling End Points
SF	Speech Frame
SFH	Slow Frequency Hopping

SG	Study Group
SID	System Identification
SIF	Signaling Information Field
SIM	Subscriber Identity Module
SIO	Service Information Octet
SIR	Signal-to-Interference Ratio
SLS	Signaling Link Selection
SMAP	Systems Management Application Process
SME	Short Message Entity
SMF	Service Management Function
SMFA	System Management Function Areas
SMG	Special Mobile Group
SMI	Structure of Managed Information
SML	Service Management Layer
SMS	Short Message Service
SMSCB	Short Message Service Cell Broadcast
SMSCH	Short Message Service Channel
SMSMC	Short Message Service Message Center
SNDCP	Subnetwork Dependent Convergence Protocol
SNMP	Simple Network Management Protocol
SNR	Serial Number
SNR	Signal-to-Noise Ratio
SP	Signaling Points
SPC	Stored Program Control
SRAT	Secondary Rate Adaptation Service
SRES	Signed Response (Authentication)
SRF	Specialized Resource Function
SS	Spread Spectrum
SS7	Signaling System 7
SSC	Single-System City
SSF	Service Switching Function
SSN	Subsystem Numbers
SSP	Switching System Platform
STP	Signaling Transfer Point
SU	Subscriber Unit
SUB	Subaddressing
T	Terminating
TA	Terminal Adaptor
TAC	Type Approval Code
TACAF	Terminal Access Control Agent Function
TACF	Terminal Access Control Function
TACS	Total Access Communication System (UK Analog)
TC	Technical Committee
TCAP	Transaction Capabilities Application Part

TCH	Traffic Channel
TCH/F	Traffic Channel Full Rate
TCH/H	Traffic Channel Half Rate
TCP	Transport Control Protocol
TDD	Time-Division Duplex
TDM	Time-Division Multiplex
TDMA	Time-Division Multiple Access
TE	Terminal Equipment
TE-TA	Terminal Equipment-to-Terminal Adapter
TFO	Tandem Free Operation
TIA	Telecommunications Industry Association
TIMF	Terminal Identification Management Function
TINA	Telecommunication Information Networking Architecture
TL	Transmission Link
TMN	Telecommunications Management Network
TMSI	Temporary Mobile Subscriber Identity
TP	Terminal Portability
TRAU	Transcoder Rate Adapter Unit
TRUP	Transparent Unprotected Service
TUP	Telephone User Part (SS7)
TXPWR	Transmit Power
UDI	Unrestricted Digital Information
UDP	User Datagram Protocol
UHF	Ultrahigh Frequency
UIMF	User Identification Management Function
UK	United Kingdom
UMTS	Universal Mobile Telephone Service
UP	User Part
UPT	Universal Portable Telephone
USO	Universal Service Order
UUS	User-to-User Signaling Supplementary Service
v	volt
VAD	Voice Activity Detector
VBS	Voice Broadcast Service
VDA	Viterbi decoding algorithm
VGCS	Voice Group Call Service
VHF	Very-High Frequency
VLR	Visitor Location Register
VSELP	Vector Self Excited Linear Predictor

W	Word
WARC	World Administrative Radio Conference
WATS	Wide Area Telephone Service
WCDMA	Wideband Code Division Multiple Access
WLAN	Wireless Local Area Network
WLL	Wireless Local Loop
WRC	World Radio Conference
WS	Work Station

LICENSE AGREEMENT AND LIMITED WARRANTY

READ THE FOLLOWING TERMS AND CONDITIONS CAREFULLY BEFORE OPENING THIS SOFTWARE PACKAGE. THIS LEGAL DOCUMENT IS AN AGREEMENT BETWEEN YOU AND PRENTICE-HALL, INC. (THE "COMPANY"). BY OPENING THIS SEALED SOFTWARE PACKAGE, YOU ARE AGREEING TO BE BOUND BY THESE TERMS AND CONDITIONS. IF YOU DO NOT AGREE WITH THESE TERMS AND CONDITIONS, DO NOT OPEN THE SOFTWARE PACKAGE. PROMPTLY RETURN THE UNOPENED SOFTWARE PACKAGE AND ALL ACCOMPANYING ITEMS TO THE PLACE YOU OBTAINED THEM FOR A FULL REFUND OF ANY SUMS YOU HAVE PAID.

1. **GRANT OF LICENSE:** In consideration of your payment of the license fee, which is part of the price you paid for this product, and your agreement to abide by the terms and conditions of this Agreement, the Company grants to you a nonexclusive right to use and display the copy of the enclosed software program (hereinafter the "SOFTWARE") on a single computer (i.e., with a single CPU) at a single location so long as you comply with the terms of this Agreement. The Company reserves all rights not expressly granted to you under this Agreement.

2. **OWNERSHIP OF SOFTWARE:** You own only the magnetic or physical media (the enclosed disks) on which the SOFTWARE is recorded or fixed, but the Company retains all the rights, title, and ownership to the SOFTWARE recorded on the original disk copy(ies) and all subsequent copies of the SOFTWARE, regardless of the form or media on which the original or other copies may exist. This license is not a sale of the original SOFTWARE or any copy to you.

3. **COPY RESTRICTIONS:** This SOFTWARE and the accompanying printed materials and user manual (the "Documentation") are the subject of copyright. You may not copy the Documentation or the SOFTWARE, except that you may make a single copy of the SOFTWARE for backup or archival purposes only. You may be held legally responsible for any copying or copyright infringement which is caused or encouraged by your failure to abide by the terms of this restriction.

4. **USE RESTRICTIONS:** You may not network the SOFTWARE or otherwise use it on more than one computer or computer terminal at the same time. You may physically transfer the SOFTWARE from one computer to another provided that the SOFTWARE is used on only one computer at a time. You may not distribute copies of the SOFTWARE or Documentation to others. You may not reverse engineer, disassemble, decompile, modify, adapt, translate, or create derivative works based on the SOFTWARE or the Documentation without the prior written consent of the Company.

5. **TRANSFER RESTRICTIONS:** The enclosed SOFTWARE is licensed only to you and may not be transferred to any one else without the prior written consent of the Company. Any unauthorized transfer of the SOFTWARE shall result in the immediate termination of this Agreement.

6. **TERMINATION:** This license is effective until terminated. This license will terminate automatically without notice from the Company and become null and void if you fail to comply with any provisions or limitations of this license. Upon termination, you shall destroy the Documentation and all copies of the SOFTWARE. All provisions of this Agreement as to warranties, limitation of liability, remedies or damages, and our ownership rights shall survive termination.

7. **MISCELLANEOUS:** This Agreement shall be construed in accordance with the laws of the United States of America and the State of New York and shall benefit the Company, its affiliates, and assignees.

8. **LIMITED WARRANTY AND DISCLAIMER OF WARRANTY:** The Company warrants that the SOFTWARE, when properly used in accordance with the Documentation, will operate in substantial conformity with the description of the SOFTWARE set forth in the Documentation. The Company does not warrant that the SOFTWARE will meet your requirements or that the operation of the SOFTWARE will be uninterrupted or error-free. The Company warrants that the media on which the SOFTWARE is delivered shall be free from defects in materials and workmanship under normal use for a period of thirty (30) days from the date of your purchase. Your only remedy and the Company's only obligation under these limited warranties is, at the Company's option, return of the warranted item for a refund of any amounts paid by you or replacement of the item. Any replacement of SOFTWARE or media under the warranties shall not extend the original warranty period. The limited warranty set forth above shall not apply to any SOFTWARE which the Company determines in good faith has been subject to misuse, neglect, improper installation, repair, alteration, or damage by you. EXCEPT FOR THE EXPRESSED WARRANTIES SET FORTH ABOVE, THE COMPANY DISCLAIMS ALL WARRANTIES, EXPRESS OR IMPLIED, INCLUDING WITHOUT LIMITATION, THE IMPLIED WARRANTIES OF MERCHANTABILITY AND FITNESS FOR A PARTICULAR PURPOSE. EXCEPT FOR THE EXPRESS WARRANTY SET FORTH ABOVE, THE COMPANY DOES NOT WARRANT, GUARANTEE, OR MAKE ANY REPRESENTATION REGARDING THE USE OR THE RESULTS OF THE USE OF THE SOFTWARE IN TERMS OF ITS CORRECTNESS, ACCURACY, RELIABILITY, CURRENTNESS, OR OTHERWISE.

IN NO EVENT, SHALL THE COMPANY OR ITS EMPLOYEES, AGENTS, SUPPLIERS, OR CONTRACTORS BE LIABLE FOR ANY INCIDENTAL, INDIRECT, SPECIAL, OR CONSEQUENTIAL DAMAGES ARISING OUT OF OR IN CONNECTION WITH THE LICENSE GRANTED UNDER THIS AGREEMENT, OR FOR LOSS OF USE, LOSS OF DATA, LOSS OF INCOME OR PROFIT, OR OTHER LOSSES, SUSTAINED AS A RESULT OF INJURY TO ANY PERSON, OR LOSS OF OR DAMAGE TO PROPERTY, OR CLAIMS OF THIRD PARTIES, EVEN IF THE COMPANY OR AN AUTHORIZED REPRESENTATIVE OF THE COMPANY HAS BEEN ADVISED OF THE POSSIBILITY OF SUCH DAMAGES. IN NO EVENT SHALL LIABILITY OF THE COMPANY FOR DAMAGES WITH RESPECT TO THE SOFTWARE EXCEED THE AMOUNTS ACTUALLY PAID BY YOU, IF ANY, FOR THE SOFTWARE.

SOME JURISDICTIONS DO NOT ALLOW THE LIMITATION OF IMPLIED WARRANTIES OR LIABILITY FOR INCIDENTAL, INDIRECT, SPECIAL, OR CONSEQUENTIAL DAMAGES, SO THE ABOVE LIMITATIONS MAY NOT ALWAYS APPLY. THE WARRANTIES IN THIS AGREEMENT GIVE YOU SPECIFIC LEGAL RIGHTS AND YOU MAY ALSO HAVE OTHER RIGHTS WHICH VARY IN ACCORDANCE WITH LOCAL LAW.

ACKNOWLEDGMENT
YOU ACKNOWLEDGE THAT YOU HAVE READ THIS AGREEMENT, UNDERSTAND IT, AND AGREE TO BE BOUND BY ITS TERMS AND CONDITIONS. YOU ALSO AGREE THAT THIS AGREEMENT IS THE COMPLETE AND EXCLUSIVE STATEMENT OF THE AGREEMENT BETWEEN YOU AND THE COMPANY AND SUPERSEDES ALL PROPOSALS OR PRIOR AGREEMENTS, ORAL, OR WRITTEN, AND ANY OTHER COMMUNICATIONS BETWEEN YOU AND THE COMPANY OR ANY REPRESENTATIVE OF THE COMPANY RELATING TO THE SUBJECT MATTER OF THIS AGREEMENT.

Should you have any questions concerning this Agreement or if you wish to contact the Company for any reason, please contact in writing at the address below.

Robin Short
Prentice Hall PTR
One Lake Street
Upper Saddle River, New Jersey 07458

ABOUT THE DISK

Included with the *Principles and Applications of GSM* is a 3-1/2" floppy disk with two directories on it. The matlab directory includes several Matlab programs that were used to generate some of the modulation figures in chapter 12. The traffic directory contains the necessary files to install WErlangB.exe on your Windows NT-4 or Windows 95 computer.

In chapter 12, we present information on various modulators used to generate DECT, GSM, and PWT signals. We generated the plots using Matlab 5, a scientific programming language from The MathWorks, Inc. A student version of Matlab is available from Prentice-Hall and will run most of the included files.

The file figure 12_2.m generates a plot of the probability density function for two Gaussian random variables used in the detection of binary signaling. In the detection of binary signals, we set a detection threshold to a signal level "A." If we set the value of A so that the horizontal shaded area and the vertical shaded area are equal, then the error rate will be minimized.

The file figure 12_6_9.m generates the data, in-phase (I), quadrature (Q) and transmitted signals for QPSK and OQPSK. When this program is run, figure 1 is QPSK and figure 2 is OQPSK for the same data signal. In the program, NSPB is thenumber of samples per bit and Nbits is the number of data bits. The data is stored in an array called "data." By changing the value of Nbits and the data stored in the array, the reader can see the signals generated for different data sequences.

The file figure12_12.m generates the data, in-phase (I), quadrature (Q), phase angle and transmitted signals for MSK. In the program, NSPB is thenumber of samples per bit and Nbits is the number of data bits. The data is stored in an array called "data." By changing the value of Nbits and the data stored in the array, the reader can see the signals generated for different data sequences.

The file thetafx2.m is used by figure 12_12.m to correct the phase angle when the phase crosses the 2p boundary. Without thetafx2.m, the plotted phase angle will be generated modulo 2p. The file figure12_14.m plots the power spectral density of MSK and QPSK signals.

In the traffic directory, the program setup.exe will install WErlangB.exe on your computer and will install an icon on the Start Menu. To install the program:

- ☞ Place the floppy disk in the A drive of your computer.
- ☞ Click on the "Start Menu" with the left mouse button and then click on "Run."
- ☞ In the Run window, select "Browse."
- ☞ In the Browse window, select the down arrow to the right of "Look in:"
- ☞ Move the mouse to "3 1/2 Floppy (A:)" and click on it.
- ☞ Double-click on the "Traffic" folder.
- ☞ Select "Setup.exe" with the mouse and click on "Open."
- ☞ In the Browse Window, click on "OK."
- ☞ The Install Shield Wizard will then start.
- ☞ Read the contents of the Welcome Window and click on " Next>."
- ☞ The Install Shield Wizard will install the program in folder "C:\Program Files\Prentice Hall PTR." If you want to install it in a different window, click on "Browse" and select the new folder name; otherwise click on "Next>."
- ☞ The Install Shield Wizard will install the program and install its icon on the Start Menu.

To run WErlangB, click on the "Start Menu"; Select Programs and then Select WErlangB. In the program Window, enter the blocking probability as the value for B(s,a) and the number of servers; then click on calculate. The program will calculate the offered load. The program has an upper limit of about 200 servers. By changing the number of samples per bit, you can see the effects on the plots and the run time of the program by using more or less samples per bit. The same NSPB and Nbits variable are used in this program.